THE LIBRARY
ST. MARY'S COLLEGE OF MARYLAND
ST. MARY'S CITY. MARYLAND 20686

S0-CIE-879

A GUIDE TO THE CHEMICAL BASIS OF DRUG DESIGN

A GUIDE TO THE CHEMICAL BASIS OF DRUG DESIGN

ALFRED BURGER
Professor Emeritus of Chemistry
University of Virginia, Charlottesville

A Wiley-Interscience Publication
JOHN WILEY & SONS
New York   Chichester   Brisbane   Toronto   Singapore

Drugs discussed here to which patent or trademark
rights may exist are unaffected as to that patent or trade
status by any statement or discussion of said drugs
which may appear in these pages.

Copyright © 1983 by John Wiley & Sons, Inc.

All rights reserved. Published simultaneously in Canada.

Reproduction or translation of any part of this work
beyond that permitted by Section 107 or 108 of the
1976 United States Copyright Act without the permission
of the copyright owner is unlawful. Requests for
permission or further information should be addressed to
the Permissions Department, John Wiley & Sons, Inc.

Library of Congress Cataloging in Publication Data:
Burger, Alfred, 1905–
   A guide to the chemical basis of drug design.

   "A Wiley-Interscience publication."
   Bibliography: p.
   Includes index.
   1. Chemistry, Pharmaceutical. 2. Pharmacy. I. Title.
RS403.B78   1983   615'.19   83-3575
ISBN 0-471-86828-0

Printed in the United States of America

10 9 8 7 6 5 4 3 2 1

PREFACE

The design of the chemical structure of biologically active compounds is guided by the techniques and working hypotheses of the field of medicinal chemistry. It is not yet a reliable enough discipline to permit absolutely dependable predictions of the profile of biological activity of a test compound, but a number of experiences and valid experimental pathways have narrowed down the choices of procedure from the randomness of earlier approaches.

This book singles out these experiences, primarily those that have advanced drug design, but makes no apologies for the failures and continuing uncertainties that beset medicinal planning. Foremost among the reasons for these uncertainties is our limited appreciation of the totality of the effects of chemical and physical properties on biological activity. Second, experimental biologists and medicinal chemists work on different facets of the relationships between chemical structure and biological activity, the underlying principle of drug design. Even in the best cooperative teams the two types of scientists emphasize different aspects of their procedures and ideas, aspects that are difficult to reconcile in a complete program of drug design. Although drug design has thus remained a compromise between different working hypotheses and approaches, it has reached a stage of being an overall guide to the activity, potency, and potential pharmacological utility of chemical compounds. The role of chemists in such joint research efforts is stressed in this volume.

This book was written in the course of one year. One cannot write during a crowded schedule and read the current literature at the same time. Therefore the reporting is up-to-date only to about 1980 overall, with a few additions of more current events.

One of the first casualties of retirement from active teaching duties is the loss of secretarial help. I was fortunate in getting the rough long-hand manuscript typed by the staff of the Word Processing Center, American Hoechst Corporation, Somerville, New Jersey, D. Merriman, Supervisor, and J. Van Elk, Coordinator. The final manuscript was typed at the University of Virginia under a grant generously furnished by John Wiley and Sons, my publishers. I am grateful for this grant and for the assistance of the typists.

Many ideas and examples quoted in this volume are results of a lifetime in medicinal chemistry. A recent source of information for data in this book has been the three-volume treatise on the subject edited by M. E. Wolff (John Wiley and Sons, 1979–1981). Grateful acknowledgment is made to this important publication.

ALFRED BURGER

Charlottesville, Virginia
July 1983

CONTENTS

INTRODUCTION   1

1. HISTORY OF MEDICINAL CHEMISTRY   5

    1.1. Evolution of the Discipline, 5
        1.1.1. Relation to Pharmacology, 6
        1.1.2. The Role of Natural Products, 7
        1.1.3. The Role of "Lead" Compounds, 8
    1.2. Historical Development of Medicinal Science, 10
        1.2.1. Foundations of Modern Therapeutics, 10
        1.2.2. The Concept of Drug Receptors, 12
        1.2.3. Organization of Drug Research, 14
    1.3. Systematic Drug Development and Rational Research Methods, 15
        1.3.1. Pro-Drugs, 15
        1.3.2. Dyestuffs, 15
        1.3.3. Minor Analgetics, 16
        1.3.4. Local Anesthetics, 17
        1.3.5. Biogenic Amines, 19
        1.3.6. Structure–Activity Relationships, 20
        1.3.7. Sedative Hypnotics, 22
        1.3.8. Vitamins and Hormones, 23
    1.4. The Crowded Years of Innovation and Expansion, 24
        1.4.1. Steroid Hormones, 25
        1.4.2. Penicillin, Sulfanilamide, and Other Antimicrobial Agents, 26
        1.4.3. Bioisosterism, 28
        1.4.4. Pharmacodynamic Drugs, 30
        1.4.5. Researches During World War II, 30
        1.4.6. Early Antitumor Agents, 31
        1.4.7. Muscle Relaxants, 32
        1.4.8. Vasopressor and Vasodepressor Drugs, 32
        1.4.9. Abstinence Drugs, 33
        1.4.10. Drugs for Mental Disorders, 34

2. RECENT RESEARCH 37

- 2.1. Antihypertensive Agents, 38
- 2.2. Antiinflammatory Agents, 42
  - 2.2.1. Nonsteroidal Drugs, 42
  - 2.2.2. Antiinflammatory Corticosteroids, 46
- 2.3. Antiviral and Antitumor Agents, 47
- 2.4. Histamine (H_1 and H_2) Receptor Antagonists, 52
- 2.5. Drugs that Counteract the Effects of Acetylcholine, 55
- 2.6. Drugs for Relief from Pain, 58
  - 2.6.1. Opiate Receptors, 64
  - 2.6.2. Centrally Acting Antitussives, 65
- 2.7. Antihyperglycemic Agents, 67
- 2.8. General Aspects of Drug Design, 69
  - 2.8.1. Screening, 70
  - 2.8.2. Molecular Modification, 70
  - 2.8.3. Drug Metabolism, 71
  - 2.8.4. Transition-State Analogs, 72
  - 2.8.5. Suicide Enzyme Inhibitors, 73
  - 2.8.6. Active-Site-Directed Irreversible Inhibitors, 74
- 2.9. Molecular Modification, 76
  - 2.9.1. Separation of Biological Properties, 76
  - 2.9.2. Methodology of Molecular Modification, 78
  - 2.9.3. Bioisosterism, 84
- 2.10. Quantitative Structure–Activity Analysis, 87
- 2.11. Conformational Analysis, 90

3. SELECTED EXAMPLES OF DRUG DESIGN 92

- 3.1. Thyroid Hormones and Thyromimetics, 92
- 3.2. The D Vitamins, 95
- 3.3. Peptides and Proteins, 96
  - 3.3.1. Calcitonin, 98
  - 3.3.2. Peptide Analogs, 98
- 3.4. Anticoagulants, 100
  - 3.4.1. Polysaccharides, 100
  - 3.4.2. Oral Anticoagulants, 101
- 3.5. Cholinergic, Curariform, and Related Agents, 102
  - 3.5.1. Acetylcholine, 102
  - 3.5.2. Neuromuscular Blocking Agents, 105
- 3.6. Anticholinergics and Histamine-1 Receptor Antagonists, 107
  - 3.6.1. Anticholinergics, 107
  - 3.6.2. Antihistaminics (H_1 Receptor Antagonists), 109
- 3.7. Local Anesthetics, 113

- 3.8. Neuroleptic Agents, 116
  - 3.8.1. Reserpine, 117
  - 3.8.2. Phenothiazines, 118
  - 3.8.3. Other Tricyclic Systems, 121
  - 3.8.4. Butyrophenones and Related Compounds, 123
- 3.9. Antidepressants, 125
  - 3.9.1. Discovery of MAO Inhibitors, 125
  - 3.9.2. Discovery of Tricyclic Antidepressants, 126
  - 3.9.3. Biochemical Hypotheses, 127
  - 3.9.4. Test Methods, 128
  - 3.9.5. Monoamine Oxidase Inhibitors, 128
  - 3.9.6. Tricyclic Antidepressants, 130
- 3.10. Antianxiety Agents, 131
  - 3.10.1. 1,3-Propanediols, 131
  - 3.10.2. Benzodiazepines, 132
- 3.11. Sedative Hypnotics and Anticonvulsants, 134
  - 3.11.1. Hypnotics, 135
  - 3.11.2. Anticonvulsants, 139
- 3.12. Analgetics, 141
  - 3.12.1. Mechanisms of Action, 141
  - 3.12.2. Endogenous Analgetic Peptides, 142
  - 3.12.3. Opioid Analgetics, 143
  - 3.12.4. Simplified Structures, 145
- 3.13. Antiinflammatory Agents, 148
  - 3.13.1. Steroidal Drugs, 148
  - 3.13.2. Nonsteroidal Antiinflammatory Agents, 150
- 3.14. Steroidal Hormones and Their Analogs, 154
  - 3.14.1. Androgenic–Anabolic Compounds, 154
  - 3.14.2. Estrogens, 156
  - 3.14.3. Progestational Agents, 158
- 3.15. Antiinfectious Agents, 159
  - 3.15.1. Topical Antibacterials and General Antimicrobials, 159
  - 3.15.2. Antimicrobials Used Topically and Systemically, 167
- 3.16. Chemotherapeutic Antibacterials, 170
  - 3.16.1. Sulfanilamides and Diaminosulfones, 170
  - 3.16.2. Bis(4-Aminophenyl) Sulfone, 177
- 3.17. Antimycobacterial Agents, 178
  - 3.17.1. Antimycobacterial Antibiotics, 179
  - 3.17.2. Synthetic Agents, 180
- 3.18. Antibiotics, 183
  - 3.18.1. Introduction, 183
  - 3.18.2. β-Lactam Antibiotics, 184
  - 3.18.3. Aminoglycoside Antibiotics, 190

- 3.18.4. Peptide Antibiotics, 191
- 3.18.5. Tetracyclines, 192
- 3.18.6. Chloramphenicol, 192
- 3.18.7. Other Antibiotics, 193
- 3.18.8. Antitumor Antibiotics, 194
3.19. Antiprotozoal Agents, 196
- 3.19.1. Antitrypanosomal Drugs, 196
- 3.19.2. Drugs for Coccidiosis, 202
- 3.19.3. Drugs for Trichomoniasis, 203
- 3.19.4. Drugs for Leishmaniasis and Histomoniasis, 203
- 3.19.5. Antiamebic Agents, 205
- 3.19.6. Other Protozoan Infections, 209
- 3.19.7. Antimalarials, 210
3.20. Anthelmintics, 220
- 3.20.1. Drugs for Schistosomiasis, 220
- 3.20.2. Other Trematode Infections, 222
- 3.20.3. Drugs for Filariasis, 224
- 3.20.4. Drugs for Cestode Infestation, 226
- 3.20.5. Broad-Spectrum Anthelmintics, 227
3.21. Antiviral and Antineoplastic Agents, 230
- 3.21.1. Natural Compounds, 231
- 3.21.2. Immunostimulants, 232
- 3.21.3. Antimetabolites as Antineoplastic and Antiviral Agents, 234
- 3.21.4. Alkylating Agents, 240
- 3.21.5. Miscellaneous Drugs, 241
3.22. Orphan Drugs, 243

References 245

Index 289

A GUIDE TO THE CHEMICAL BASIS OF DRUG DESIGN

INTRODUCTION

My interest in the chemistry of medicinal agents originated when, in 1928, I did volunteer work in the Vienna laboratory of Professor Sigmund Fränkel, whose treatise *Arzneimittel-Synthese* had then reached its 6th edition. In 1929 I became a research associate in the Drug Addiction Laboratory of the National Research Council at the University of Virginia. There, in cooperation with the pharmacologist Nathan B. Eddy of the University of Michigan, our unit studied molecular modifications of morphine and analogs of structural fragments of this opium alkaloid, with the avowed purpose of separating analgesia and dependence liability in the resulting compounds. We prepared one series of compounds after another based on reactions of the allylic alcohol system of morphine and on hydrophenanthrene and similar ring systems containing the amino alcohol functions of this alkaloid. While running rather similar reactions hundreds of times in such serial syntheses myself, I worried about the lack of reasoning underlying molecular modification in those days. Therefore the early (1932) publications by Hans Erlenmeyer (1) on the application to drug design of Langmuir's (2) and Grimm's concepts of isosterism and Hinsberg's (3) ideas of ring equivalents came as a welcome revelation. Here was the first suggestion of a rationale for molecular modification, including a prophecy that molecular shape would come to be considered a prime condition for analogy of biological behavior. A few years later Schaumann (4) demonstrated the importance of steric similarity when he explained the analgetic* activity of meperidine on the basis of its structure, which represented an unorthodox fragment carved out of the skeleton of morphine.

Biochemistry entered the picture of medicinal chemistry—a designation unalterably associated with biochemical explanations of the mechanism of action of medicinal agents—with the demonstration in 1940 by Woods and Fildes (5) that many drugs act as antagonists to biosynthetic substrates. This idea caused a virtual revolution in the selection of "lead" compounds, since an apparently unending supply of biosynthetic substrates emerged from the

---

* The *Oxford English Dictionary* points out that the term *analgetic* for insensibility to pain is a better formation than *analgesic* and more parallel with *anesthetic*, total insensibility.

improved analytical identification of intermediary metabolites. Although the repeated discovery of odd structures as drugs during random screening continued to accompany the more logical selection of "leads" on biochemical grounds, the latter raised the level of confidence in the discovery of medicinal agents and offered an intellectually rewarding approach to work in this field.

By the 1860's pharmacologists (6) had staked out claims for the conception and actual laboratory preparation of experimental drugs, a preoccupation for which they were little suited by temperament or chemical experience. The split of pharmacology from chemistry came when the design of pharmacological experimentation and the explanation of physiological mechanisms was placed on a firmer theoretical foundation and absorbed the time, energy, and curiosity of experimental biologists.

As medicinal chemistry became more biochemically oriented, the borderline with experimental biology became less distinct. Medicinal chemists should have stopped at this point because they, in turn, were ill-prepared to and not inclined to carry out biological experimentation. Instead, very many medicinal chemists became parlor pharmacologists, that is, they began to talk and think about pharmacological experiments at the expense of efforts they should have spent perfecting chemical experiments and chemical working hypotheses. As more of them fell into this pattern that should have been reserved for conversations during coffee breaks, medicinal chemical symposia and lectures followed suit. It became customary at such research conferences to invite clinicians and physiological pharmacologists to report on the problems they encountered during pharmacotherapy. The chemists in the audience were barely familiar with biological and medical terminology. They had been brought up to master organic, physical, analytical, and biological chemistry, and only few had had a formal course in biology or pharmacology (this educational shortcoming has been ameliorated in the last decade). But these chemists felt that on top of perfecting chemical, statistical, and instrumental experiences with a direct bearing on chemical thought and experimentation, they should be able to participate in biological decision-making. It is indeed necessary for medicinal chemists to be able to understand their biological teammates, but they should not permit their valuable time to be dominated by problems to which they, as chemists, could not make a systematic contribution.

Pharmacologists have every obligation to talk and write about the responses of cells and tissues to drugs, to describe drug metabolism, and to devise the best laboratory tests to evaluate the activity and toxicity of a drug. Chemists listening to or reading about such discussions usually feel discouraged. What does one do to one's "lead" compound if it produces a given side effect? Should one introduce chemical substituents? Or make the molecule more rigid, or less rigid, or homologize it? Even if one only wishes to make a competitive drug with virtually the same activity profile as the prototype, discussions at the cellular or tissue level—let alone behavioral data in animals or humans—will not serve as guides for chemical planning, design, or understanding. The only common meeting ground is the explanation of the mechanism of action of a drug in biochemical terms, in studies of the inhibition of enzyme systems, antagonism to substrates, and other chemical reactions. Such reactions might then serve as models for chemical work on the drug. This means that medicinal chemists should participate more actively in the work of biochemists and give it direction toward the study of drugs in normal and abnormal physiological environments.

One of the most intriguing questions in medicinal chemistry is, How did an original discoverer of a drug get the idea that a given chemical structure might have pertinent biological activities? Since this problem arises every time a disease entity is singled out for pharmacotherapeutic study, this volume collects at least some of the historically more important aspects of drug discovery and selected cases of unusual circumstances that led to prototype compounds with therapeutic properties, provided they teach us what to look for in similar circumstances.

In treatises on medicinal chemistry, the space allotted to the description of the clinical and experimental biological background data has grown, as it should, with the advancing scope of knowledge in these fields. But in accord with the increasing interest of medicinal chemists in biological experiments, many sections of these treatises now describe biological events that cannot be translated into biochemical or medicinal–chemical data. Therefore a voice should be raised in defense of medicinal chemistry and to point out the features that make it an interdependent and yet independent science.

In this book I have no intention of reviewing the whole field of medicinal chemistry. No attempt whatsoever is made to provide complete coverage. This has been done authoritatively by 82 experts in the fourth edition of *Burger's Medicinal Chemistry* (7). Instead, the book is limited to underscoring what chemical work has been done and still needs to be done in medicinal science. When necessary for an understanding of what the chemists can do, references are made to biological background and experimentation, but these references are as brief as possible and restricted to quotations from the published literature.

Much medicinal chemical work is a repetitive and parallel effort. Such studies are dictated primarily by commercial competition as well as by the hope that persistent modification will sharpen selectivity and reduce toxicity of almost every type of drug. Except in systematic surveys, they do not teach much. Therefore, this volume concentrates on the chemical rationale of medicinal science. It should give the readers an idea of what they can expect to encounter in the field and should guide them to existing texts, reviews, and treatises for more complete and systematic information. Many references are made to the three-volume treatise edited by Wolff (7). Reading the chapters in those books will provide the systematic and biological details that the present volume does not deal with.

The first chapter arranges the field of medicinal chemistry by historic sequence of observations. From 1950 on, however, this arrangement becomes unfeasible for the whole field because of the explosive simultaneous developments in many laboratories. Consequently, in Chapter 2 the most important events are described according to disease entities, and then the various new aspects of drug design are examined. This scheme is expanded in Chapter 3 so that a few instructive topics may be studied in somewhat greater detail. Subjects still in a purely empirical stage—such as drugs to control aging—are not treated because they cannot teach any chemical rationale, at least not yet. The experimental pharmacological details that a biologist must know are restrained, because medicinal chemists should not pose as parapharmacologists.

Throughout the book it is assumed that the reader, a medicinal scientist or student of medicinal chemistry, is familiar with many aspects of this subject. Therefore detailed explanations of facts, nomenclature, and well-known data that form the backbone of medicinal science are not given. Chemical formulas are shown where necessary for a quick assessment of a structural type, but they are not shown where a chemical name in the text is self-explanatory to a trained chemist or graduate student.

1

HISTORY OF MEDICINAL CHEMISTRY

1.1. EVOLUTION OF THE DISCIPLINE

Medicinal chemistry is the branch of chemistry that deals with the discovery, design, and development of therapeutic chemical agents for use in clinical and veterinary medicine. It deals with the relationships between chemical structure and biological activity, the identification of drug metabolites, and the biochemical explanation of the transport and actions of prophylactic, therapeutic, and curative medicines. The chemical concepts of medicinal chemistry have also permeated the fabric of biology, genetics, medicine, toxicology, and pest control.

The term "medicinal chemistry" evolved hesitatingly in the United States in the 1920's and even more slowly in other countries. Previously, colleges of pharmacy and research departments in the pharmaceutical industry had called their chemical sections "pharmaceutical chemistry," and a few conservative departments still adhere to this practice. It was recognized that "pharmaceutical chemistry" might be confused with "pharmaceutics," which deals with the formulation and the coating and finishing operations of a drug product in the apothecaries and in the industry. In addition, the ever closer ties of drug design with biochemistry and biochemical explanations of drug action made a change in designation advisable. With the appearance since the 1940's of several text and reference books, major journals (*Journal of Medicinal Chemistry, European Journal of Medicinal Chemistry*), several monograph series (e.g., *Progress in Drug Research*) and the *Annual Reports in Medicinal Chemistry*, the name "medicinal chemistry" has become firmly established. The last vestiges of confusion with medical chemistry, that is, analytical and diagnostic chemical procedures, were removed in Europe when frequent national and international symposia on medicinal chemistry were initiated there. The term "pharmacochemistry" has not found application in English-speaking countries (8).

This book was conceived as an overview of the mission of *chemists* in all the processes that lead up to the discovery and development of drugs.

Close cooperation between chemists and biologists is of the essence for this mission. However, the chemical identification, characterization, and preparation of biologically active materials—and most of all, the ultimate decision about structural changes that promise greater potency or a more acceptable spread between activity and toxicity—rest with the chemist.

Most accounts of the history of medicine give full credit for the discovery of drugs to experimental and clinical pharmacologists. This is still done in current reports on medicinal discoveries, both in medical journals and in the news media. When a pharmacologist tests an *existing* drug for a new activity unrelated to its established use and finds one in such trials, he is indeed the discoverer of this new application. But when a new substance is submitted for specified biologic tests, the chemists who dreamed up and prepared the material are recognized for their intellectual acumen and technical skill. Since the screening of large numbers of chemicals is time-consuming, expensive, and wasteful, any chemical or physical method that shortens this tedious process should be credited prominently as a contribution to the ultimate success in the search for medicinal agents.

It is not uncommon to delve into antiquity or the Middle Ages for the roots of medicinal chemistry, or iatrochemistry as it was called. However, the inventive and practical input into drug discovery and development practiced in those early days did not approach our present concept of the field, and the explanation of drug action was relegated to supernatural beliefs. The description of crude botanical drug powders and extracts is part of what has been called pharmacognosy in modern times. The use of natural materials and inorganic substances reported by alchemists at the end of the Middle Ages represented essentially the dawn of pharmacology and toxicology but harbored little chemistry. It would stretch the imagination if one classified the Swiss alchemist Paracelsus (1493–1541) (9), the Dutch van Helmont (16th century) (10) or Sylvius (17th century) (11) as ancestors of biochemistry and thereby as forerunners of those who try to explain the mode of action of drugs. The few chemical operations of alchemists that could be considered to bridge the transition to more modern science were mostly restricted to the use of inorganic elements (As, Au, Sb) and compounds.

1.1.1. Relation to Pharmacology

In 1876, the pharmacologist Rudolf Buchheim of the University of Giessen wrote that "the mission of pharmacology is to establish the active substances within the [natural] drugs, to find chemical properties responsible for their action, and to prepare synthetically drugs that are more effective" (6). This definition would now be applied unhesitatingly to medicinal chemistry. Pharmacology occupies itself with Buchheim's further statement that it "study the changes by the drug in the organism and then explore the possible influence of such changes upon pathological conditions." This dichotomy is more than semantics. Experimental biologists are ill-prepared by experience or practical inclination to search for new chemicals, let alone to synthesize them or prove their structures; but without test chemicals they cannot verify the validity of test methods they have devised for a given disease entity.

Chemical thought processes have evolved slowly over the decades of therapeutic discovery. These ideas have at their center two requirements of medicinal chemistry: One is to find prototype ("lead") compounds that can provide an entry into a given area of drug research. The second is to rationalize and pinpoint molecular modification and avoid the almost senseless empirical choice of candidate compounds that characterized molecular modification before 1935 and gave it a low rating in the minds of some scientists. In both approaches considerable refinements and progress have been made by applying modern theoretical considerations of organic and physical chemistry to medicinal problems. This minimizes having to make assumptions in biology because, in spite of all efforts to unify physical and biological sciences, the increasing complexity of these sciences has separated them further in solving problems common to both. To be sure, all biology is based on chemistry, but the ladder reaching from fundamental biochemical metabolism to the structural explanation of macromolecular biological events has not yet been found in most cases.

Empiricism has characterized all efforts of medicinal science. As an example, one might read logic into the discovery of iodine (12) and its use in the treatment of goiter. But iodine was used first not for thyroid diseases but as a topical antiseptic (13), a use for which it has survived since 1839. The later connection of iodine and goiter was based on chance clinical observations. So much for iodine as an essential trace element; others such as chromium (14) have not yet even found a biochemical niche, except, perhaps, in insulin resistance.

1.1.2. The Role of Natural Products

The many methods practiced by medicine men and tribal doctors to treat illnesses in all parts of the precivilized world do not qualify as pharmacology or any other biomedical designation. In fact, many "drugs" conveyed to us by medicinal folklore had little to do with their alleged therapeutic purpose, because diseases were often misdiagnosed. Nevertheless, they provided important sources of natural products that centuries later could be extracted, purified, and identified. Natural products have remained valuable sources of potential drugs. They must have arisen from metabolites very similar to those found in the mammalian organism and therefore should be recognized more easily by mammalian biosites than some unrelated, totally unnatural chemicals. In addition, their structures are often novel and unexpected and are apt to arouse chemists out of their traditional thought patterns. Even so, natural products chemistry is organic chemistry, often at its very best, but it is not medicinal chemistry, which is motivated by biological activity.

One has to be a confirmed teleologist to assume that natural products have been placed in plants or animals as sources of therapeutic agents for other animal species including humans. The majority of natural products probably represent biosynthetic intermediates or end products, often stored away in tissues where they are least in the way of the fundamental metabolic processes of the organism. Some natural products are toxic to animals, such as dicoumarol which causes the sweet-clover hay disease in foraging cattle, or the toxins of the barracuda which exert their action when the fish is ingested by humans. Similarly, the toxic action of the fungus *Claviceps purpurea* has been known for centuries to lead to epidemics of miscarriages and psychotic episodes. Some of these poisons, when purified and administered in judicious doses, have been used as therapeutic agents, but only after considerable chemical manipulation.

Some natural plant products serve as repellents to insects and other predators, but at least in one case a biogenetic relationship has become known. Some insects synthesize 6,7-dihydro-5H-1-formylpyrrolizine as a pheromone and use pyrrolidine alkaloids from *Crotalaria, Senecio,* or *Eupatorium* as biosynthetic sources for this purpose (15). Less causative relationships are seen in tannins that are toxic to insect herbivores. Tomato and potato plants produce proteinase inhibitors when attacked by chewing insects; these inhibitors affect the digestive processes of the insect and are accumulated by a putative plant wound hormone called proteinase inhibitor-inducing factor. 2-Tridecanone from a wild tomato inhibits feeding by the tobacco hornworm. Other defense agents against insects include cadinene, a sesquiterpene, and myrcene, a diterpene, which protect Douglas firs from budworm attack. A naphthoquinone derivative from the tropical medicinal shrub *Plumbago capensis* inhibits molting in several lepidopterous agricultural pests and inhibits an enzyme involved in the biosynthesis of chitin. Polyacetylenes from Compositae (daisies, black-eyed susans, marigold, fleabane) are toxic to mosquito larvae. One of these, α-terthienyl, is more potent than DDT in insects exposed to light. Leaf damage or UV irradiation of plants brings about the synthesis of antifungal and antibacterial phytoalexins; in soybeans these agents also deter feeding on the plant by the Mexican bean beetle, but in high concentrations the phytoalexins are very detrimental to the soybeans ("suicide response") (16).

1.1.3. The Role of "Lead" Compounds

Before synthetic organic chemistry hit its stride, natural products were the sole source of experimental medicinal materials. Their activities were almost always recognized by pharmacologists before chemists attempted to purify the active principles. At least one major American pharmaceutical company based its operations on this principle until 1950. They did not investigate the chemistry of any natural product—alkaloids, steroids, vitamins, hormones—unless pharmacologists had established the usefulness of the substance in medicine. Only then did organic chemists undertake the purification, structural studies, and synthesis that made the compound economically available.

This had not always been so. In the declining days of Renaissance men, with their broad though superficial expertise in many fields, specialization as we know it today had not yet become a prerequisite to solving interscience problems. As structural theory and methodology of organic chemistry expanded, it became possible to undertake the determination of the structure of many natural products—with preference to biologically interesting ones— and of complex synthetic compounds. In selected instances, synthesis followed suit, enhancing the availability of the product for biological study. Synthesis represents an intellectual triumph, but at first it often yields only minute amounts of product as a museum sample and as a confirmation of the structure and configuration of the compounds. If chemical development improves the accessibility and yield of a complex substance, the synthetic sequence may be of help in the preparation of derivatives and analogs needed to select a material with optimal medicinal properties for a given purpose. Thus the purification of medicinal drugs, whether natural or synthetic, and the perfection of analytical and synthetic methods have been a prerequisite for the emergence of medicinal chemistry. In recent years, the methodology of searching for "lead" compounds has been based more and more on biochemical considerations. Whenever it is possible to connect pathogenesis with a metabolic or nutritional excess or deficiency, the chemicals that cause such disturbances will be identified. With the introduction of sensitive spectroscopic equipment over the past three decades, many previously unrecognized chemicals have been associated with new disease entities. If the disease is caused by a deficiency, one can try to replace the missing material by direct administration or incorporation in an "enriched" diet. If an enzymic reaction has gone amiss because of excess of a substrate or a metabolically altered substrate, one can attempt to interfere with the damaging reaction by administering a structural analog of the pathogen. This line of approach has been very rewarding. Although in a way it is only molecular modification of a substrate, it introduces unaccustomed structures of biochemicals as prototypes for further modification. It also satisfies our pride in overcoming pathogenic obstacles set up by nature with severe consequences for the health of the community.

The influence of biochemistry on medicinal thinking dates back only five decades, when metabolite antagonists were first recognized and designed. Much more recently, experiments were initiated that might furnish new "leads" based on chemistry alone, without biological input. They have led to analogs of the transition states of chemicals, which could be derived from organic chemistry precepts. Some of these transition-state analogs can be formed by enzymes, which then are efficiently and often irreversibly inactivated. These chemicals have been called "suicide enzyme inhibitors"; they are discussed in Section 2.8.5. Because they have made their appearance so recently, their impact on drug design has not yet reached its crest.

1.2. HISTORICAL DEVELOPMENT OF MEDICINAL SCIENCE

If we are to follow the development of medicinal reasoning, we must look at it from a historic viewpoint. In the early stages we encounter more empirical pharmacology than chemistry, because chemistry on the whole had a slow start in the first half of the 19th century. As time advanced, chemistry took more effect in shaping drug discovery and development.

1.2.1. Foundations of Modern Therapeutics

It is customary to begin the history of modern therapeutics with digitalis because it was mentioned by Welsh physicians in 1250, described and named by Fuchsius in 1542, and introduced for the treatment of "dropsy" (congestive heart failure) by Withering in 1785 (17). The active principles of selected species of this plant, digitoxin (from *Digitalis purpurea* L.) and especially digoxin, a secondary glycoside (from *D. lantana* Ehrh. or *D. orientalis* Lam.), are still manufactured by extraction after 200 years. The chemical pharmacists (18) who first carried out these separations were influenced by the commercial incentive of preparing purified drugs, and they also undoubtedly dreamed of casting light on the biological organization and physiological function of the plant organisms they extracted.

In 1805, the young German pharmacist Sertürner purified the main opium alkaloid, morphine (19), and unwittingly started the course of events that gave us semisynthetic alkaloids (codeine, 20) and alkaloidal analogs such as hydrogenated morphine and codeine derivatives. It also gave us heroin, first prepared in 1874 (21, 22). Atropine was isolated in 1832 by Geiger and Hesse (23), papaverine by Georg Merck in 1848 (24), the same year that van Heyningen purified quinidine. More significant was the isolation of quinine in 1820 by Pelletier and Caventou (25) and the isolation of cocaine in 1859, by Niemann (26) and by Wöhler (27). This list could be extended to the present day; natural products chemistry has had a revival since the isolation of reserpine from *Rauwolfia serpentina* (28, 29), and several medicinally interesting substances have made their appearance in these studies. However, the majority of the hundreds of alkaloids, terpenes, steroids, and so forth, obtained in these researches presented fascinating structural revelations but were of no interest as therapeutic agents.

This in itself was enough to raise the question why some compounds are biologically active while others are not. Similar questions had already puzzled the alchemists, such as why some elements are soluble in acids while others cannot be dissolved, or why some compounds taste sweet whereas others taste sour. This latter problem touches directly upon the relationship of biological activity to chemical properties. Then there are the reactions on the retina evoked by different colors; these colors are due to different emitted or absorbed wavelengths of visible radiation. Such colors were seen in dyestuffs, some of which had just been synthesized in the 1860's. The chemical structures of these dyestuffs were known to contain chromophoric, bathochromic, and auxochromic groups, which are responsible for their color and colorfastness. These groups could be varied, and after some trial and error their effects on dyestuff properties became predictable. If one wanted to produce a certain tint, a number of functional groups and structural moieties could be expected—although not guaranteed—to yield the desired color qualities. Chemists soon recognized that there was an interdependence of the effects of the three types of structural determinants, and that changing one of them mandated certain changes in the others if the same or nearly the same color and dyestuff characteristics were to be preserved. Other properties of the dyestuffs also depended on the structure of the dye, particularly their affinity for cotton (cellulose), wool or silk (protein), or later for synthetic fibers and plastics. Resistance to actinic bleaching and to detergents also became associated with structural features of dyestuff molecules.

The stage was thus set for studying the effects of chemical structure on other properties of various substances. The hair and feathers of animals had been dyed for millennia, long before they were recognized as being composed of protein fibers. Muscles also responded to chemicals and could be stained by dyestuffs, as could other protein-type tissues. Muscles also responded to nondyeing chemicals by contracting and relaxing and obeying impulses from their innervation that could be affected by chemicals. One of these is curare (*ourari*, from *uria*, "bird," and *eor*, "kill"); it was used by South American Indians as an arrow poison on their hunting and tribal warfare expeditions (30) and had been brought to the attention of Western science by Humboldt in 1805. Claude Bernard (31) tested curare (a mixture of many alkaloids) as a neuromuscular blocking agent in a study of the muscular paralysis caused by this poison. The chemical properties of the separated curare alkaloids did not become known until almost a century later, but the relative ease of isolation of *d*-tubocurarine made this alkaloid more amenable to structural studies. *d*-Tubocurarine is a monoquaternary bis-isoquinoline base, too complicated structurally to assign a special significance to molecular segments, but an attention-getter because of its quaternary ammonium character. This ammonium grouping was suspected, although far from being proved, in 1868 when A. Crum-Brown and T. R. Fraser (32) in Edinburgh recognized the curariform neuromuscular blocking action of the methiodides of several other unrelated alkaloids, and especially of the simple quaternary ammonium ion, tetraethylammonium. They then tested other tetraalkylammonium ions for the same activity and observed graded increasing action as one ethyl group after another was replaced by methyl, tetramethylammonium being 125 times more active (i.e., more toxic) than tetraethylammonium. (For a summary of these researches, see ref. 33.) Thus, the two Scottish scientists established for the first time that there is a connection between chemical structure and biological activity. A similar observation was made by Richardson shortly afterwards, when the hypnotic (cell-depressant) activity of aliphatic alcohols was shown to be a function of their molecular weight (34). This pointed to the importance of physcial properties for biological manifestations.

To take advantage of such discoveries and convey medicine from the Middle Ages to modern concepts, many practical problems had to be solved almost in one swoop. This left little time for causative thinking or groping for explanations of physiological phenomena. The term "biochemistry" did not become established for another 40 years (35), and instrumentation to deal with submilligram quantities of reagents in a mixture of ill-defined biochemicals did not exist. Kühne invented the term "enzyme" (36) as a contribution to fermentation experiments, relating the old German *enzymos*, "ferment," to zymase (37).

1.2.2. The Concept of Drug Receptors

Curious chemists questioned the mode of action of nutrients and drugs; they are chemicals and as such must react with other chemicals in cells and tissues. Even a resulting biological response must be the final stage of chemical reactions. In 1878, Langley studied the antagonistic effects of atropine and pilocarpine and suggested that "there is some receptor substance with which both atropine and pilocarpine are capable of forming compounds . . . according to some law in which their relative mass and chemical affinity for the substance are factors" (38). The concept that the mode of action of drugs was probably subject to the law of mass action was developed further by Clark (39), who emphasized the idea of active enzyme sites with which substrates and drugs should interact.

At about the same time (1885), the German immunologist Paul Ehrlich began to advance theories of drug receptors (40). Ehrlich, who was to become the undisputed founder of modern medicinal science and of the overall methodology of medicinal chemistry, visualized receptors as cellular bodies equipped with "side chains" that could interact with substrates and drugs. He wrote:

> The protoplasm is equipped with certain atomic groups, whose special function consists of fixing to themselves certain food stuffs important to cellular life. In accord with the nomenclature of organic chemistry, these groups may be designated as side chains. As they have the function of attaching themselves to certain nutrients, the nutrients must have atomic structures themselves, every group uniting with the combining group of the side chain. The relationships of the corresponding groups of the two reactants must be specific. The groups must fit each other, e.g. as a male and female screw (Pasteur) or as a lock and key (Emil Fischer). Toxins possess a haptophore (binding) group corresponding to that of the nutrient. In addition to these haptophoric groups which effect the union of the toxins to the protoplasm, the toxins also possess a second "toxophoric" group which, as far as the cell goes, is not only useless but actually injurious. (41)

Ehrlich regarded the interaction of alkaloids (e.g., the antagonists atropine and pilocarpine) and their cell receptors as "loose and reversible." He felt such antagonists must have similar anchoring groups and that differences in their toxophoric groups accounted for their different physiological effects. Although he never stated this explicitly, Ehrlich compared the "side chains" with auxochrome groups in dyestuff molecules, even calling the complementary functions of drug molecules "pharmacophores." He thus equated cellular and molecular dimensions, making cellular portions react with drugs as would functional groups of a molecule. This was a confusion on a $1:10^6$ scale of molecular sizes; nevertheless there was an anticipation of modern ideas of receptor structure.

Hypothetical speculations, if not supported by convincing evidence, may side-track scientific thought from target problems rather than guide them to a solution of such problems. Innumerable attempts have been made to outguess the chemical properties of receptors but reliable data have not been furnished. An excellent compilation of current ideas concerning receptors may be found in three recent reviews (42). See also (42a).

The prevalent view of drug receptors is that they consist of a cascade of cell-membrane-bound lipoproteins and/or glycoproteins and/or cellular nucleic acids. There must be a primary macromolecular region to which the biologically active compound fits sterically; the disturbance caused by this primary act of ligation is then amplified by such agents as cyclic purinosyl phosphates; steroid, adrenergic, and other hormones; and numerous coordinated molecules and cofactors whose significance is only beginning to emerge. Since the picture of receptors thus remains too hazy to be of practical use, drug design based on receptor structure has remained wishful thinking. The situation has been presented in an article (43) whose introductory paragraphs are quoted here because of their aptness.*

> The methodology of drug design could be greatly improved if receptors and their mode of interaction with active substances were known in precise molecular detail. Such information could be used, for example, to design conformationally defined structures in which pharmacophoric groups are oriented in the proper spatial arrangement for optimal receptor interaction. Compounds with this three-dimensional complementarity to a receptor should show greater potency, exhibit higher specificity, and have fewer effects at other receptors than conformationally flexible structures. Aspects of this ideal of drug design have been approached in the synthesis of enzyme inhibitors starting from the solid-state structures of enzyme–substrate complexes obtained through x-ray crystallography and in the preparation of intercalating substances on the basis of the Watson–Crick model of DNA and the crystal structures of DNA model fragments.

---

* Reprinted with permission from *J. Med. Chem.*, **24**, 1026 (1981). Copyright 1981 American Chemical Society.

The molecular features of important pharmacologic receptors, however, are presently unknown beyond their pharmacological and biochemical classification and their description as membrane-bound proteins. For these systems, receptors must be characterized and differentiated not by their own architecture but by what is known about compounds exhibiting receptor activity. . . . Consequently, the existing models for the receptors are abstract and are far removed from our understanding of the chemical structure and dynamics of proteins and their mode of interaction with small molecules.

1.2.3. Organization of Drug Research

Ehrlich's contributions do not rest only on his theories of drug action, drug metabolism, and drug resistance and his breakthrough researches on antisyphilitic and antitrypanosomal agents—he also perceived and organized medicinal research (44) in a manner that has not changed materially to the present day. He realized the need for teamwork, assigning specialized efforts to experts in chemistry, biology, and medicine and coordinating these researches in an institutional setting. The great medicinal research institutes all over the world—whether supported by private foundations, government departments, universities, or the pharmaceutical industry—have been patterned on Ehrlich's rather modest establishment in Frankfurt. As biomedical doctrines grew in complexity, they had to be subdivided beyond the early aspirations, but the team effort, exemplified best in the industrial research setting, has remained the foundation of medicinal research.

Each branch of science rises and falls cyclically over the decades. Periods of intense activity are followed by quieter years in which stock is taken of successes and failures, scientific gains are consolidated, and the foundation is laid for renewed progress. Ehrlich's lifetime (1854–1915) coincided with a period of a general upturn in drug research, and although he led the way in chemotherapy—a term he coined in 1891 for the classical cure of infectious diseases (44, 45)—he was by no means the only leader in pharmacotherapy. First, the continuing isolation and purification of botanical alkaloids and other natural products provided pharmacologists with ample materials for biological testing. Then the invention of general anesthesia (46) and the concomitant revolution in surgery and obstetrics was an incentive to expand the use of antiseptics and topical antibacterial phenols that had been introduced by Lister (47). In all these cases organic chemists furnished a limited selection of compounds to experimental biologists, who then used the compounds for therapeutic purposes in the laboratory or clinic. Medicinal chemists entered the picture later by trying out molecular variants of the "lead" drugs with the hope of improving the actions of the original prototypes. This was the case, for example, with divinyl ether (48, 49) in the series of general anesthetics, and many phenolic compounds, di- and trihydric phenols, bis(hydroxyphenyl) alkanes, and so forth (50–52) among the topical antiseptics.

1.3. SYSTEMATIC DRUG DEVELOPMENT AND RATIONAL RESEARCH METHODS

The pattern of accidental or semiplanned drug discovery (53) followed by systematic variation of the original "lead" has been repeated time and again. Individual success stories of such research projects have greatly advanced therapeutic art and have given us many if not most of the drugs now at the disposal of the physician. Because of this decisive role of enriching our armamentarium of drugs and the occasional inspired thoughts that rose over the sameness of uninspired work procedures, the most representative stories of systematic drug development are told in the sections of this book devoted to drug types according to the diseases they help to manage. However, these repeat performances can teach us but so much. Therefore, as we look back at the intellectual awakening of medicinal science, only those moments are recalled that laid a firm foundation to thoughts and procedures that have survived the fashions of temporary therapeutic studies.

1.3.1. Pro-Drugs

Much has been made in recent years of pro-drugs as a means of administering medicinal compounds in the form of derivatives that can be converted metabolically to an active agent. This sequence is to provide for the transport of the disguised drug past metabolic barriers and get the active compound to the physiological vicinity of the tissues where its action is desired. In wide use in medicine today are both biochemical precursors (e.g. dopa for dopamine) and derivatives with protecting ester or acyl (amide) groups from which the active drug can be recovered metabolically.

The medicinal chemists who proudly announce such "drug design" usually overlook the history of these ideas. A century ago, the Swiss biochemist Nencki (54) wished to obtuse some of the side effects of salicylic acid by converting this compound to readily hydrolyzable esters; the phenyl ester called salol was introduced for this purpose. Thirteen years later, acetylsalicylic acid was synthesized by Felix Hoffman and introduced by Dreser as aspirin as an analgetic and antiinflammatory drug (55).

1.3.2. Dyestuffs

Nowadays we label bioreagents with ^3H, ^{14}C, ^{13}C, or fluorescent probes to locate them at components of cells and tissues and follow their fate. Before the advent of these more sophisticated labeling methods and the concomitant autoradiographic or spectroscopic read-out, the more conventional biological staining of tissues had to suffice. Some dyestuffs achieved selective staining and thereby gave rise to the hope that they might be selectively toxic to some cells. This opened the door to experiments in the chemotherapy of infectious diseases and to less corrosive antiseptics than the phenols.

Methylene blue (56, 57), gentian violet (58), and acriflavine (59) are examples of such dyestuffs. Other dyes were found to be trypanocidal, especially afridol violet (60), trypan red (61) and trypan blue (62). Some of these dyes were not as selective in staining tissues as could be hoped, and this prompted substitution of nonchromophoric groups for azo, quinone, and similar moieties. Replacement of these chromophores by amide and urea groups led to "colorless dyestuffs," some of which retained or even improved upon the cytostatic properties of the parent dyes. This was the road to suramin (63–65) in 1920. It took another decade before it was recognized that metabolic components, at least of certain azo dyes, could be the carriers of their antibacterial activity, as shown for sulfanilamide (66).

Another sequel to the assumption that selective adsorption of dyestuffs on cells was conducive to antimicrobial chemotherapy led to the introduction of toxic atoms into analogs of dyestuff molecules, particularly arsenic with its tradition as a systemic poison. The antitreponomal drug arsphenamine represents an example of this concept: the azo nitrogens of an azo dye were replaced by arsenic and gave medicine the first effective antisyphilitic agent (67, 68). Again, it turned out that the analogy of arsenoso to azo compounds was fortuitous, because metabolic oxidative cleavage of arsphenamine was needed for bioactivation. In due time the active oxidation product, oxophenarsine, took the place of earlier dyestuff-like drugs (69).

1.3.3. Minor Analgetics

Parallel with the emergence of some rationalizations in the design of chemotherapeutic agents, endocrinology and pharmacology began to follow more sensible patterns during the last two decades of the 19th century. The prerequisite for this development was the isolation of hormones and other biochemicals, and the increasing availability of synthetic organic chemicals for biological experimentation. Synthetic substances began to undergo screening in the relatively primitive pharmacological tests of that time. Any discovery of biological activity thus remained heuristic, but soon synthetically accessible analogs were prepared with the hope they might constitute improvements over the earliest prototype compounds. For example, the antipyretic action of acetanilide was discovered when a nurse erroneously dispensed this compound to a patient (70). The compound had to be abandoned because of its toxicity, and p-aminophenol was tried instead in the belief that aniline was oxidized to it in the body. Toxicity remained high, but loath to abandon the "lead," derivatives of p-aminophenol were prepared and tested. Of these, acetophenetidine (phenacetin, p-ethoxyacetanilide) was introduced (71) in 1887. Acetaminophen (p-hydroxyacetanilide), first prepared in 1878 (72), was not recognized as the major metabolite of acetanilide and phenacetin until much later, in 1949; it has become the standard analgetic and antipyretic in this series. This is an example of the roundabout fate of structurally very simple compounds whose biochemical fate had to wait for an explanation for many years. Besides acetaminophen, aspirin has maintained itself as an over-the-counter analgetic and antipyretic in spite of extended and intense efforts to find additional non-narcotic analgetics in virtually every structural class of compounds. The belated suggestion in 1971 that aspirin and some other nonsteroid antiinflammatory agents owe their activity to an inhibition of cyclooxygenase, which is needed in the biosynthesis of prostaglandins (PGF 2α), has reaffirmed the need for close coordination of medicinal chemistry with biochemistry.

1.3.4. Local Anesthetics

One of the problems occupying medicinal chemists at the turn of the century was the improvement of local anesthetics. The prototype alkaloid of such agents, cocaine, had been isolated in 1859 by Niemann (26) and by his professor, Wöhler (27), the founder of organic chemistry. It occurs to the extent of 0.6–1.8% in the leaves of the coca shrub, *Erythroxylon coca*, native to the high Andes Mountains. Its numbing action on the tongue (73) and the discovery of its local anesthetic action in other tissues (74, 75) led to its clinical introduction by Köller in 1884 (76) at the suggestion of his friend Sigmund Freud. Even at that early date chemists recognized the multiple pharmacological activities of cocaine and especially noted its troublesome CNS activity, which can lead to dependence liability and which Freud had great difficulty with in his practice. An account of the pharmacological and clinical history of cocaine may be found in "Goodman and Gilman" (77).

The difficulties in procuring coca leaves in Europe in the 1890's and the hope that other compounds might segregate the CNS, local anesthetic, and vasoconstrictor activities encountered in cocaine led to a search for other, more specific, local anesthetics. Two alkyl esters of aminobenzoic acids were the first screened (by self-experimentation on the tongue, and by anesthesia of the rabbit cornea) out of a large number of diverse compounds tried. They were ethyl p-aminobenzoate (benzocaine) (78–80) and orthocaine (methyl 3-amino-4-hydroxybenzoate, Orthoform) (81, 82). These compounds are not highly soluble and are therefore used only as topical anesthetics and not in underlying tissues. Einhorn, a pioneer medicinal chemist, was struck by the aromatic ester groups found in these agents as well as in cocaine. When the molecule of cocaine was stripped of its functional groups and the tropane ring was simplified by ring opening, the significance of the carbomethoxyl group, the tropine ester, and the bicyclic moiety for the local anesthetic activity could be assessed. From these experiments resulted several dialkylaminoalkyl esters containing piperidine and other monocyclic systems such as the eucaines (83), amylocaine (called stovaine, an English wordplay on the surname of its inventor, Ernest Fourneau, which in French means "stove") (84), and procaine (85). In procaine the p-amino group was introduced in analogy to the same function in benzocaine. Although procaine does not provide the desirable vasoconstrictor properties of cocaine, it is devoid of the unwanted CNS activities of the coca alkaloid. It became the prototype of local anesthetics for two generations of surgeons and dentists, in spite of its brief duration of action. Its ester group is hydrolyzed too readily by serum cholinesterase, and the components, diethylaminoethanol and *p*-aminobenzoic acid, are inactive as anesthetics. In spite of this drawback, thousands of ester analogs have been synthesized in the industry and by candidates for Master of Science in Chemistry degrees. About two dozen analogs became clinical drugs in the competitive and profitable field of local anesthetics.

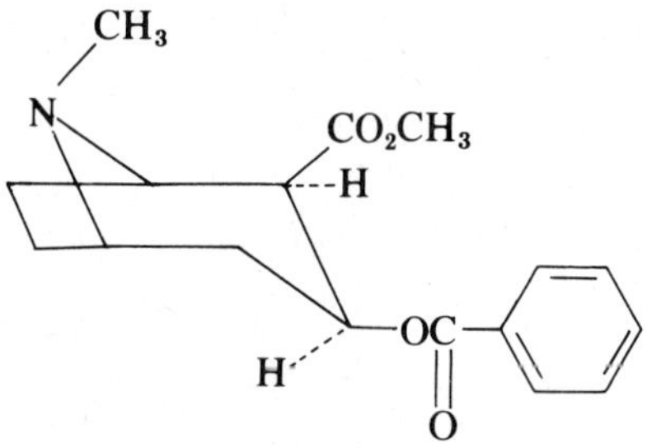

Cocaine

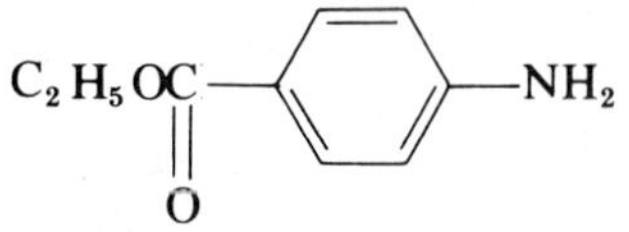

Benzocaine

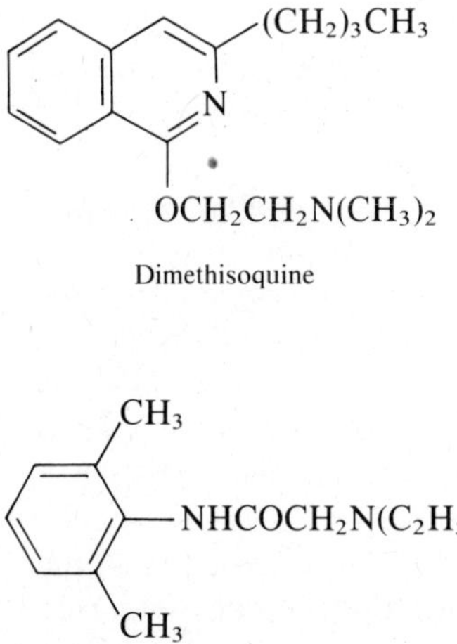

Dimethisoquine

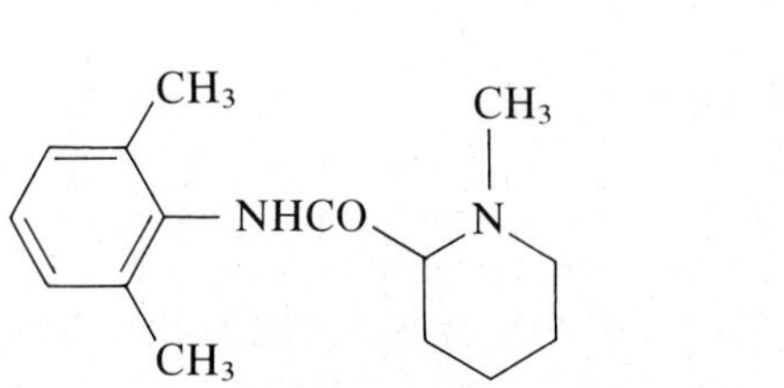

Procaine

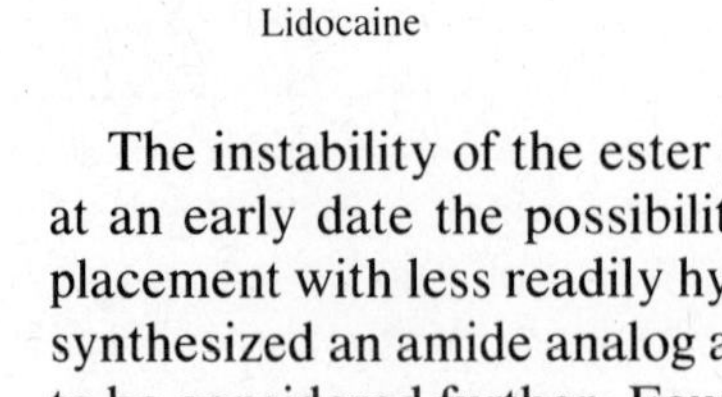

Lidocaine

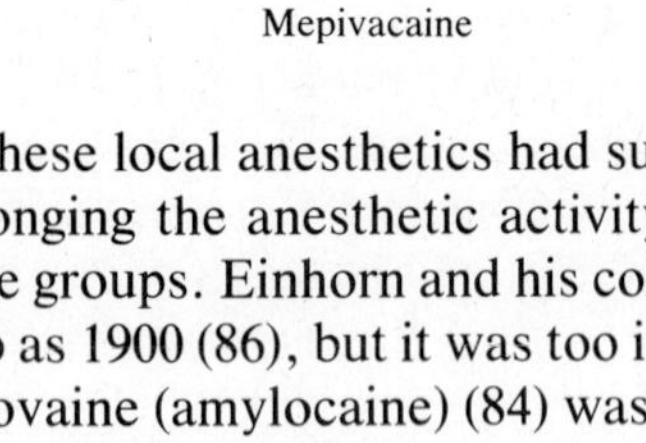

Mepivacaine

The instability of the ester group of these local anesthetics had suggested at an early date the possibility of prolonging the anesthetic activity by replacement with less readily hydrolyzable groups. Einhorn and his colleagues synthesized an amide analog as long ago as 1900 (86), but it was too irritating to be considered further. Fourneau's stovaine (amylocaine) (84) was heavily

SYSTEMATIC DRUG DEVELOPMENT AND RATIONAL RESEARCH METHODS

branched and therefore less hydrolyzable. In the mid-1930's interest in these compounds was revived, but from a different angle. Von Euler and Erdtman worked on the determination of the position of the side chain of gramine [3-dimethylaminomethylindole] and in the course of this study synthesized its 2-isomer (87). An intermediate in this synthesis was 2-dimethylamino-2-acetotoluidide, which in a chance test produced numbing on the tongue. Erdtman and his student N. Löfgren expanded this "lead" and synthesized other anilides for tests as local anesthetics (88). The drug lidocaine was the crowning result of this effort (89, 90). It contains two sterically hindering *ortho* methyl groups that protect the anilide linkage from premature hydrolytic cleavage. Of drugs of this kind, lidocaine (91) and mepivacaine (92) have remained the most widely used local anesthetics, with prolonged duration of action. The principle of prolonging action by designing hard-to-hydrolyze functions that connect the basic side chain was advanced further in dimethisoquine (93), which contains an ether group at that point. However, one such structural change, even though highly conducive to hydrolytic stability, does not guarantee pharmacological or toxicological advantages.

1.3.5. Biogenic Amines

The history of the adrenergic hormones and analogous synthetic amines contains an account of logical events made possible by the ever-increasing sensitivity of analytical chemical methods. A pressor effect exerted by adrenal extracts was observed in 1895 by Oliver and Schäfer (94) and the apparent principle that caused this effect was called epinephrine by Abel (95). The material turned out to be an amine that gave crystalline salts, and this facilitated isolation (96–98) and purification in three widely separated laboratories, one of them in the industry where the compound was trade-named Adrenaline. Its relatively simple molecule lent itself to rapid structural determination (99), and this was followed almost immediately (in 1903) by total synthesis of the racemate by Stolz (100) and by Dakin (101). Stolz's colleague, F. Flächer, completed the task by optical resolution (102) and by identifying one of the enantiomers with the natural hormone. Since epinephrine is a secondary methylamine, the homologous primary amine, norepinephrine, or Noradrenaline (*nor* means "no R"), was also synthesized from a common intermediate (103). Not much attention was paid to this synthetic compound until Cannon suggested its occurrence in the natural adrenal hormonal conglomerate (104). Another 25 years passed before norepinephrine was isolated from nervous tissue (105–107) and recognized as a neurotransmitter (104, 108). This sequence of events is not without parallel. Acetylcholine (ACh) was synthesized in 1867 (109) but its pharmacology had to wait until the 20th century (110, 111) and the recognition of its role as a neurotransmitter until 1921 (112, 113). Similarly, histamine was synthesized (114) a dozen years before its pharmacology was first investigated (115) and even longer before its occurrence in animal tissues was discovered (116).

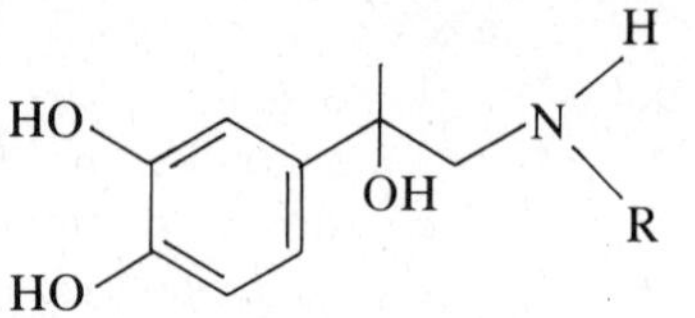

Norepinephrine: R = H
Epinephrine: R = CH_3
Isoproterenol: R = $CH(CH_3)_2$

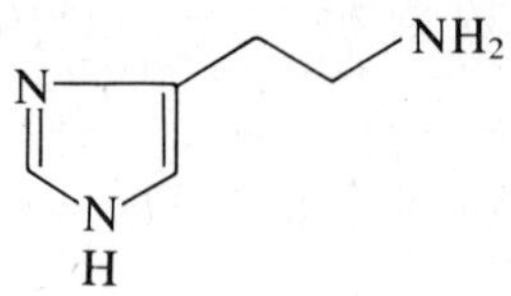

Histamine

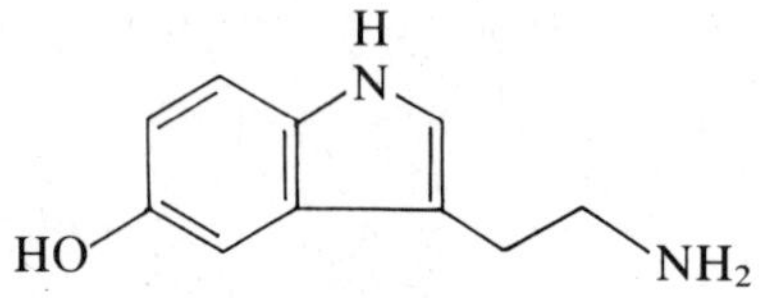

Serotonin
(5-HT)

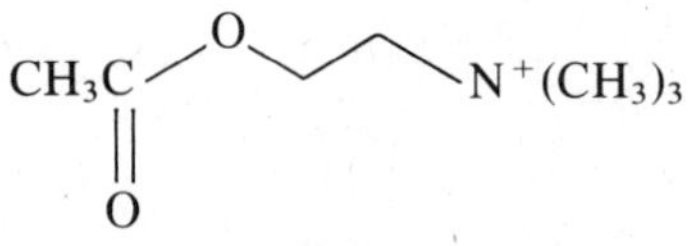

Acetylcholine
(ACh)

All these compounds contain the grouping —C—C—N connected to an aromatic or quasi-aromatic heterocyclic ring (in ACh an electron-attracting acetoxy group). Barger and Dale (117), in a classical study of structure-activity relationships (SAR), summarized these connections for naturally occurring biogenic amines as well as synthetic analogs. Among similar amines and alkamines that fall into this pattern are (i) the ephedrine alkaloids from *Ephedra sinica* (118), whose pharmacological study (119) uncovered CNS stimulatory as well as vasopressor activities, and (ii) amphetamine, synthesized in 1887 (120), in which such CNS properties predominate. When medicinal chemists gathered the courage to change the natural *N*-methyl group of epinephrine, a series of new SAR was unearthed, culminating in the *N*-isopropyl homolog isoproterenol (121) with maximal α-adrenergic properties.

1.3.6. Structure–Activity Relationships

A similar act of changing the traditional *N*-methyl group of alkaloids led earlier in the series of morphine alkaloids to *N*-alkylnorcodeine derivatives, especially *N*-allylnorcodeine (122), which antagonized morphine-induced respiratory depression. It became the prototype for *N*-allylnormorphine (nalorphine) (123) and naloxone (17-allyl-4,5α-epoxy-3,14-dihydroxymorphinan-6-one) (124, 125), powerful narcotic antagonists.

It is possible that these terminally unsaturated derivatives act as suicide enzyme inhibitors in the biosynthesis of endorphins. Unexpectedly, other alkyls such as *N*-phenethylnormorphine and its analogs are more potent analgetics (126) than the natural *N*-methyl alkaloids.

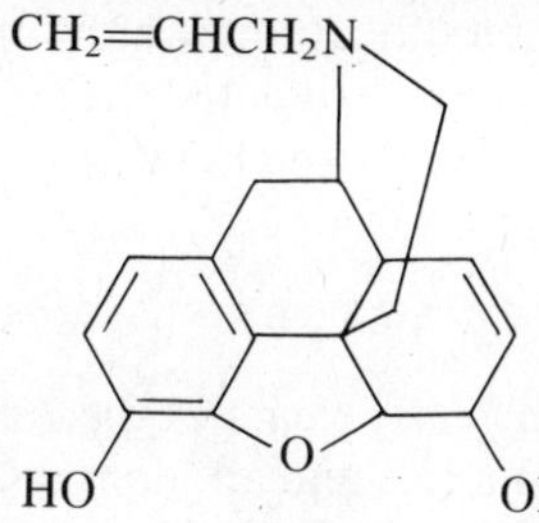

SYSTEMATIC DRUG DEVELOPMENT AND RATIONAL RESEARCH METHODS

Nalorphine

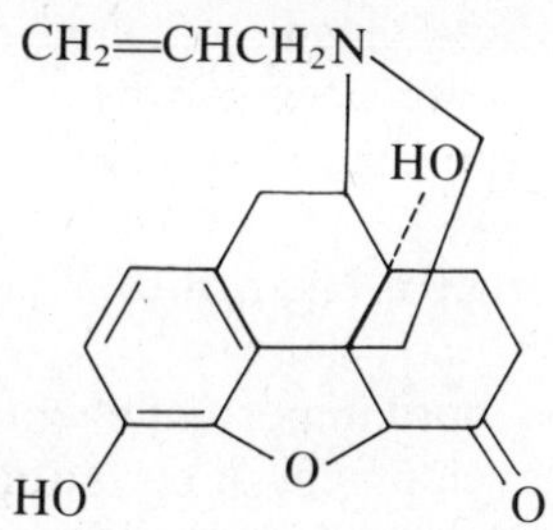

Naloxone

The picture of SAR as painted so far appears somewhat disorganized. In a series of structurally closely analogous compounds, some regularities seem to be traceable but there are always so many exceptions that one cannot rely on the apparent regularities. In some cases, the medicinal chemist is to blame for not thinking a problem through. For example, it is widely accepted that substitution of an aromatic ring by chlorine in the molecule of a biologically active compound "deepens" activity. This has been explained by invoking the Hammet substituent constant of chlorine and the push–pull electron-density changes brought about by the substituent. It may be, however, that the chlorine atom blocks aromatic hydroxylation and thereby prevents metabolic removal of the active species. Although such speculations anticipate more contemporary ideas of drug action, the problems existed 85 years ago and seemed to resist solution. In 1893 the French physiologist Charles Richet tried to formulate SAR mathematically but this procedure had to wait another six years when Hans Horst Meyer (127) and E. Overton (128) published studies in which narcotics were ranked according to ratios of their solubility in blood and lipids (i.e., water and organic solvents). These theories of a relationship of the physical properties of a compound and its biological actions were taken up by others and revived decisively by Corwin Hansch in 1964 (129); they are reviewed in Section 2.10. The lipid theory was followed by ideas concerning other physical properties such as ionization (130, 131) and a proposal by Linus Pauling (132) that narcotics interfere with clathrate formation of ice crystals in the CNS and thereby prevent normal nervous transmission (of pain impulses).

These proposals, whether right or wrong, were but rare attempts to rationalize an increasingly unruly accumulation of therapeutic data that defied explanation. But, driven by the need for new drugs, research had to go on with or without rationalization on the molecular level. The period from 1890 to 1920 saw the elucidation of the structure of many additional natural products, some of them with useful or at least pharmacologically interesting activities. It also led to those great chemotherapeutic and pharmacotherapeutic discoveries described on the preceding pages. Other drug discoveries, made by chance, did not add to knowledge in medicinal chemistry, but each new empirical "lead" initiated molecular modification, which in turn terminated in a few needed drugs. A chronologically tabulated account of year-to-year happenings may be found in a recent treatise (133).

1.3.7. Sedative Hypnotics

One of the more important events during the early years of the 20th century was the development of barbiturates as hypnotics and anticonvulsants. Linear halogenated ureides had been used as sedative–hypnotics, following the recognition of the anesthetic (i.e. CNS depressant) properties of halocarbons such as chloroform. Cyclic ureides of the barbiturate type were in line to be tried next, especially if they contained two alkyl substituents in position 5. A precedent for studying cyclic ureides could be seen in paraldehyde (134), a powerful though habit-forming hypnotic with an oxygen-heterocyclic structure. Since the starting materials for 5,5-disubstituted barbiturates were diethyl 2,2-dialkylmalonates, and since it was by far easier to substitute malonic ester with two like than unlike alkyl groups, the earliest dialkylbarbiturates were symmetrically substituted. A typical example was barbital (135). A new synthesis of aromatically 5-substituted 5-alkylbarbiturates made possible the preparation of phenobarbital (136, 137) which, in addition to its hypnotic activity, was found to be an excellent anticonvulsant (138).

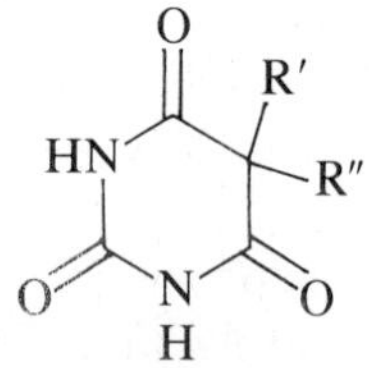

Barbital: $R' = R'' = C_2H_5$
Phenobarbital: $R' = C_2H_5, R'' = C_6H_5$
Amobarbital: $R' = C_2H_5, R'' = (CH_3)_2CHCH_2CH_2$
Secobarbital: $R' = CH_2=CHCH_2, R'' = CH_3(CH_2)_2CH(CH_3)-$

A method of stepwise substitution of diethyl malonate then opened the floodgates to other nonsymmetrical barbiturates (139), such as amobarbital and secobarbital, which are superior to symmetrical derivatives in many pharmacological ways. This is an early instance in which clever organic chemistry made possible medicinal design and inexpensive drug production.

1.3.8. Vitamins and Hormones

Two major classes of biological cofactors made their appearance from 1910 on: the various endocrine hormones and the vitamins. The term "vitamine" was coined by Casimir Funk in 1911 (140). Hormones (from ορμάω, *hormao*, "I move"), compounds secreted by endocrine glands, nerves, and other tissues, travel to different anatomical targets where they transfer biochemical information by participating in reactions of the cell nucleus or cytoplasm and probably at drug receptors. Although some of their functions overlap with those of vitamins and some vitamins behave like hormones in animal species that can synthesize them, there are differences between the two classes of biochemicals, especially from the point of view of medicinal chemistry. Many hormones occur in multiple forms with naturally modified structures and minor differences in their biological mission. Their structures can be modified, often quite considerably, in the chemical laboratory with retention of various facets of their activity spectrum. By contrast, vitamins are usually unique and appear to constitute enzyme cofactors designed evolutionally to optimize certain biochemical functions. In a few cases, closely related forms can be interconverted or interchanged, for example, pyridoxamine, pyridoxine, and pyridoxal; the A, K, and D vitamins; thiamine and codecarboxylase; and niacin and niacinamide. But in most cases only one form is suitable as an enzymic cofactor. The unnatural retinoic acid, which can promote the growth of vitamin A-deficient animals but lacks other vitamin A functions, is of value in the treatment of psoriasis. Other analogs of the vitamins with unnatural functional structures are of no medicinal interest except that some analogs of thiamine, niacinamide, and other B vitamins have shown antagonistic biochemical activities. For all practical pur-

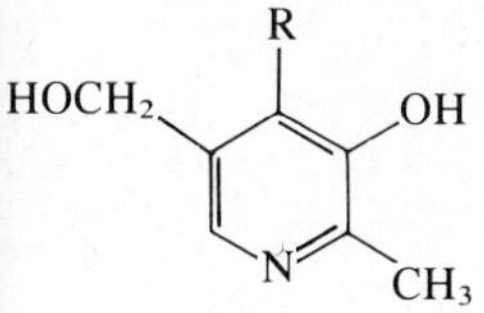

Pyridoxine: R = CH$_2$OH
Pyridoxamine: R = CH$_2$NH$_2$
Pyridoxal: R = CHO

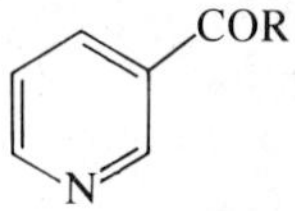

Niacin: R = OH
Niacinamide: R = NH$_2$

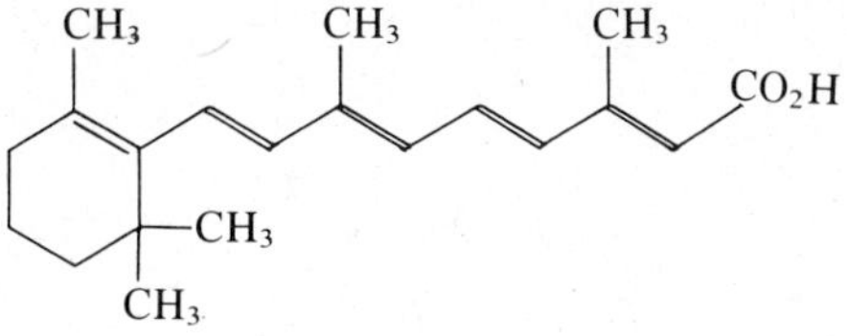

Retinoic Acid

poses, vitamins are nutritionally essential natural products whose synthesis and commercial manufacture have been developed with cost-reduction in mind. They have revolutionized human and animal health care. The history of the structural proof of vitamins and their syntheses is brilliantly studded with Nobel price awards to organic chemists and biochemists. The hormones, which are less unique physiologically and can be subjected to extensive structural modification, have been a fertile field of medicinal investigation. Many synthetic analogs are more potent, specific, and pharmacologically acceptable than the natural hormones.

The isolation and structural proof of various hormones and their eventual synthesis paralleled the complexity of their chemistry. The early hormones yielded their structural secrets to methods that most organic chemists now would scorn. Separations without chromatography, often requiring the concentration of ten-thousands of liters of biogenic liquids or the extraction of equally many glands from slaughterhouses, preceded manual crystallizations and elementary analysis without the benefit of spectroscopic methods. Because at least 10 to 50 mg of hormone were needed for these tasks, hormones that occur in extremely small amounts had to wait until instrumental refinements for their isolation became available. Some structural classes, such as the steroids in the 1940's and polypeptide hormones in the 1950's, were surrounded by an aura of technical inaccessibility. One was supposed to have been a steroid or protein chemist from way back before one could dare to tackle structural and synthetic problems in these series. All this changed when the peculiarities of these structures became routine knowledge. The often inexplicable reactivities of axial and equatorial steroids yielded to conformational analysis, and the tiresome and awkward stepwise synthesis of polypeptides was speeded up immeasurably by Merrifield's automated solid-phase procedures.

The thyroid hormones became amenable to structural proof (141–143) and synthesis (144) in the 1920's. This opened the way to preparing and testing congeners (145–148). Antithyroid drugs (149, 150) were first found

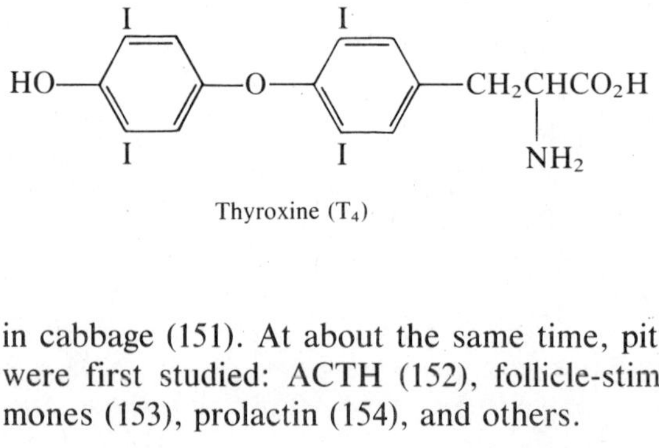

Thyroxine (T_4)

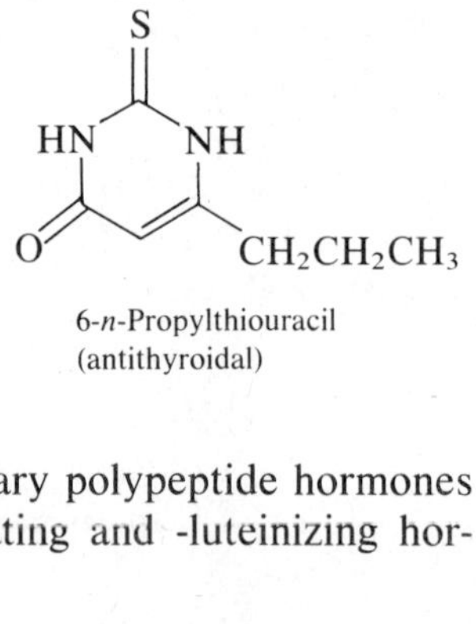

6-n-Propylthiouracil
(antithyroidal)

in cabbage (151). At about the same time, pituitary polypeptide hormones were first studied: ACTH (152), follicle-stimulating and -luteinizing hormones (153), prolactin (154), and others.

1.4. THE CROWDED YEARS OF INNOVATION AND EXPANSION

The years 1928–1962 have been called the golden age of medicinal science. Even World War II could not interrupt the exuberant developments in medicinal chemistry and in the experimental biochemical and biomedical sciences, with a concomitant revolutionary upturn in all branches of medicine. Although the period will prove not to have been unique, it certainly constituted an unprecedented emergence of new methods and new drugs. Medicinal chemistry contributed to these developments by furnishing hundreds of thousands of these compounds for biological evaluation. The most active areas were the antimalarials, antituberculous agents, sulfonamide drugs, antihistaminics, antipsychotic drugs, centrally acting analgetics, antihypertensive agents, oral antidiabetics, steroids, and antibiotics. Towards the end of

THE CROWDED YEARS OF INNOVATION AND EXPANSION

the golden-age period, nonsteroidal antiinflammatory agents, antidepressants, antianxiety agents, semisynthetic penicillins, and the first potent anticancer drugs became the leaders in research efforts. Details of some of these bewildering numbers of discoveries and expansions are furnished in the chapters devoted to individual drug types. Here only the principal innovations of medicinal-chemical thinking, planning, and performance are summarized.

The golden age of medicinal chemistry was also characterized by the unprecedented growth of the pharmaceutical industry. Many firms that had only a handful of medicinal chemists in their laboratories increased their numbers by a factor of 10 or even 50 during this period. Diversification of fields of interest became possible; companies that had specialized in pharmacotherapeutics took up chemotherapeutic research, and vice versa. In all phases of chemical drug investigation, one clinically applicable discovery and development almost drove the next one off the status of highest current interest. One important drug after another came off the ever-lengthening assembly line, and both the industry and the public daydreamed that this would go on forever.

1.4.1. Steroid Hormones

Wherever extraction of plant or animal tissues gave biologically active materials, the typical methodology and modus operandi of the natural-products chemist came into play. This occurred with steroid hormones of which estrogens (155, 156), progesterone (157–160), and androsterone (161–163) were

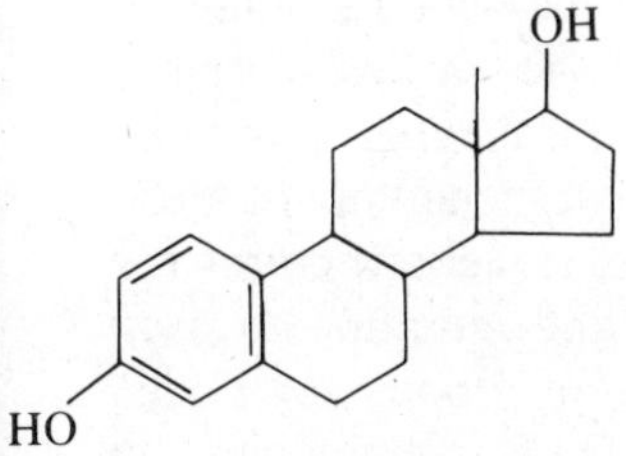

Estradiol 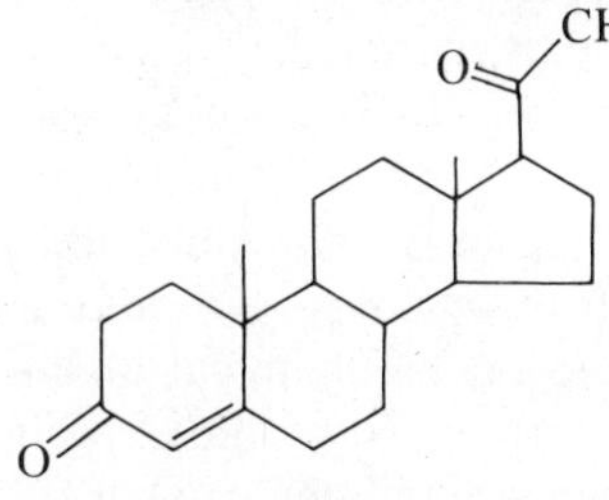   Progesterone 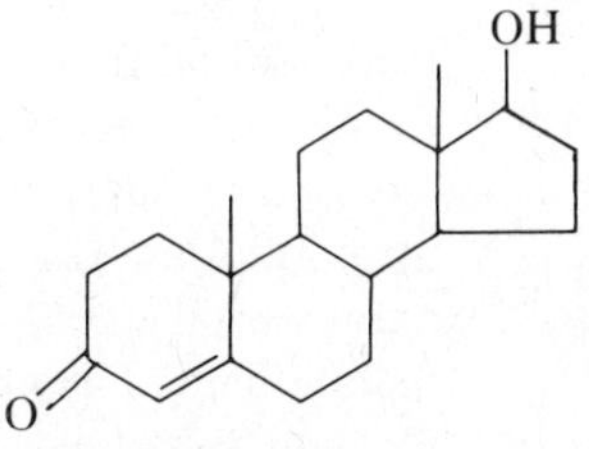   Testosterone

the first. There was fierce competition for priority between four or five laboratories. Isolation and concentration from enormous amounts of biological materials, structure elucidation based on a few milligrams of hormone, and ultimately partial synthesis (164–168) made the active substances accessible for further work in these fascinating areas. Now, after 45 years, it is difficult to reconstruct the awe and wonder with which these events were greeted. The greatest excitement was still to come in the chemistry (169–171) and total synthesis of the corticosteroid hormones (172, 173).

1.4.2. Penicillin, Sulfanilamide, and Other Antimicrobial Agents

The late 1920's witnessed the discovery of penicillin by Alexander Fleming (174) but the full impact of this observation was not realized until ten years later when chemical studies, chemical manufacturing, and biomedical applications of antibiotics got under way. Moreover, Fleming's descriptions

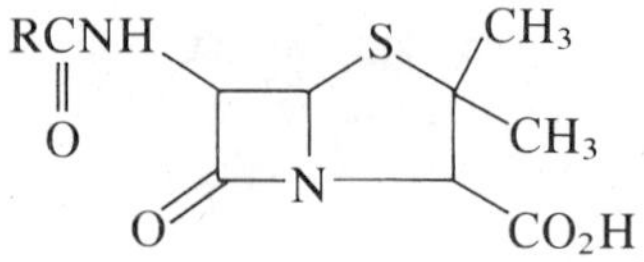

Penicillins
(general formula)

were overshadowed by the development of the bacteriostatic sulfonamide azodyestuffs (175). This work represented the last application of visible biological staining to chemotherapy. Both the working hypothesis and the medicinal use of these dyestuffs were obliterated when sulfanilamide was found to be the "activated" drug metabolite that exerted the bacteriostatic activity (176). Molecular modification of sulfanilamide resulted in over 10,000 reported compounds and untold thousands of additional analogs that lie buried in the files of pharmaceutical manufacturers. From this vast number, a few dozen reached clinical utility and carried civilians as well as combat troups through World War II and the dangers of gram-positive bacterial infections. The war years also reactivated the search for antimalarials after the Japanese occupied Malaysia and other quinine-producing areas. An estimated 500 American chemists were engaged in the wide-ranging molecular modification of the cinchona alkaloids and the earlier German, French, and Russian "lead" drugs. Additional structural types were discovered by screening (177). In Britain, Curd and Rose branched out of these traditional approaches to new drugs (178) and devised novel structures for antimalarial agents based on biochemical analogies and attention to drug metabolism. They were influenced by the experiments of Woods and Fildes (5), which had established sulfanilamide as an antagonist to *p*-aminobenzoic acid, a reagent needed in the biosynthesis of the essential growth factor, folic (pteroylglutamic) acid. Their theory of regarding drugs as metabolite

p-$H_2NC_6H_4CO_2H$
  p-Aminobenzoic
  Acid

p-$H_2NC_6H_4SO_2NH_2$
  Sulfanilamide

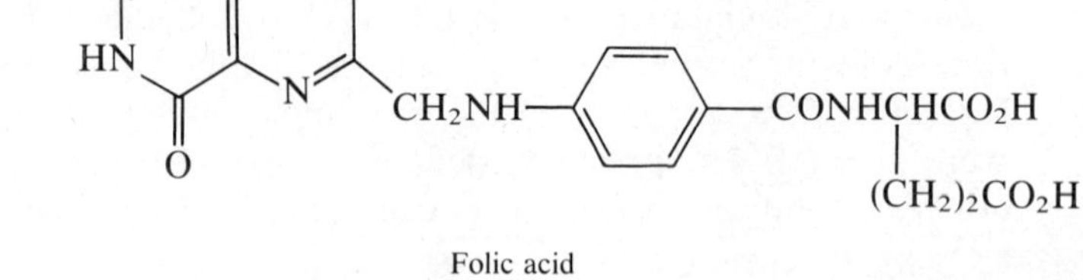

Folic acid

antagonists had a profound influence on drug design and gave medicinal chemistry the reasoned biochemical background that elevated it from a hit-and-miss art to a defendable science. Even though a biochemical relationship between a drug and a metabolite has often been established only post-factum—as is usually the case with antibiotics—guidelines for drug design have been enriched greatly by this line of thought.

On the heels of antimalarial chemotherapy, attention turned to antituberculous agents. The first discovery in this area was streptomycin (179), but with the exception of reducing its aldehyde group to CH_2OH in dihydrostreptomycin, medicinal chemistry did not contribute concretely to the SAR of this antibiotic. The first logical discovery of a tuberculostat based on the metabolite antagonist theory was that of p-aminosalicylic acid (180). It combined the observation of the involvement of salicylic acid in the oxidative metabolism of virulent mycobacteria (181) with that of the antibacterial action of some substituted p-aminobenzoic acids. At about the same time, niacinamide was reported to be antituberculous in large doses (182) whereas niacin itself was inactive. Thus the tuberculostatic effect was not related to vitamin activity. Of the many analogs of niacinamide tested, pyrazinamide (183) and ethionamide (2-ethylthioisonicotinamide) (184–186) have remained accepted antituberculous agents.

The hope that such biochemical analogies might unearth further antituberculous drugs was disappointed by the discovery of isoniazid, one of the most active and pharmacologically acceptable agents. This drug was found by one of the most lowbrow serendipitous procedures, namely as a synthetic intermediate in the course of the preparation of another compound on which hopes of attaining activity had been pinned (187–189). The fact that three separate laboratories arrived at isoniazid by the same experimental mistake does not improve the history of this drug.

$$H_2N-\langle\rangle-SO_2-\langle\rangle-NH_2$$

Dapsone

Sulfanilamide derivatives had shown considerable although not clinically useful antimycobacterial properties but their potential was fulfilled only when a benzene ring was interspersed between the SO_2 and NH_2 portions of the sulfonamide group. The resulting vinylogue of sulfanilamide, dapsone, suppressed rat leprosy (190) and has become the agent of choice in human leprosy (191).

The societal effects of the antituberculous drugs were seen dramatically in the fate of the many tuberculosis hospitals that dotted health resorts in the first part of the 20th century. A few years after the introduction of these drugs, especially isoniazid, many of these hospitals closed their doors because patients had become ambulatory and were cured by the new agents.

A similar decline in patient population took place a little later in the overcrowded mental health facilities when psychoactive drugs released many schizophrenic and endogenously depressed patients from the constraints of these asylums. The alleviation of these two health problems contributed much to the public's appreciation of medicinal science.

1.4.3. Bioisosterism

Many of the drug discoveries already described and some that will be mentioned later were accidental or serendipitous observations that did not teach improvements in strategic design procedures. Three problems confront the medicinal chemist who wants to follow a "lead." One is, to discover what molecular features of the lead compound are involved in causing the biological activity; two, to find the relationship between the compound and a biochemical metabolite, enzymic substrate, or perhaps even an enzyme that could explain its activity. The third problem is to delineate those molecular segments or stereochemical conditions that can be altered gradually while retaining the biological activity or altering it in a desirable direction. These questions are addressed by medicinal chemists and biochemists. If the biological activity does not show optimally expected limits and the compound's toxic manifestations are unacceptable, molecular modification offers a great incentive to improve its activity and toxicity profiles. Guidelines for such molecular changes are provided by the rules of bioisosteric replacements.

Isosterism originally meant, according to Langmuir (2), a close similarity or even likeness in the electron arrangements, shapes, and sizes of related atoms that could explain the analogies or actual identity of physical constants of molecules formed by such atoms (N_2 vs. CO, etc.). Erlenmeyer applied this concept to biologically active molecules (1). It has been reviewed (192). The ground rules proposed by Erlenmeyer were too restrictive; two compounds were termed isosteric only if they formed mixed crystals. In most cases identical boiling points and other physical characteristics were postulated that were difficult to attain for compounds with similar biological activity. The researches on the sulfanilamide derivatives and antihistaminics in the early 1940's were guided by the concept of isosteric replacement (193). Later, Friedman lifted the restrictions in the definition of isosterism by introducing the term "bioisosterism" for compounds that "fit the broadest definition for isosteres and have the same kind of biological activity." This includes agonists as well as antagonists with activity in the same test system. A survey of bioisosteric groups, both classical (194) and nonclassical (195), offered medicinal chemists a wider choice of groups and functions that could be exchanged, with a good prognosis that the compounds containing these groups would exhibit similar or antagonistic biological activity. Application of these expanded rules greatly lessened the chance that the compounds designed on this basis would have no biological activity or only unrelated activity. The present state of bioisosterism has been reviewed (196); this review includes some new and interesting cases and references.

The first applications of classical isosterism, before it was even defined, may be found in ring equivalents (3) as they are encountered in the pairs benzene and thiophene or the pairs pyridine and thiazole. When sulfapyridine was found curative in pneumococcus infections but fairly toxic clinically, isosteric replacement of pyridine by thiazole, pyrimidine, and so on, led to a number of useful sulfonamide bacteriostats. Similarly, the series of histamine-1 receptor antagonists (antihistaminics) and anticholinergic agents (antispasmodics) was developed from early blocking structures by replacing benzene by thiophene, >CH— by >N—, —CH$_2$ by O or S, and so on. The sulfur atom of the phenothiazine ring system of neuroleptic drugs was replaced by —CH=CH— or —CH$_2$CH$_2$—, leading to the azepine ring drugs that opened up the field of tricyclic antidepressants. Of these, the first example was imipramine with a standard >N(CH$_2$)$_3$NMe$_2$ side chain. By classical isosteric exchange of >N— with >C=, amitriptyline with the side chain >C=CH(CH$_2$)$_2$NMe$_2$ was obtained. The success of bioisosterism in molecular modification has become so well established that most medicinal chemists now apply it to their problems routinely. It has become so common that novelty of invention has been questioned in at least two cases of patent court decisions when a new drug was based on the structure of an existing agent and the "novelty" claimed consisted only of one bioisosteric replacement. In drug design where so much depends ultimately on luck and intuition, the predictive rules of bioisosterism, though by no means infallible, have been a welcome guiding light. Recently, Ganellin and his colleagues based the design of histamine-2-receptor antagonists on bioisosteric considerations (197). Two of their candidate drugs, derived from 4-methylhistamine, contained a side chain terminating in a thiourea group, —NHC(=S)NHR, but

Cimetidine

they had to be abandoned after extensive clinical trials because of severe side effects. These side effects are seen frequently in thiourea derivatives, and therefore a bioisosteric substitute for the thiourea group was sought. It was found in a direct comparison of thiourea itself with electronegatively substituted guanidines such as H$_2$NC(=NNO$_2$)NH$_2$ or H$_2$NC(=NCN)NH$_2$. Cyanoguanidine itself was found to be a true isostere of thiourea in several physical respects, and the —NHC(=NCN)NHR group was therefore introduced into the side chain of the H$_2$ receptor antagonist structure. This process yielded cimetidine, the prototype antisecretory and antiulcer agent. The most important lesson to be learned from this success story is that proper chemical and biochemical thinking can solve problems in drug design without the tangible input of pharmacology.

1.4.4. Pharmacodynamic Drugs

Hand in hand with the rapid development of chemotherapeutic agents during the 1930's and 1940's, outstanding progress was made in pharmacodynamic agents. Antihistaminic drugs (H_1 receptor antagonists) were introduced in France (198–199) and the United States (200–202) and then tried in the treatment of psychoses (203), motor agitation (204), and as adjuncts to anesthesia (205). Anticonvulsant drugs were investigated extensively, led on by diphenylhydantoin (206), a drug that had been synthesized in 1908 (207). The first totally synthetic potent morphine-like analgetic, meperidine (pethidine),

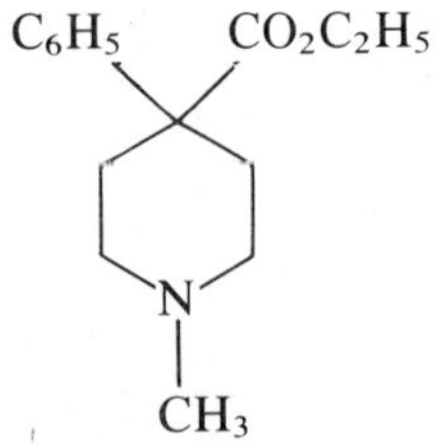

Meperidine
(Pethidine)

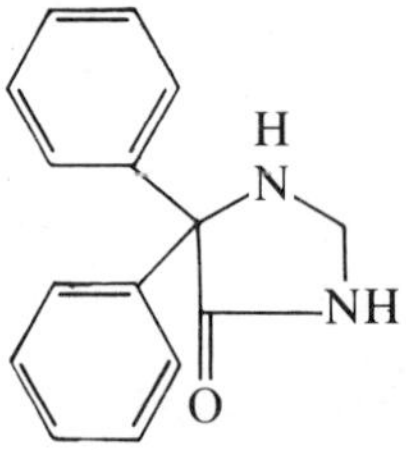

Phenytoin
(Diphenylhydantoin)

became known (208, 209). It would be nice to be able to report that it was designed by dissection and analogy of the morphine molecule, but that was not so; it was to be a "reversed" antispasmodic structure, and its analgetic properties were observed during pharmacological workup (4). N-Allylnormorphine was recognized as a morphine antagonist in a series of other N-alkylnormorphine derivatives (123); its methyl ether, N-allylnorcodeine, had already been known 28 years earlier to antagonize morphine-induced respiratory depression (122).

1.4.5. Researches During World War II

The years after 1943 witnessed many diverse important drug discoveries, for example, that of the hallucinogen LSD (210) and the wartime crash-program completion of antimalarial research that yielded not only traditional aminoquinolines (211–216) but also biochemically based pyrimidines (217–

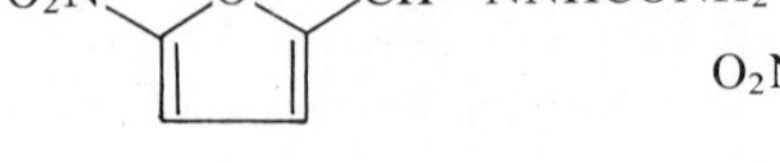

Nitrofurazone

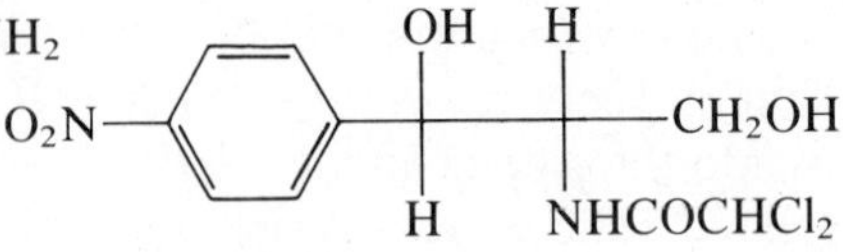

Chloramphenicol

219). Antibiotics research forged ahead with streptomycin (179) and bacitracin (220). The useful antimicrobial activity of nitrofurfural derivatives was recognized (221), although mental reservations against the much maligned nitro compounds were not allayed; even when the first broad-spectrum antibiotic was found in chloramphenicol (222, 223), its nitro group was blamed for the manageable bone marrow depression caused by this compound.

At the end of World War II, most of the wartime research was declassified and published. The *Journal of the American Chemical Society* had to add two extra volumes to its usual schedule to accommodate five years of pent-up antimalarial work alone. At the same time, antituberculous thiosemicarbazones, although quite toxic, made therapy in war-ravaged Europe possible as a stopgap (224), soon to be replaced by *p*-aminosalicylic acid (180) and other more effective and safer agents.

1.4.6. Early Antitumor Agents

Another WWII discovery was the antileukemic action of the simplest nitrogen mustard, HN2. This arose from clinical pharmacological observations based on a battlefield accident. A ship loaded with mustard gas (sulfur mustard) was bombed in an Italian invasion harbor and the chemical was spread on the waves. Military personnel rescued from the ocean showed an unusually low white-blood-cell count. The leukopenic action of mustard gas had been recorded in the literature (225) previously. This suggested to Gilman and Philips (226) an application in leukemia where leukocytes multiply in an uncontrolled fashion. The high toxicity of sulfur mustard made such trials impossible and therefore a less toxic "alkylating agent" was sought and found in HN2 [$(ClCH_2CH_2)_2NCH_3$]. Other molecular modifications fol-

$S(CH_2CH_2Cl)_2$ $H_3CN(CH_2CH_2Cl)_2$
Mustard gas   Mechlorethamine (HN2)

lowed; they are mentioned in the section on antitumor agents. The replacement of S by N in these compounds was in line with the thought processes that led to the synthesis of dimercaprol as a ligating agent for use in metal poisoning (227).

The scientific study of cancer began about 1915 with the discovery that certain chemicals administered to rodents resulted in the appearance of specific tumors. The alkylating agents were the first to point a way to cancer chemotherapy. From the beginning it was clear that the ill-defined metabolic differences between malignant and communal cells would raise the greatest obstacles to attaining reasonable specificity of any therapeutic agent in this field. This held especially for compounds that might block the biosynthesis of nuclear monomers containing the purine and pyrimidine bases of nucleic acids. When the participation of tetrahydrofolic acid in the one-carbon transfer in this biosynthesis became known, an incentive was given to alter its structure and to let such putative metabolic antagonists interfere with purine biosynthesis. Aminopterin and methotrexate were the first such compounds (228, 229). A decade later, 5-fluorouracil initiated the large number of structural analogs of purines and pyrimidines with antineoplastic properties (230); some of them inhibit enzymes needed in the biosynthesis of the bases (231, 232), others are incorporated into faulty nucleotides and prevent their utilization in cellular multiplication.

1.4.7. Muscle Relaxants

Also before 1950, the first muscle relaxant other than curare, called mephenesin, was found by pharmacological observation (233). A more acceptable version of such drugs, with antianxiety properties as well, was developed by extensive molecular modifications. The drug, meprobamate

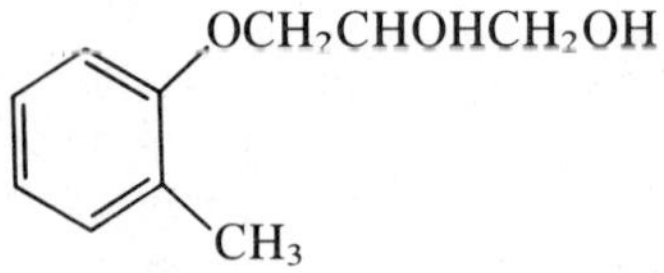

$$\text{Mephenesin}$$

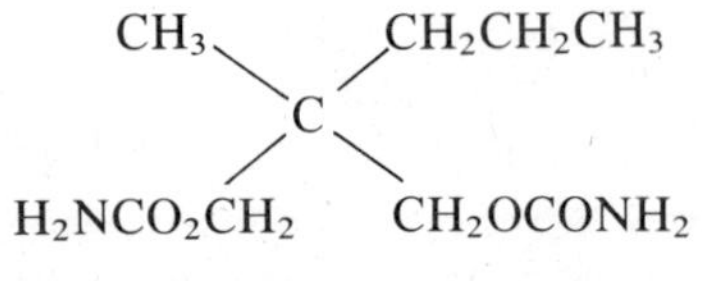

$$\text{Meprobamate}$$

(234, 235), became the pharmacological prototype, with a characteristic test pattern by which newer antianxiety agents could be recognized.

1.4.8. Vasopressor and Vasodepressor Drugs

A breakthrough in the field of vasopressor and vasodepressor agents, of potential value in regulating blood pressure, occurred with the publication by Ahlquist in which two different types of adrenoreceptors, α and β, were postulated (236). Agonists and antagonists to these pharmacologically distinguishable receptors soon made their appearance, but the early examples lacked specificity. Adrenoreceptors are not easily separated from other autonomic receptors functionally as well as anatomically, and drugs which, for example, block nervous transmission at ganglia usually produce cholinergic blockade in cholinergic and adrenergic ganglia although in unequal measure. Many ganglionic blocking agents are quaternary onium ions, mostly quaternary ammonium ions. They can reduce hypertension. Paton described how he discovered the prototype of these agents, hexamethonium, while making a systematic search for a competitive antagonist to decamethonium, which he hoped to develop as a muscle relaxant for anesthesia. Instead, hexamethonium turned out to be the first effective drug for essential hypertension (237). Other ganglionic blocking agents did not require qua-

THE CROWDED YEARS OF INNOVATION AND EXPANSION

ternary ammonium groups so long as they contained large sterically hindering moieties, for example, mecamylamine (238) and chlorisondamine (239, 240).

$(CH_3)_3N^+(CH_2)_6N^+(CH_3)_3$

Hexamethonium

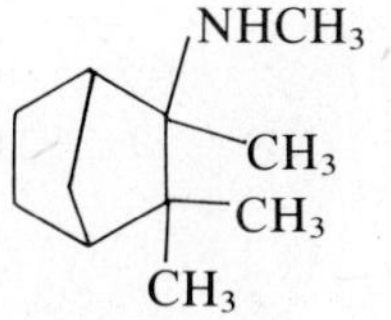

Mecamylamine

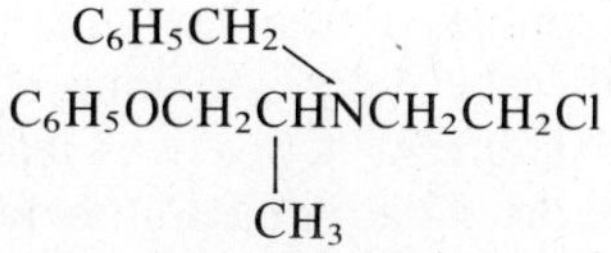

$C_6H_5CH_2$
$C_6H_5OCH_2CHNCH_2CH_2Cl$
$|$
CH_3

Phenoxybenzamine

The use of these agents, and with it medicinal–synthetic interest in this field, has declined since other types of antihypertensive drugs—the main use of the hexamethylene bisquaternary agents—have become available. However, succinylcholine has remained a valuable cholinergic blocking and depolarizing agent for relaxation during anesthesia. It is now difficult to realize that these compounds were at the center of interest of competitive medicinal research around 1950.

One thinks of alkylating agents primarily as antitumor agents that tie together two segments of nucleic acids and thereby inhibit separation of the double helix. However, alkylating agents—such as phenoxybenzamine, with one chloroethylamino group that can cyclyze to an ethylenimmonium ion—block α-adrenoceptors and act as peripheral vasodilators and antihypertensive agents (241). The active species responsible for α-adrenergic blockade is postulated to be a carbonium ion formed by opening of the aziridinium ring (242). The prototype of these agents was Dibenamine [N,N-dibenzyl-N-(β-chloroethyl)amine], which was investigated by Nickerson and Goodman (243) as an antileukemic nitrogen mustard. The interest of these pharmacologists in autonomic phenomena led to the incidental tests of Dibenamine as a potential adrenergic blocking agent. Medicinal chemistry entered the picture during molecular modification to breed toxicity out of Dibenamine. The phenoxy group of phenoxybenzamine suggested itself based on the similar arrangement in choline ethers.

1.4.9. Abstinence Drugs

The potent, orally active analgetic methadone had been synthesized before 1943 (244), but the clinically most interesting pharmacological property of the (−)-enantiomer was not recognized until later (245). Methadone is useful in the treatment of narcotic abstinence syndromes in heroin abuse by substituting for morphine derivatives. Similarly, in chronic alcoholism, disul-

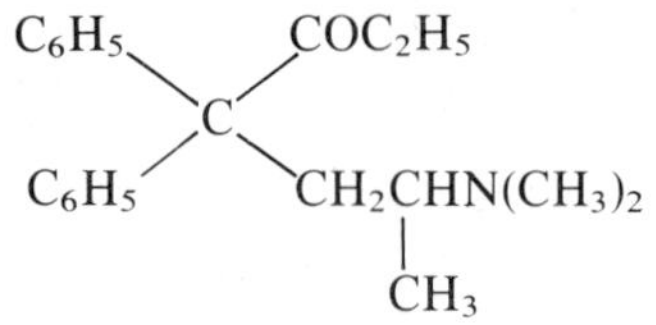

$$\underset{C_6H_5}{\overset{C_6H_5}{>}}C\underset{CH_2CHN(CH_3)_2}{\overset{COC_2H_5}{<}}$$
$$\qquad\qquad\qquad |$$
$$\qquad\qquad\qquad CH_3$$
Methadone

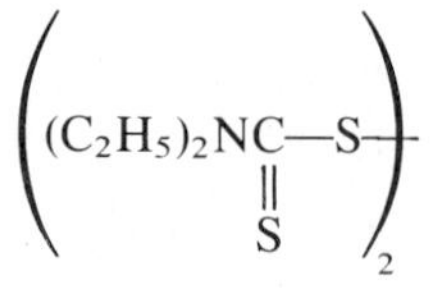

$$\left((C_2H_5)_2N\underset{\underset{S}{\|}}{C}-S-\right)_2$$
Disulfiram

firam, like its lower homolog tetramethylthiuram, causes hypersensitivity to ethanol (246) and thereby deters drinking of alcoholic beverages (247). The discovery of this deterrent action was made by two Danish pharmacologists who took the drug—a reputed anthelmintic and industrial antioxidant—and became ill at a cocktail party. They recognized that disulfiram had altered their response to alcohol. A similar sensitization occurs by industrial exposure to cyanamide (248) or ingestion of animal charcoal (249) or the fungus *Coprinus atramentarius* (250).

1.4.10. Drugs for Mental Disorders

Antihistaminics had been seen in France to alter psychotic episodes in patients treated for allergies (203) and motor agitation (204), and in the case of promethazine to potentiate clinical anesthesia (205). Molecular modification of promethazine, especially lengthening the side chain to C_3NR_2, emphasized the effects on anesthesia and artificial hibernation (252) but soon revealed another property, antipsychotic (neuroleptic) activity (253). The apparently optimal structure for this purpose was recognized in chlorpromazine (254). Simultaneously, the antihypertensive rauwolfia alkaloid reserpine was found to deplete stores of biogenic amines (255), regarded as necessary for the maintenance of "normal" mental and emotional performance; thus reserpine was tried and used in psychotic patients. In both series, extensive molecular modification probed structure–activity relationships, more so in the phenothiazine field because of synthetic obstacles for indole alkaloids. Every kind of related ring system, substituent, and other pertinent structural change was studied, and several drugs with more antiemetic or modified neuroleptic activity progressed to clinical use.

A second group of antipsychotic agents arose from a Belgian screening program that had a reasonable foundation of CNS-active structures. These compounds belong in a series of butyrophenones (256) and are represented clinically by haloperidol and other derivatives. Again, medicinal chemistry contributed a very large number of analogs and variations.

This was also true for muscle relaxant and antianxiety agents (234, 235). The scoop in this field came unexpectedly when dozens of laboratories screened assorted compounds for a meprobamate-like profile. Among these

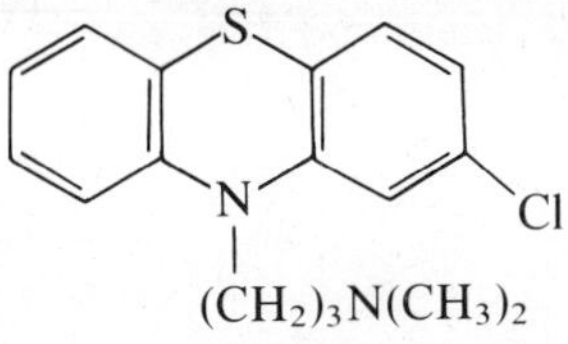

Chlorpromazine

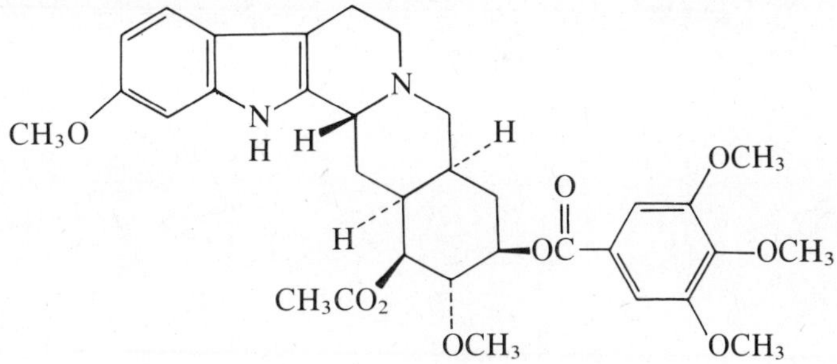

Reserpine

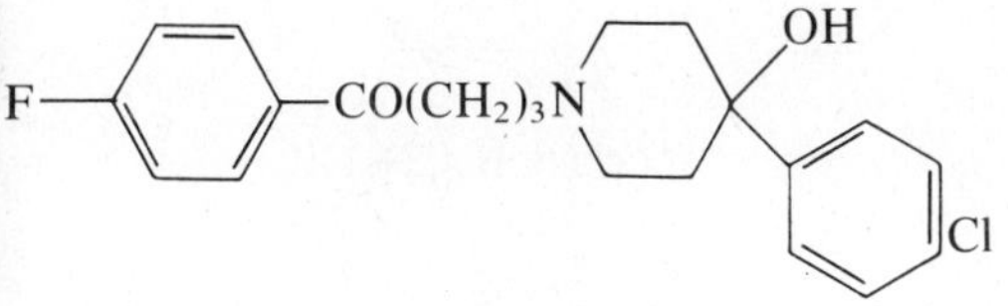

Haloperidol

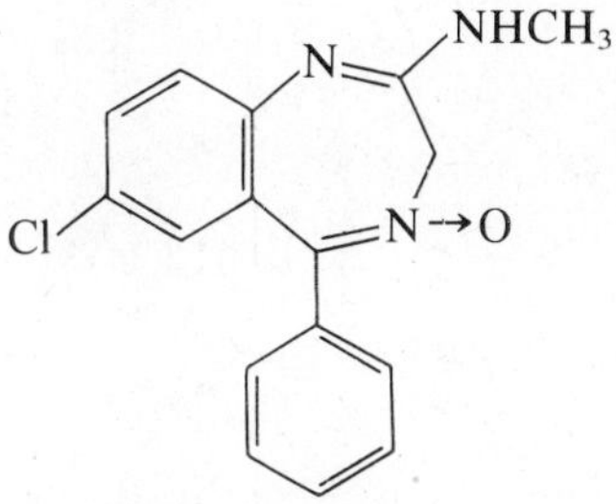

Chlordiazepoxide

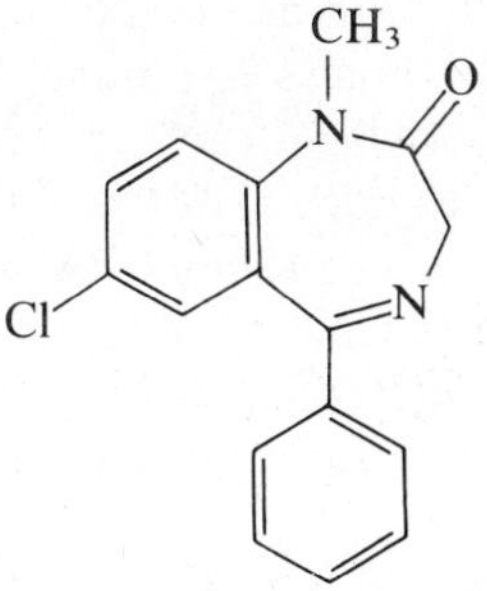

Diazepam

were some benzodiazepines that responded beautifully in such tests (257). At that time their structure was not even understood, and chemical studies had to catch up with pharmacological development (258). The first drug in this series, chlordiazepoxide, was followed by diazepam (259), which for many years became the most widely prescribed of all medicinal agents. In all this, little if any medicinal–chemical rationale was involved until the synthesis of a large number of structural analogs permitted standard deductions to be drawn about structure–activity relationships and bioisosteric replacements (260).

Endogenous depressions are believed to be due to low concentrations of various biogenic amines and other modulators of nervous transmission at neuronal receptors. This may be caused by enzymatic destruction of such amines or by the inability of these compounds to reach the receptors if they are taken up by the cells and stored there. Two types of drugs have been

Iproniazid

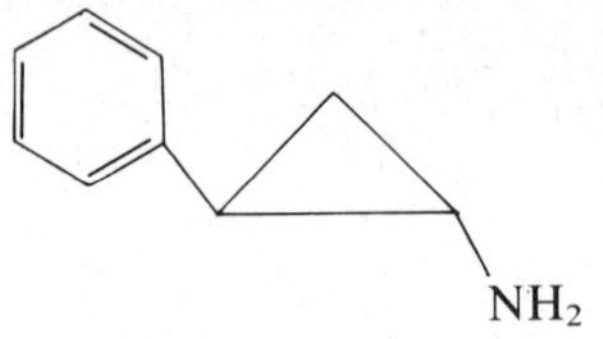

Tranylcypromine

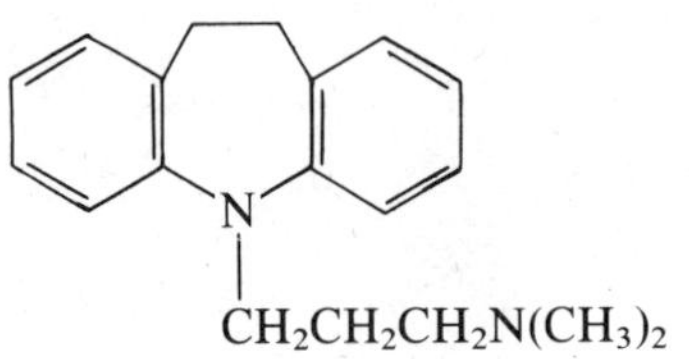

Imipramine
CH₂CH₂CH₂N(CH₃)₂

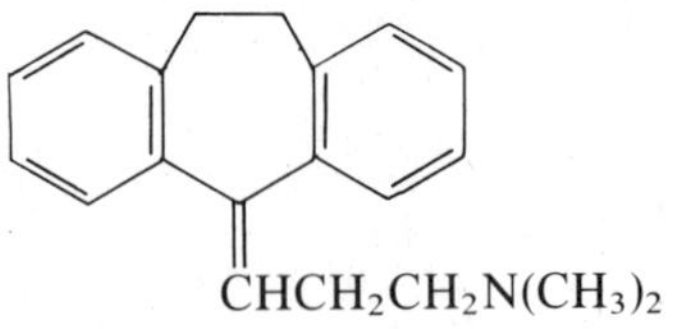

Amitriptyline
CHCH₂CH₂N(CH₃)₂

found to prevent these processes. Inhibitors of the amine-degrading enzyme, monoamine oxidase (MAO), prevent deamination and thereby maintain the concentration of the biogenic amines. A number of hydrazine derivatives led by (the now abandoned) iproniazid (261), as well as cyclopropylamines (tranylcypromine) (262, 263), are both MAO inhibitors and antidepressants. The amine uptake inhibitors are represented by imipramine (264), synthesized as an isostere of a phenothiazine neuroleptic (265) but found clinically and serendipitously to be an antidepressant. Other "tricyclic" antidepressants followed suit, for example, an isostere of imipramine called amitriptyline (266, 267).

In manic disorders, lithium ion, long known as a uricosuric agent, was found to control the manic phase of manic depression (251).

2
RECENT ACTIVE RESEARCH AREAS

The accelerating progress of drug development reached its full stride around 1950. The industry renovated its laboratories and staffed them more fully; many colleges of pharmacy added or expanded their graduate departments, and led by the National Institutes of Health the great governmental and private research organizations had taken the leading role in medicinal science, unmatched anywhere. Public financial support for these institutions had multiplied manyfold, both for their intramural investigations and for extramural support of medicinal and allied studies in the universities. As a result of these events, a bewildering array of new drug developments took place in many different study areas. It is not the purpose of this book to relate these events as they occurred year by year; this has been done in tabular form (133). Even in this introduction to the ensuing sections it is necessary to group topics according to study areas rather than historic dates in order to retain the continuity within a given disease-oriented direction of research. As an example, the highlights of work on a given area of drugs over the past 30 years are described, and then the discussion returns to about 1950 to pick up a given pharmacodynamic area, and so forth. It is as if the reader were walking slowly through an art museum, looking at Dutch masters in one room and at Italian paintings of the same period in another room rather than viewing the pictures according to the year they were finished and regardless of the tradition and environment that affected their creation.

As in art, where romanticism, impressionism, cubism, expressionism, and other fashions rise and fall, therapeutic research has changed its emphasis and explored different areas in the course of time, returning to a dormant problem when new fundamental knowledge makes possible a new attack upon it. Furthermore, the proclamation of national health goals (abolishment of stroke, heart disease, cancer), whether realistically attainable or not, has often poured research funds into particular areas—and cut them off when tangible results are not realized within 5 or 10 years. Similar conditions prevail in the pharmaceutical industry. The most broadly conceived research program is abandoned if no new and profitable drug arises from it within a given time. During the past 35 years several waves of interest have crested in the biomedical–therapeutic area. Four of these fields have held the center of attention almost continuously. They are work on drugs for hypertension, arthritis, cancer, and infectious disease. Although some forms of hypertension can be controlled—though not cured—with drugs, much remains to be done in this highly active and competitive area. The reason for the low success rate is the lack of knowledge of the biochemical causes of different forms of hypertension. This holds even more for arthritic diseases and most types of malignancies where every cause, from viral etiology to disturbances of the immune system, has been proposed. In the case of infectious diseases, control by antibiotics is reliable in many cases but the constant emergence of resistant strains of pathogens sparks an indefinite research effort to find additional ("third-generation") antiinfectious agents.

A similar condition pertains to drugs for the treatment of psychiatric disorders, where ignorance of their etiology and the lack of curative agents has begun to slow down the once overwhelming research effort. Other major areas that have made less progress in recent years are antihyperglycemic and antiviral agents. These relative slumps in productivity have prompted many investigators to try their hand at veterinary antiparasitic and similar diseases, where drugs might promise major economic and scientific if not humanitarian returns.

2.1. ANTIHYPERTENSIVE AGENTS

Hypertension, or pathologically elevated systemic arterial blood pressure, is a cardiovascular disease that occurs in 23 million Americans alone. It may be a precursor of atherosclerosis that can lead to coronary heart disease or stroke. Antihypertensive drugs have made an inroad on the mortality figures due to these conditions but the pharmacological etiology of various forms of hypertension is so complicated that definite rates of reduction of risks are not yet available. These etiological difficulties have also impeded the development of pertinent animal model tests for specific causes of hypertension. Thus the veratrum alkaloids, extracted as long ago as 1855 (268) and recognized for their antihypertensive properties four years later (269), have become unimportant because of toxic side actions and lack of specificity. The same fate befell cholinergic agents, whose vasodepressor activity has been known since the turn of the century (110). Ergot (270) was recognized as an α-adrenoceptor blocking agent in 1906 (271). Vasodilators should reduce the blood pressure (272) but their action is largely peripheral, while at least some important causes of hypertension lie in the CNS. Another working hypothesis is that rennin, an enzyme (273) from kidney extracts (274), liberates angiotensin (275) and inhibition of angiotensin-releasing enzyme counteracts hypertension. To the medicinal chemist these pharma-

ANTIHYPERTENSIVE AGENTS

cological opinions have suggested analog synthesis in a variety of structural systems ranging from alkaloidal analogs to small peptides.

An important milestone in the therapy of hypertension was the introduction of the rauwolfia alkaloids, reserpine and congeners. Not only did reserpine offer an effective treatment but its success, although beset with depressant CNS side effects, revived interest in natural-products chemistry, especially in alkaloids and drugs containing an indole nucleus. Used hundreds of years ago as a plant with varied therapeutic effects (276), the powdered root of *Rauwolfia serpentina* was worked up pharmacologically (277) and clinically (278, 279). Extraction and separation of about 60 alkaloids gave pharmacologists purer materials for testing, and fractionation led to the isolation of a weak base, reserpine, which produced most of the antihypertensive and sedative properties of the alkaloidal mixture (280).

At almost the same time, the isolation (281), characterization of vasoconstrictor properties (282), and synthesis (283) of 5-hydroxytryptamine (serotonin) (284) raised the possibility of an antagonistic interplay of reserpine and serotonin and intensified the synthesis and biological evaluation of thousands of indole compounds as potential antihypertensive and CNS-active substances.

In Switzerland, where chemical and pharmacological work on reserpine originated, other materials were also screened for effects on elevated blood pressure. Among structural types singled out were hydrazinophthalazines,

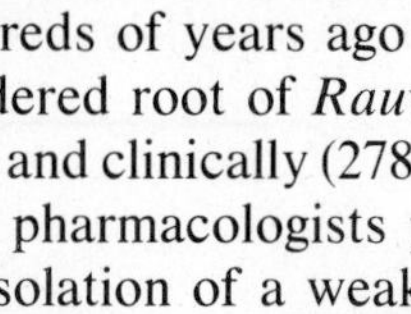

Hydralazine

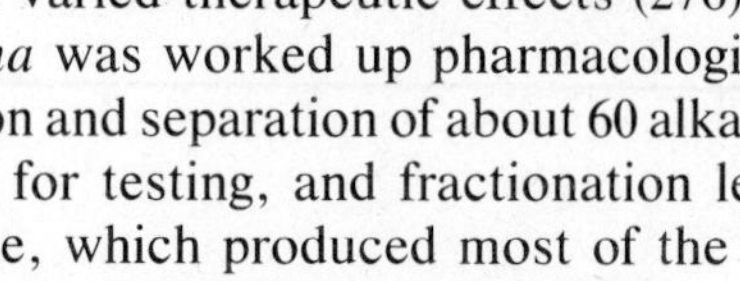

Methyldopa

of which hydralazine (285, 286) was chosen as a clinical antihypertensive agent (287). (It should be noted that the different modes of action that interested pharmacologists in research on hypertension are essentially disregarded in this survey of events.)

In the case of another antihypertensive drug, L-α-methyldopa, its mechanism of action was misunderstood to begin with. Perhaps due to delays in publication, the antihypertensive action of methyldopa was announced (288) before the synthesis (289, 290) of the compound. Yet the synthesis had been undertaken on the theoretical basis that the sterically slightly hindered α-methyldopa would inhibit aromatic amino acid decarboxylase and thereby, two steps down the line, slow down the biosynthesis of the hypertensive norepinephrine and epinephrine. What had not been anticipated was that methyldopa itself was also a (poor) substrate of the enzyme dopa decarboxylase and was partly decarboxylated to α-methyldopamine. This amine then falls prey to the other enzymes in the biogenetic route to epinephrine, probably being hydroxylated to α-methylnorepinephrine, which then competes with norepinephrine at the receptor site. Thus, like other dopa decarboxylase inhibitors, methyldopa has apparently no intrinsic effect on the blood pressure but must first be transformed metabolically to an antagonist of a biogenic amine (291).

The remaining antihypertensives have had an irrational history from a biochemical point of view, although this has not detracted from their clinical importance. Thus, chlorothiazide was designed logically as a diuretic based on the diuretic side effect of sulfanilamide (292) but its antihypertensive activity was discovered clinically (293). Several other diuretics, such as fu-

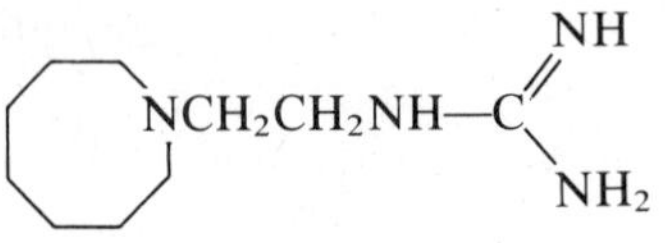

Guanethidine

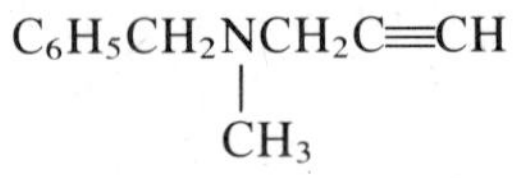

Pargyline

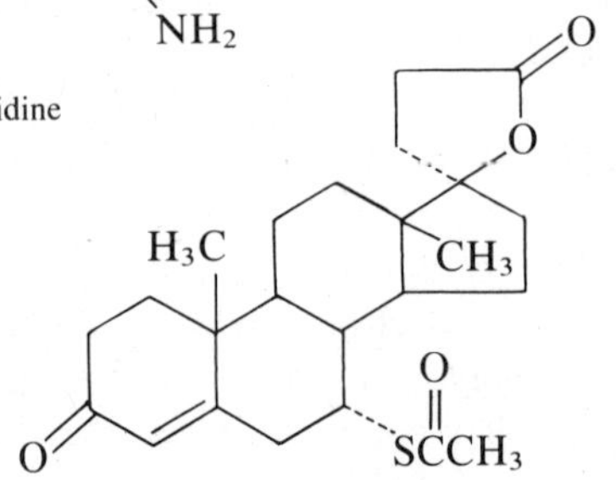

Spironolactone

rosemide (294), followed the same path. Spironolactone (295) antagonizes mineralocorticoids such as aldosterone and was studied first for the treatment of various edemas. Its antihypertensive properties were discovered by Wilkins (296). Guanethidine was patterned on the activity of a related "lead" amidoxime, (3-hexahydro-1-azepinyl)propionamidoxime (297); it produces ganglionic blockade, prevents release of norepinephrine from nerve terminals, and also depletes tissue stores of this neurohormone (298). Pargyline is the only MAO inhibitor marketed as an antihypertensive drug (299, 300); it was prepared as a potential antidepressant agent (301). In other words, the pharmacological evaluation of these drugs, which included the commercially ever-attractive effect on hypertension, triggered the main interest in these agents.

Working with adrenolytic agents, the ability of a dichloro analog of isoproterenol (poorly named dichloroisoproterenol or DCI, since the catechol moiety is replaced by 3,4-dichlorophenyl) was observed to block the myocardial stimulant and vasodilator properties of catecholic adrenergic amines (302). These effects are interpreted as a blockade of β-adrenoceptors. Unexpectedly, DCI turned out to be carcinogenic. Undaunted, chemists undertook extensive molecular modification. Many other compounds share the β-receptor blocking action, which suggested clinical utility in angina pec-

ANTIHYPERTENSIVE AGENTS

toris, cardiac arrhythmias, essential hypertension, and some other diseases. During the molecular modification of the "lead" compound, phenethanolamines and phenoxypropanolamines were studied. The latter are hybrids of typical adrenergic amines and choline ether analogs, with —OCH_2— sand-

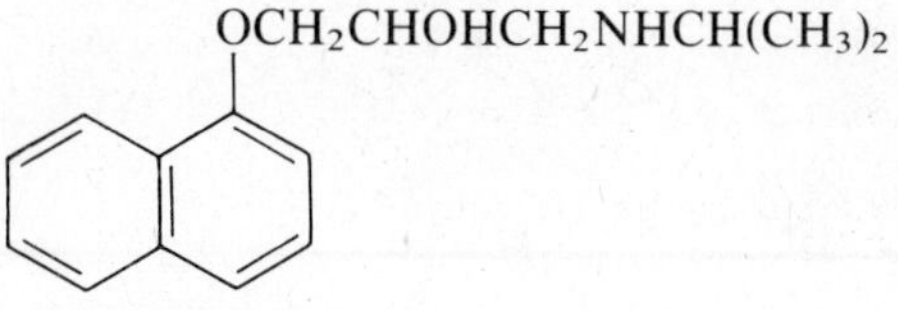

$OCH_2CHOHCH_2NHCH(CH_3)_2$

Propranolol

$OCH_2CHOHCH_2NHC(CH_3)_3$

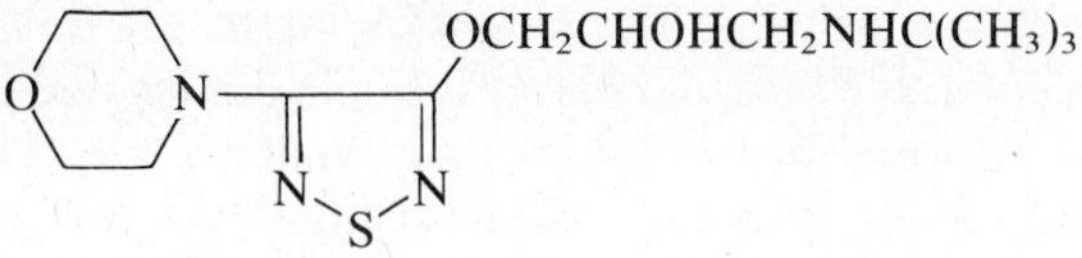

Timolol

wiched between the aromatic and ethanolamine groups. The clinically most successful drugs, chosen here only for their pharmacological acceptability, were a secondary naphthyloxy propanolamine compound, propranolol (303, 304), and sotalol (305). The latter agent contains one acidic CH_3SO_2NH group in place of the two catecholic hydroxyls of isoproterenol. It blocks β_1 adrenoreceptors a little more strongly than β_2. Timolol (306) is the newest entry with more highly claimed β_1 blocking specificity.

The latest contenders for a role in antihypertensive therapy are a group of small peptides that inhibit angiotensin-converting enzyme (ACE, kininase II). This exopeptidase catalyzes the removal of a carboxy-terminal dipeptide group from polypeptide substrates. For example, inhibition of ACE prevents the formation of the hypertensive, antinatriuretic angiotensin II and the simultaneous inhibition of the degradation of the hypotensive natriuretic bradykinin. Such inhibitors of ACE were discovered in the venom of the pit viper *Bothrops jararaca* (307), which consisted of at least 15 bradykinin-potentiating peptides. The most potent of these is BPP_{5a} (<Glu-Lys-Trp-Ala-Pro) but its duration of action *in vivo* was too short since it is also a substrate of ACE. A nonapeptide called teprotide (5-oxoPro-Trp-Pro-Arg-Pro-Gln-Ile-Pro-Pro) is a competitive inhibitor of ACE. The active site of ACE appeared to be similar to that of carboxypeptidase A (308–310) and it

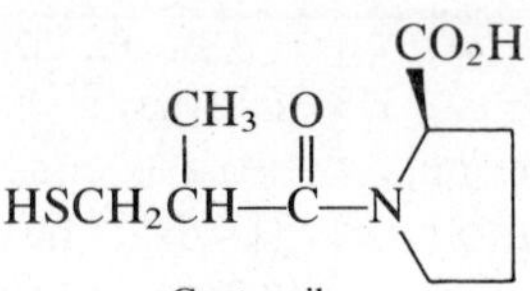

Captopril

became possible to deduce from models what kinds of peptides would occupy the (positive) carboxy-terminal binding site of the enzyme. Of many peptides tried, D-3-mercapto-2-methylpropanoyl-L-proline (captopril) was chosen as a specific competitive inhibitor of ACE. It is effective in stroke-prone patients and several types of hypertension although ineffective in DOCA (desoxycorticosterone acetate)-salt hypertension.

2.2. ANTIINFLAMMATORY AGENTS

2.2.1. Nonsteroidal Drugs

No other disease syndrome has been researched so persistently—and as yet so inconclusively—as arthritis. The long-term interest in this condition may stem from the deplorable fact that barely a single human and animal escapes some degree of arthritis, particularly with advancing age. This debilitating syndrome usually leads to painful inflammation, tissue degeneration, and associated diseases arising from the immobilizing effects of arthritis. The cry for relief from the overt symptoms of arthritis is mirrored in the abundant use of over-the-counter analgetics and the concomitant sale efforts in the news media. For the pharmaceutical industry, antiinflammatory drugs represent a profitable and ongoing venture of research and production. For the medicinal scientist they offer the challenge of seeking new structural types that will improve palliative treatment and perhaps remove some of the natural causes of the diseases called arthritis and rheumatic conditions. Biochemical pharmacologists will want to improve basic knowledge of mechanisms of action of the diseases themselves and of the drugs used to treat them. The fact that it took 71 years to even guess the mode of action of aspirin should stimulate work in this area of research. Since aging politicians and company presidents are victims of degenerative diseases, they are more prone to allocate research funds for these conditions than for less pressing health problems.

It is therefore not surprising that some of the earliest medicinal-chemical investigations were directed at antiinflammatory agents. Only they were not called that but instead were spoken of as analgetic–antipyretic drugs. Among them was salicylic acid, which soon had to be modified because of side effects. The acid, found in oil of Gaultheria in 1844, had previously been prepared oxidatively from salicin by Piria in 1838; salicin is a glucoside of salicyl alcohol (Leroux, 1827) and was found in the bark of the willow *Salix alba*. Salicylic acid was first used in rheumatic fever by MacLagan in 1877 (311) and has maintained its supremacy in this condition to the present day. A few years later Knorr (312, 313) had some pyrazolones tested as antipyretics–analgetics since he mistakenly associated the structure of quinine with this ring system. Aminopyrine (314, 315) has survived as a therapeutic agent for 100 years, particularly in Europe and in spite of side effects such as blood dyscrasias. Perhaps because there were no better "leads" for a long time, drugs in both these series have been modified and used time and again. Of the salicylates, <u>esters such as aspirin (55)</u> have certainly not been

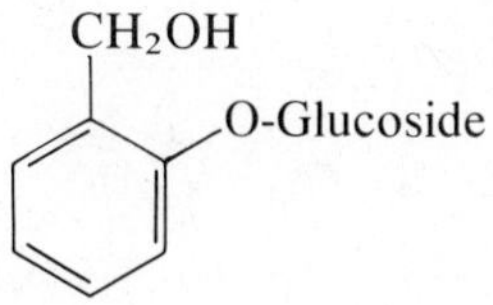

Salicin

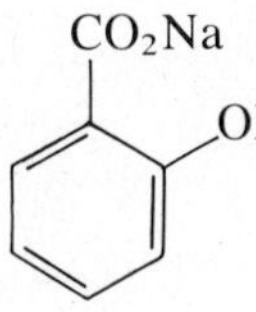

Salicylic acid Na salt

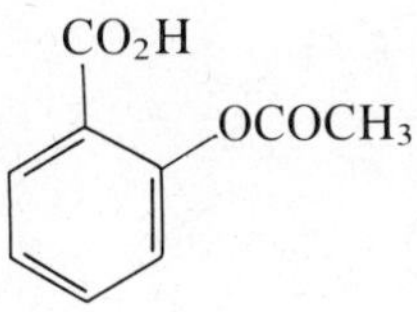

Aspirin

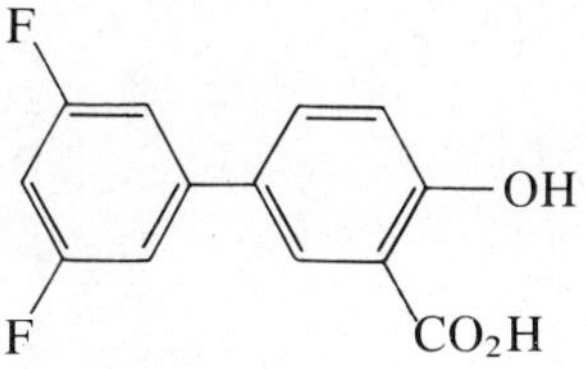

Diflunisal

replaced; 5-(2,4-difluorophenyl) salicylic acid (diflunisal) (316, 317) is a recent entry to the salicylate drug field. It was chosen from 500 candidate compounds by screening. Among pyrazolidinediones, phenylbutazone (318) is still used as an antiinflammatory and uricosuric; its active metabolite, oxyphenbutazone (319), is somewhat less toxic (320).

Salicylic acid is not the only prototype, nonsteroid aromatic acid that exerts antiinflammatory activity. Both arylacetic and arylpropionic acids with homocyclic and heterocyclic aromatic nuclei produce this aspirin-like

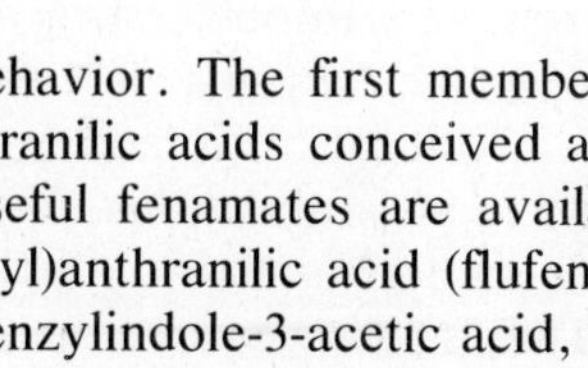

Phenylbutazone

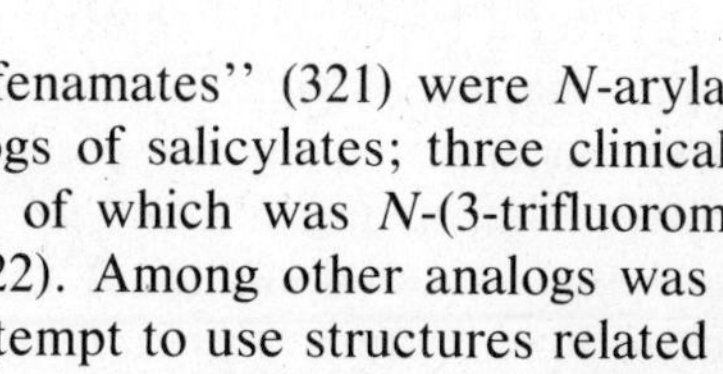

Oxyphenbutazone

behavior. The first members of these "fenamates" (321) were N-arylanthranilic acids conceived as amine analogs of salicylates; three clinically useful fenamates are available, the first of which was N-(3-trifluoromethyl)anthranilic acid (flufenamic acid) (322). Among other analogs was 1-benzylindole-3-acetic acid, made in an attempt to use structures related to serotonin metabolites for antiinflammatory activity. As many as 350 indole derivatives were screened and two were tested clinically; one of these was indomethacin (323). A further medicinal improvement was found in an indene isostere, sulindac, substituted by fluorine in the indene and by methylsulfoxide in the phenyl group (324–326). These are only two examples of very many molecular variations leading to about 20 clinically useful representative agents. This holds particularly for substituted arylpropionic acids of which ibuprofen [p-$(CH_3)_2CHCH_2C_6H_4CH(CH_3)CO_2H$] (327) may be cited as an

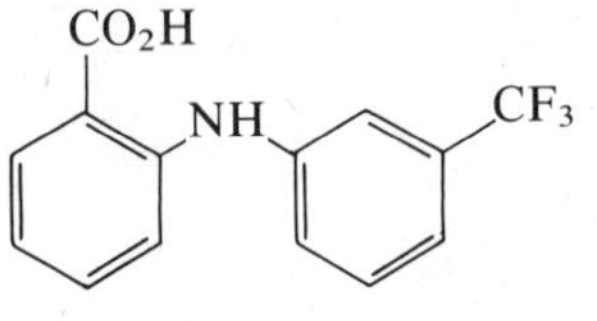

Flufenamic Acid

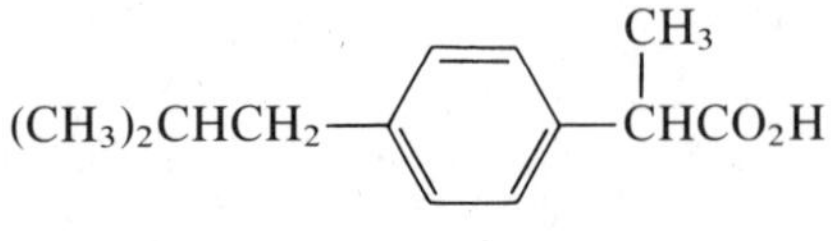

Ibuprofen

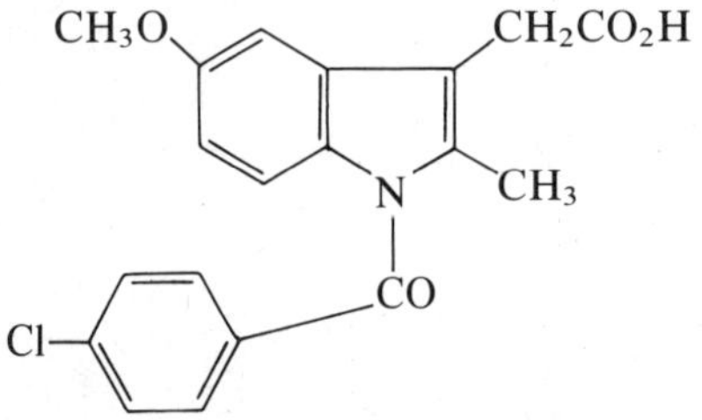

Indomethacin

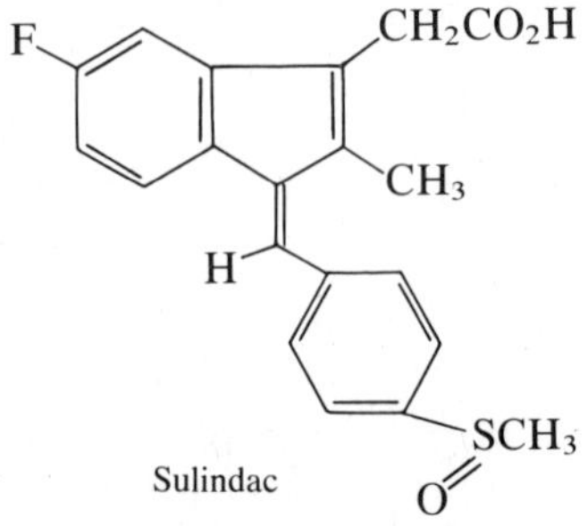

Sulindac

example of a widely used aspirin-like antiarthritic and antidysmenorrheal drug. As a rule, the (S)-(+) isomer of the asymmetric acids is more potent, but there are exceptions. The mechanism of action of these compounds, including salicylates, involves inhibition of several pathways associated with the arachidonic acid cascade, including cyclooxygenase and prostaglandin synthetase (328). The subject has been reviewed (329).

Research activity in antiinflammatory drugs has intensified to the point that most major pharmaceutical companies have a current test program going on in this field. The success of their efforts depends largely on the validity of their pharmacological test methodology (329, pp. 1213–1226). In these companies, chemicals available from this and any other research effort are usually run through the antiinflammatory screen; consequently, a potpourri of structural types has been found to respond positively to one test or another. Pharmacological explanations of underlying mechanisms of action have concentrated on the pathogenesis of rheumatoid arthritis and have helped in improving both *in vitro* biochemical and *in vivo* inhibition assays. They involve immunological responses, generating an immune complex (rheumatoid factor) of an autoimmune nature. This is followed by complement fixation and responses of the lymphocytes and macrophages, which cause release of inflammatory prostaglandins and lysosomal hydrolases. Several types of antirheumatic drugs with a slow onset of action have been discovered by tests founded on immune-based bioassays.

ANTIINFLAMMATORY AGENTS

Among these drugs are thiogold compounds. Gold had been noted by Robert Koch in 1890 to inhibit *Mycobacterium tuberculosis in vitro* and this led to trials in arthritis and lupus erythematosus, which were then mistakenly associated with tuberculous manifestations. Inorganic gold sodium thiosulfate (330) was the first derivative tried as an antirheumatic. Aurothioglucose was recommended in 1927 (331) for the relief of pains in the joints, and the

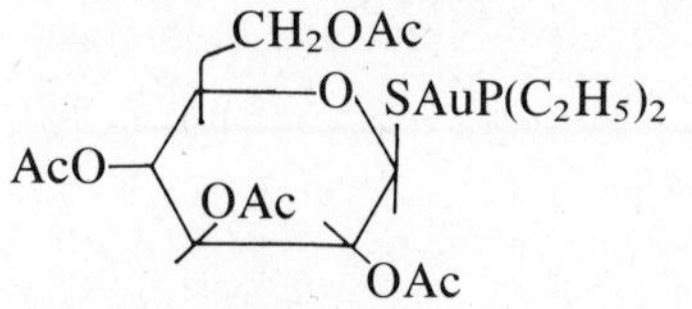

Auranofin

British Empire Rheumatism Council gave radiological evidence in 1961 for the slowing of joint disease by sodium aurothiomalate. A more recent revival of this old theme is found in auranofin (332).

Other pharmacological "shots-in-the-dark" are antimalarials of the chloroquine group and D-penicillamine (333), which have antirheumatic activity but are limited by their toxicity. Penicillamine had been developed originally as a chelating agent for copper ions, which must be removed in Wilson's disease.

These attempts to find antiinflammatory agents in diverse areas of drug research emphasize the uncertainties in understanding the etiology of inflammation (see 334–338).

Both immunostimulants and immunosuppressants have been tried in rheumatoid diseases because an immune system defect is apparently involved in their etiology. The most important immunostimulant for this application is the broad-spectrum anthelmintic levamisole, which has delayed but significant antirheumatic activity (339). Structure–activity relationship studies extended this phenomenon to related anthelmintics and their metabolites. The occurrence of the two unrelated activities in these structures has not been explained.

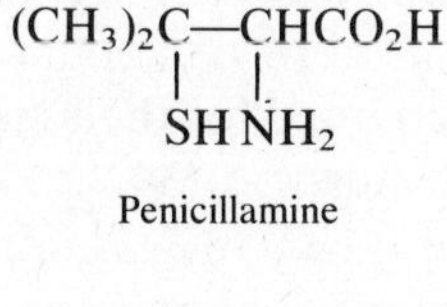

$(CH_3)_2C$—$CHCO_2H$
     | |
     SH NH_2

Penicillamine

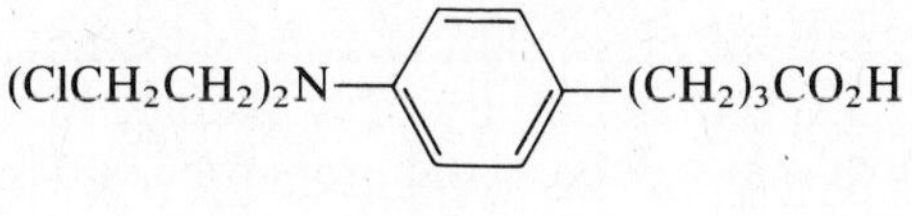

$(ClCH_2CH_2)_2N$—⟨ ⟩—$(CH_2)_3CO_2H$

Chlorambucil

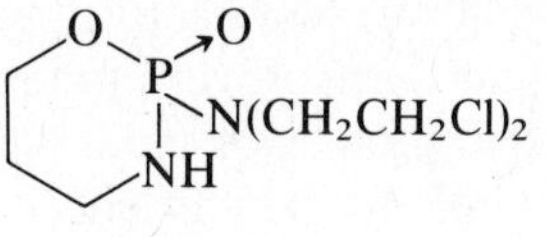

Cyclophosphamide

Among immunosuppressants, alkylating agents such as chlorambucil (4-[bis(2-chloroethyl)amino]benzenebutanoic acid) (340) and cyclophosphamide (341) have been used in rheumatoid arthritis (342) but their side effects hinder continued therapy with these agents. The same holds for one of the oldest antiinflammatory drugs, the alkaloid colchicine. Its botanical source, the autumn crocus (meadow saffron), *Colchicum autumnale*, came originally from Colchis in Asia Minor. The plant was known to Dioscorides to be poisonous but was not recommended for gout until Baron von Störck used it for this purpose in 1763. Benjamin Franklin who himself suffered from

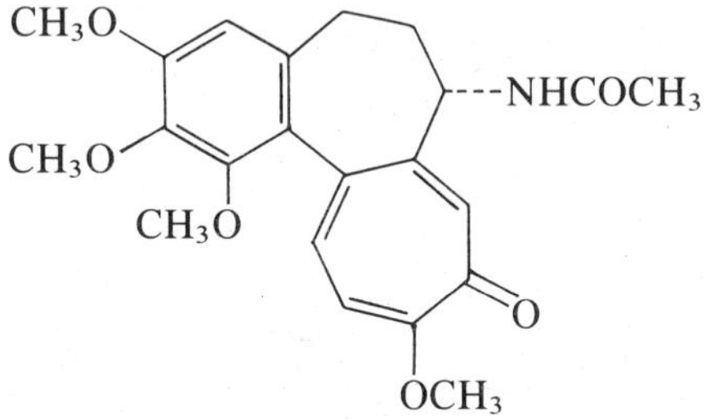

Colchicine

gout is said to have brought *Colchicum* to the United States. Extracted from *Colchicum* in 1820 (343), colchicine became established principally as an antigout medicine. Structure–activity relationships of derivatives and analogs have been reviewed (344). The primary mode of action of colchicine is inhibition of microtubule assembly, which in turn blocks a number of immunological and humoral mediator functions.

2.2.2. Antiinflammatory Corticosteroids

The cortex of adrenal glands has yielded about 50 substances on extraction by natural-products chemists (169–171, 345). Some of these compounds were active in alleviating the symptoms of rheumatoid arthritis (346); they turned out to be pregnane derivatives substituted by alcohol or keto groups in position 11 and a dihydroxyacetone group at position 17. The glucocorticoid activity of these natural hormones, notably of cortisol and cortisone, was recognized as the desirable therapeutic requirement of these compounds. Their mineralocorticoid properties, although minor, accounted for their side effects, which have been summarized as Cushing's syndrome (347). The sheer number of glucocorticoids suggested that structure–activity relationships should exist, but for half a decade no synthetic corticoid was found that was more active than cortisone. This led to the assumption that cortisone and cortisol were unique, as had been seen with most of the vitamins, and that molecular modification had no place in this series. It therefore came as a relief when Fried and Sabo (348) reported that the synthetic compound 9α-fluorocortisol was ten times as potent as cortisol. This opened up a fe-

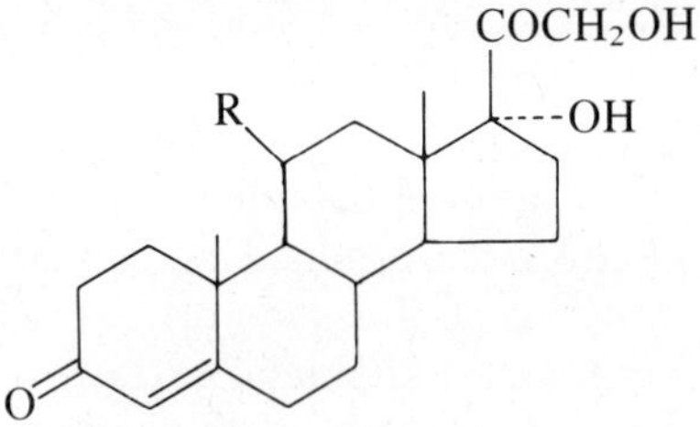

Cortisone: R = =O

Cortisol: R = H, HO

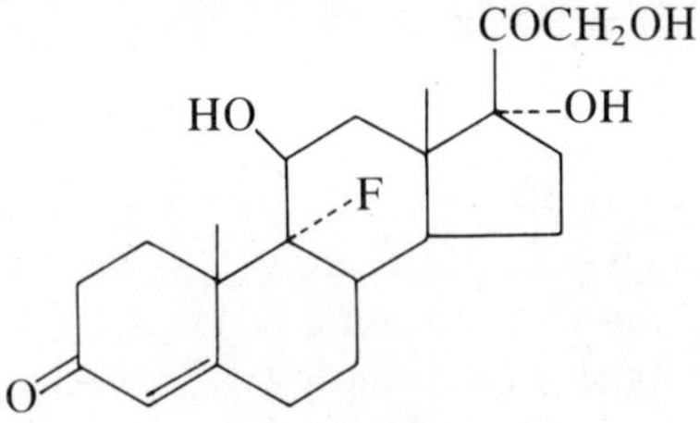

Fluorocortisol
(9α-fluorocortisol)

verish search for analogs in which medicinal and side effects would be separated more effectively. It has succeeded only to some extent. Even though some synthetic corticosteroids have high antiinflammatory activity, in some cases 1000 times that of cortisone, side effects have come to the fore so often that only relatively few analogs have been accepted clinically.

New concepts in medicinal chemistry have provided interesting guidelines in the molecular modification of corticosteroids. Fried and Borman (349) discovered that each substituent or linkage (halogen, OH, alkyl, unsaturation) affects the activity of the molecule almost independently of other groups. The effect of each substituent was assigned a numerical value called enhancement factor. The activity of the compound is calculated by multiplying the biological activity of the unsubstituted compound by the enhancement factors of the newly introduced groups or linkages. Exceptions to these regularities were noted for 2α-methyl and 9α-fluoro substituents (350). In any event, this prediction of total activity has helped to design more potent analogs, but the regulation of the ratio of glucocorticoid to mineralocorticoid activities has remained empirical. These relationships have been reviewed (351).

The mode of action of antiinflammatory corticoids has also been studied (352). It involves the inhibition of phospholipase A_2 and thus inhibition of the release of arachidonic acid; this prevents the biosynthesis of prostaglandins, slow-reacting substance, and leukotriene C, which are derived from arachidonic acid. In addition, the interaction of the corticosteroids with the steroid receptor rationalizes the thermodynamic properties of this binding process and makes the binding affinity—a determinant of potency—predictable (353). Lists of currently prescribed antirheumatic corticosteroids and their uses (topical, systemic) may be found in a review (354).

2.3. ANTIVIRAL AND ANTITUMOR AGENTS

The same uncertainties that plague the explanations of arthritis and inflammation are also encountered in the chemotherapy of tumors and viruses.

They concern the role that genetic, immunological, and metabolic events play in their etiology and the significance of these factors in devising structures of drugs directed against some phases of these diseases. At the same time, the urgency of the popular dread of cancer and the economic losses and suffering caused by even common virus infections has prodded medicinal scientists to blind searches for chemotherapeutic agents for such diseases. Random synthetic compounds, antibiotic beers, plant and animal extracts, and other mixtures have been submitted for screening by test methods that only gradually achieved a higher confidence index of therapeutic significance. The past 25 years have witnessed the injection of rational thoughts into these searches. Since both neoplasms and viruses owe their invasiveness to nucleic acid components of their nuclei, metabolite analogs in the widest sense of the word have been prepared and tested as antitumor and antiviral drugs. In many laboratories, these two types of tests have paralleled each other. Depending on the expertise of the participating biologists and their interest in viral or neoplastic phenomena, the emphasis has been greater in one or another of these two areas. Since some cancers are caused by viruses, and some virus infections may result in malignancies, the lines of demarcation have been even fuzzier. Any medicinal chemist who conceives some rationale for either antitumor or antiviral agents will be well advised to prepare his or her compounds for both test areas. In both fields, some standard tests may screen out *a priori* unlikely candidate materials, but it is impossible at present to claim that a given compound will show preference against any virus or any tumor under specific test conditions.

Conforming with other sections in this book, antiviral and antitumor test systems are not described here unless they give chemists a clue to what types of chemicals should be fed into the testing programs with a reasonable expectation of antiviral or antitumor activity. This involves inhibition of some step of nucleic acid biosynthesis, or protein biosynthesis for viral protein coats. Construction of structural analogs of the monomeric building blocks of these biopolymers can proceed along traditional experimental lines. The prime targets are analogs of purines, pyrimidines, ribosides and deoxyribosides, nucleosides, amino acids, small peptides, and perhaps the hitherto neglected lipids found in viral lipoprotein envelopes. Such analogs have been conceived, prepared and tested by the thousands. Irreversible inhibition of enzymic processes using the natural substrates has also received much attention because such inhibition targets candidates for competitive inhibition in the biosynthesis or utilization of the natural prototype monomers. Alkylating agents with carbonium ion intermediates are perceived as being able to tie branches of nucleic acid helixes together and prevent their separation. For proteins, similar steric immobilization can be visualized with such agents.

In the biosynthesis of basic components of nucleotides, ancillary contributors of one-carbon and one-nitrogen fragments are needed. Such moie-

ANTIVIRAL AND ANTITUMOR AGENTS

ties are furnished by *N*-formyltetrahydrofolate, and this compound as well as its biosynthetic precursors (folate and dihydrofolate) have been imitated by designing potential metabolite antagonists. The first were aminopterin and methotrexate (228, 229; see also Section 1.4.6).

The seminal discoveries of nucleoside analogs that turned out to be useful antimetabolites were 6-mercaptopurine (355) and 5-fluorouracil (230). In the case of 2'-deoxy-5-fluorouridine, its mechanism of action was found to involve inhibition of thymidylate synthetase (231, 232), and this set the pace

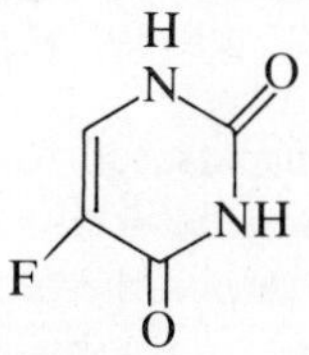

Fluorouracil

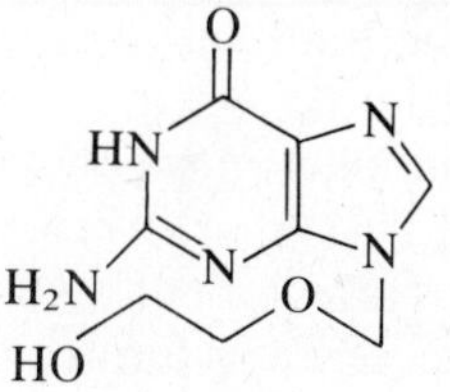

Acyclovir

for the biochemical studies in this field. At first the pentose moieties (ribose and 2-deoxyribose) remained sacrosanct, until some arabinosides were screened successfully for antiviral properties (356). Later, even the pretense of glycosidation was dropped when acyclovir [9-(2-hydroxyethoxymethyl)guanine] (357) was found to inhibit the multiplication of herpes viruses both *in vitro* and *in vivo*.

The classical alkylating agents comprise nitrogen mustards and analogous alkyl esters, alkanesulfonates, 1,2-epoxides, aziridines, and so on. These are all compounds that can react directly—or by way of intermediate charged particles—with cellular biochemicals, in contrast to metabolic analogs that interfere with enzymic reactions for which cellular biochemicals may be substrates. For the chemist, this means that other chemically reactive substances should be evaluated as anticancer agents. This is a two-edged sword: A properly reactive compound may react with nuclear materials as well and thereby initiate mutations that can lead to carcinogenicity. Even non-nuclear cell components might react and give rise to general toxicity. As an example, *N*-nitroso compounds are widely known as effective carcinogens that can hardly be completely avoided in the diet or in the environment because biochemical nitrosation takes place constantly in the animal organism itself. Yet, structure–activity relationship studies have revealed that certain *N*-nitrosoureas [RN(NO)CONHR'], where R can but must not be an alkylating group, can serve as clinically useful antitumor drugs (358). A considerable number of nitrosoureas, including naturally occurring antibiotics of this structural type (359), have been tried successfully in clinical cancers.

Another type of aggressive chemical is seen in α,β-unsaturated carbonyl compounds, which can add to sulfhydryl groups of enzymes and thereby deactivate the biocatalysts; the plural is chosen on purpose because this

$$RCOCH=CHR' + HSE \rightarrow RCOCH_2CHR'(SE)$$

is usually a nonspecific reaction, deactivating both pathogenically involved enzymes of invasive cells and enzymes of host tissues. Thus their toxicity is predictably high, but this has not tempered efforts to work with such compounds. Those that have acceptable therapeutic indexes often have sterically hindering groups adjacent to the —C=C—C=O group, which may explain why some thiol enzymes add to this group and others are slowed down in this reaction.

Carbonium ion mechanisms have been demonstrated for some unorthodox structures. 1-Aryl or 1-heterocyclically substituted 3,3-dimethyltriazenes [ArN=NNMe$_2$] (360), especially triazenoimidazoles, are N-demethylated metabolically and furnish methyl carbonium ions, which appear to be responsible for their anticancer activity (361). One of them, darbazine (DTIC), emerged as the drug of choice in the treatment of melanotic melanoma and other tumors (362). It is of interest to trace the history of the triazenoimidazoles. They were first synthesized (363) as latentiated forms of 5-diazoimidazole-4-carboxamide, a precursor of 2-azahypoxanthine, that is, as pro-drugs to a metabolite antagonist in purine biosynthesis. Their mechanism of action as methyl cation producers by a series of metabolic steps emerged only later. Similarly, several methylhydrazines, originally prepared as MAO inhibitors in adrenergic amine biosynthesis (364), were

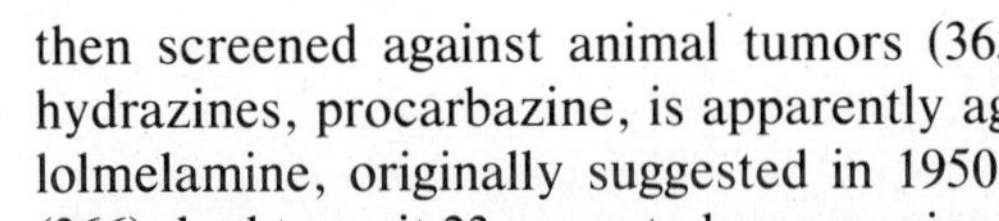

Procarbazine

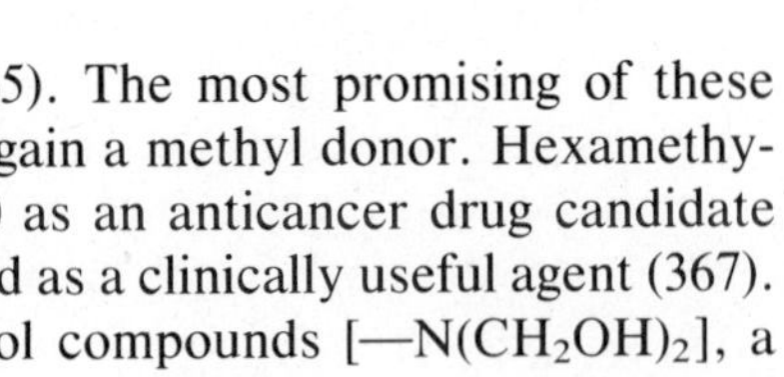

Hexamethylolmelamine

then screened against animal tumors (365). The most promising of these hydrazines, procarbazine, is apparently again a methyl donor. Hexamethylolmelamine, originally suggested in 1950 as an anticancer drug candidate (366), had to wait 23 years to be recognized as a clinically useful agent (367). It is degraded rapidly by way of methylol compounds [—N(CH$_2$OH)$_2$], a variant to methyl cation intermediates.

The moral of these observations is that chemists should search for alkyl cation donors by plausible degradative mechanisms and then have them screened as antitumor agents.

Chemical reactivity leading to attacks on DNA is apparently not restricted to organic compounds, since an inorganic substance, cis-dichlorodiammineplatinum, (H$_3$N)$_2$PtCl$_2$, first tried as an antibacterial agent (368), has

ANTIVIRAL AND ANTITUMOR AGENTS

proved to be a medicinally applicable cancer drug. It seems to form two types of bidentate bonds with all the DNA bases except thymine (369), and thus its action is similar to that of the nitrogen mustards.

Some of these mechanisms must also play a role in antiviral drugs. Steric interference seems to be a factor in amines of the cage type, such as adamantamine (370) and related compounds (371, 372), which have limited prophylactic use in influenza (373). A number of thiosemicarbazones, repeatedly tried as chemotherapeutic agents in bacterial, mycobacterial, and other infections have also been tested as antiviral drugs (374–376). They may owe their activity to the formation of coordination compounds with copper, cobalt, nickel, manganese, or zinc which are known to be needed for the biosynthesis of several enzyme systems (377). One of these derivatives, 1-methylisatin-3-thiosemicarbazone (methisazone) is clinically useful in vaccinia infections (378). Each such agent has been tested by several methods that might reveal the step or steps of viral replicative processes affected by the compound. The possible steps include attachment to the host cell, adsorption to and penetration of the cell surface membranes, removal of the viral protein

Methisazone

envelope, interference with the biosynthesis of the exposed viral RNA, DNA, or mRNA, and interference with polymerase or cellular transcriptase enzymes. Although it is important to understand these mechanisms of action, they contribute only marginally to drug design. What we need in the conception of both antitumor and antiviral drug structures is a clearer knowledge of the enzymes that catalyze the biosynthesis of nucleic acids and the associated proteins (the viral immunospecific proteins and the nuclear histones). A description of at least some features of the active sites of those enzymes would rejuvenate research ideas in these fields. So far it has only been possible to synthesize analogs of purines, pyrimidines, and amino acids, and to attack the natural aberrant nucleosides and proteins by crosslinking and similar structural maneuvers.

A survey of the structural types that have been coaxed through antiviral screening procedures leaves one bewildered. Little logic can be derived from these trials, and a success here and there in laboratory animals does not contribute to clarifying this situation. When interferon, a glycoprotein found in minimal amounts in leukocytes (379), was first shown to give rise to a chain reaction that regulates immune reactions and leads to anticancer and antiviral activity, inducers to interferon were sought. They were found for mice infected with viruses, but not for humans at the time of this writing.

Compounds that look like anticholinergics, antimalarials, and a variety of antibiotics have been tried, and finally such oddities as phosphonoacetic acid, an inhibitor of DNA polymerase induced by herpesvirus and cytomegaloviruses, have been tested with a measure of experimental success. However, the field has remained largely empirical, and defendable working hypotheses need to be proposed.

2.4. HISTAMINE (H_1 AND H_2) RECEPTOR ANTAGONISTS

Histamine, [4(5)-imidazolylethylamine], was synthesized in 1907 by Windaus and Vogt (114) because of its structural relationship to histidine and because of the occurrence of the imidazole ring in both histamine and the alkaloid pilocarpine. Only later did Dale and Laidlaw (115) observe the vascular effects of histamine, which resembled those seen after anaphylactic shock. Histamine was then found in ox mucosa (380) but its biochemical origin remained in doubt. Abel (381) confirmed that it occurred in many animal tissues and that it caused a number of toxic effects. Stimulation of the secretion of gastric acid (382) was regarded as such an action (383). It was only natural that the earliest efforts to counteract effects of histamine were directed against its most striking manifestations, especially allergic phenomena.

By the early 1930's, both *in vitro* and *in vivo* tests had come into general use to determine the effects of compounds that contract various autonomically innervated muscles that activate gastrointestinal and bronchial tissues. Likewise, drugs that counteract the contractions caused by acetylcholine (ACh) were being studied as therapeutic agents. Besides ACh, various other compounds were found to produce similar spastic conditions, among them histamine, barium chloride, 5-hydroxytryptamine (serotonin), and later, slow-reacting substance of anaphylaxis. Using these test procedures (384), new compounds could be screened as antihistaminics, anticholinergics, antiserotoninergics, and so on.

The earliest observations concerning antihistaminic effects *in vivo* were made by Bovet, who found that 2-(*N*-piperidinomethyl)-1,4-benzodioxane protected animals from bronchial spasms caused by histamine aerosol (385) or ACh (386). This compound is essentially a basic choline ether, which might explain why it also has anticholinergic properties. Other basic ethers containing a dialkylaminoethoxy chain attached to various positions of ben-

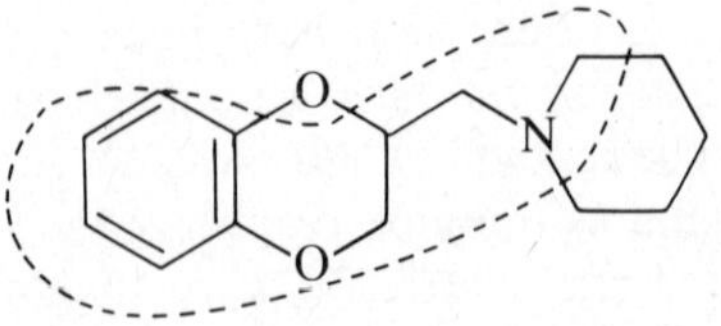

2-(*N*-piperidinomethyl)-1,4-benzodioxane

zodioxane exhibited similar properties. These *in vivo* experiments were performed at the Pasteur Institute in the same year that Domagk decided to rely on *in vivo* trials of sulfamyl azo dyes for antibacterial tests. This was a classical turn in pharmacological strategy, to achieve practical therapeutic results by direct trials in laboratory animals rather than by *in vitro* methods such as the time-honored Magnus test.

It soon turned out that blocking bulk rather than the sequence of individual atoms or the exact kinds of ring systems led to antihistaminic compounds. Molecular modification in this series began a few years after the introduction of the concepts of isosterism in medicinal chemistry (1, 192). With the exception of the sulfonamide antibacterials, the antihistaminics provided the most fertile proving ground for isosteric replacements. The first demonstration of the value of this principle was made by Staub (387); she found in 1937 that derivatives of N,N-dimethylethylenediamine [$RR'NCH_2CH_2NR''_2$] were equivalent or superior to N,N-dimethylaminoethoxy aryl analogs [$ArOCH_2CH_2NR_2$] and that a fairly good separation of antihistaminic and antiacetylcholine activities could be achieved by judicious molecular modification. Clinical studies (388) by Halpern of ethylenediamine derivatives prepared by Mosnier (389) led to the introduction of phenbenzamine [$C_6H_5N(CH_2C_6H_5)(CH_2)_2N(CH_3)_2$] into the therapy of allergic disorders.

Further bioisosteric replacements comprised the exchange of benzene rings for thiophene, furan and pyridine, both in the anilino and benzylamino groups. In the United States, Rieveschl and Huber (390, 391) arrived independently at diphenhydramine[$(C_6H_5)_2CHOCH_2CH_2N(CH_3)_2$]; this benzhydryl ether analog of the antispasmodic diphenylacetate ester, adiphenine [$(C_6H_5)_2CHCO_2CH_2CH_2N(C_2H_5)_2$] (392, 393), exhibited potent antihistaminic activity in addition to considerable anticholinergic properties. Diphenhydramine also has CNS depressant activities, a dichotomy that is shared by most other antihistaminics. Several of them are now in use as over-the-counter hypnotic–sedatives. Diphenhydramine (or its 8-chlorotheophyllinate salt, dimenhydrinate) is also effective in preventing motion sickness; the discovery of these "side effects," which in some cases (394) became "main" effects, resulted from clinical observations of patients originally treated for allergies.

A veritable flood of molecular modifications followed the discovery of these "leads". Structure–activity relationships deduced from these extensive data indicate that in general, antihistaminic activity may be expected in structures of the type where Ar may be heterocyclic or a benzyl-type group; X is N, CH, or CO; the carbon chain is mostly C_2 but if the two

$$\begin{matrix} Ar & & R \\ & XC_{2-3}N & \\ Ar' & & R' \end{matrix}$$

aromatic groups are connected (as in fluorene or by an atom or group as by S in phenothiazines), it may occasionally be C_3; and the R on nitrogen is a methyl group or other small groups but may also be connected (as in piperidyl). Many other regularities have been noted, such as preference of conformations of the C_2 chain. A recent review may be consulted for additional data (395).

As indicated above, the antihistaminic drugs exhibit multiple activities that can be observed by multipurpose screening. The most easily found effects are those on the CNS, mostly depressant activities, which have led to the development of neuroleptic agents by extending the studies of phenothiazine antihistaminics to molecular modifications in which the neuroleptic activities had become the "main" ones.

None of the thousands of "antihistaminics" counteracts the stimulation of the secretion of gastric acid and gastric motility by histamine (396). This led to the proposal of two receptors at which histamine acts. One is called histamine-1 or H_1 receptor and is responsible for the share of the manifestations of the allergic response attributable to histamine. Antihistaminics are therefore inhibitors of H_1 receptors (397). They are distinguished by structures that sterically block the approach of histamine to the H_1 receptor site.

Inhibitors of H_2 receptors have been based not on bulky blocking groups but instead tentatively on the detailed conditions of the imidazole ring present in histamine (398). Since 4-methylhistamine shows a weak but noticeable inhibition of the secretion of gastric acid, this homolog was chosen to study the proton tautomerism (hydrogen shift) in its imidazole ring as well as the steric interaction between the 4-methyl group and the α-methylene of the side chain (399). This study was one of the first to zero in on increases of specific activity of a multiactive molecule on the basis of chemical information only. The elaboration of the side chain needed for optimal H_2 receptor antagonist activity followed traditional though no less ingenious lines of molecular modification. The real quantum leap occurred in the abandonment of a thiourea group in the side chains of the earlier drugs in this series, that of burimamide [—$(CH_2)_4$NHCSNHMe] (which enabled Black to define histamine H_2 receptors) (398, 399) and metiamide [—$CH_2S(CH)_2$NHCSNHMe] in which bioisosteric replacement of one methylene by sulfur, plus substitution by 4-methyl in the ring, alters imidazole tautomerism and increases activity. These compounds had unacceptable side effects that were blamed, logically, on the thiourea group, which had caused similar toxicities in quite unrelated series of drugs. The personal experience of the investigators with guanidines suggested replacement of the thiourea group [—NHCSNHR] by a guanidine unit [—NHC(=NH)NHR] but this procedure increased basicity and reduced activity. It appeared possible that basicity would be decreased by introducing an electron-withdrawing group into the guanidine moiety, for example —NHC(=NNO_2)NHR or —NHC(=N—CN)NHR. As a preliminary step to such operations, cyanoguanidine [H_2NC(=NCN)NH_2] was compared directly with thiourea [H_2NC(=S)NH_2]; the two compounds turned out to be classical isosteres in every measurement of their physical properties. Following this "lead," the side chain —CH$_2$SCH$_2$CH$_2$NHC(=N—CN)NHCH$_3$ was introduced into 4-methylimidazole. The resulting drug, cimetidine, has become a successful and almost trouble-free drug for hyperchlorhydria and the therapy of peptic ulcers.

Throughout this work, paramount emphasis was placed on the proton tautomerism of imidazole which, if attained in the right proportion, was deemed essential for H$_2$ receptor antagonism. However, potent H$_2$ receptor antagonism has since been observed in compounds that do not contain an imidazole ring but instead a furan, thiophene, or even benzene ring substituted by a basic dimethylaminomethyl group (400). Such compounds cannot exhibit the kind of proton tautomerism seen in imidazole, and this put the original reasoning in question. Moreover, further bioisosteric modification of the guanidino group in the side chain, from —CH$_2$S(CH$_2$)$_2$NHC(=NCN)NHMe to —CH$_2$S(CH$_2$)$_2$NHC(=CHNO$_2$)NHMe, resulted in ranitidine, a drug that is about five times more active than cimetidine, more selective for H$_2$ receptors, and acceptable in extensive clinical trials. See also ref. 400a.

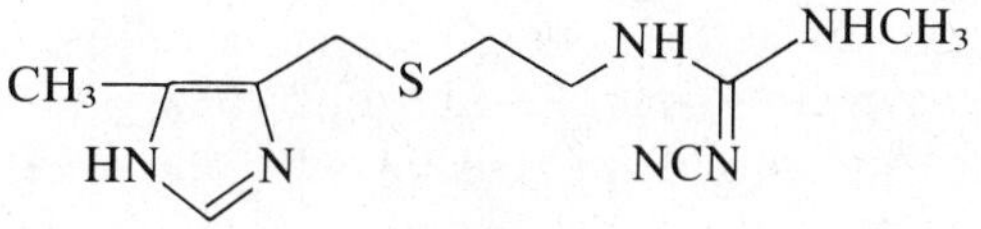

Cimetidine

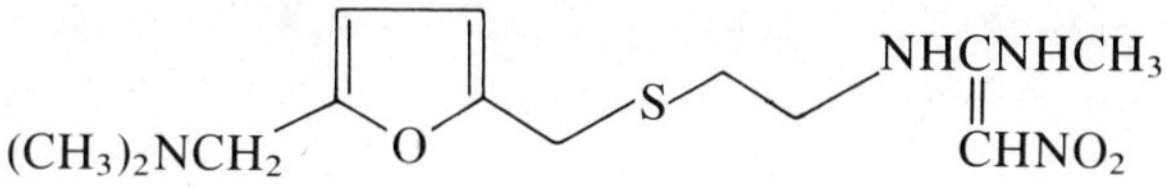

Ranitidine

The finding that such quasi-aromatic derivatives can have potent effects jeopardizes the earlier explanation (401) that for H$_2$ receptor antagonists, receptor recognition is determined by the imidazole ring. As in so many other cases of medicinal biochemistry, the factual data have remained untouched but their theoretical rationale must await further support (402). For pyridyltriazole H$_2$ receptor antagonists, see ref. 402a.

2.5. DRUGS THAT COUNTERACT THE EFFECTS OF ACETYLCHOLINE

Like histamine, the neurohormone acetylcholine (ACh) and its cholinergic analogs exert a variety of physiological actions. Interestingly, two types of cholinergic receptors were recognized at an early date, aided by the fact that two different alkaloids produce responses similar to those observed for ACh. The alkaloid (2S, 3R, 5S)-(+)-muscarine acts at postganglionic parasympathetic neuroeffector sites, primarily on smooth (involuntary) muscles and secretory glands. ACh actions at these sites have been called muscarinic; the receptor is also called muscarinic; these actions include cardiac inhibition, peripheral vasodilation, contraction of the pupils of the eyes, increased secretion by most secretory glands, and stimulation of gastrointestinal and urinary tract contractions (peristalsis).

The second type of ACh actions parallel those of $(-)$-nicotine and are called nicotinic, as is the second ACh receptor. They involve sequential stimulation followed by blockade of end plates, autonomic ganglia, and skeletal muscles. The muscarinic and nicotinic receptors are believed to differ chemically, sterically, and physiologically; they may be widely differing complex proteins but their nature is not yet understood. Extensive investigations of the electric organs of several marine animals that are rich in ACh receptors and partial purification of fractions containing these receptors seem to indicate that the receptors are proteins. The nicotinic ACh receptor from fetal calf muscle consists of four to six high-molecular proteins, perhaps associated with actin (402b).

It is not surprising that agents which counteract the many muscarinic and nicotinic effects of ACh must differ in structure and selectivity of action. They do not prevent ACh from being released at nerve endings but may compete with it for receptor sites. Fortunately, some prototype antagonists of ACh have been known to pharmacologists for a long time and have provided "leads" for the very extensive molecular modification in the series of anticholinergic drugs. This survey would not include such agents if they had not occupied the attention of medicinal chemists and pharmacologists for decades to an almost unseemly degree. The reason for this lasting interest lies in the three major clinical applications of anticholinergics: antisecretory, antispasmodic, and mydriatic. Occasionally there is a fair separation of these activities, but more often than not they overlap a great deal. Before the recent advent of histamine-2 receptor antagonists, the antisecretory anticholinergics were the principal therapeutic agents for excessive gastrointestinal motility and peptic ulcers. Naturally, one did not wish to cause dry mouth or pupil dilation with the same agents, and the trend of research was dictated by separation of these properties. This also means that not all anticholinergics are antispasmodics or mydriatics. The overlap between anticholinergic properties is aggravated by overlap with antihistaminic activities, and also with antiserotoninergic behavior, especially where gastrointestinal peristalsis is concerned. Even the time-honored prototypes atropine and scopolamine, specific anticholinergics, suffer from lack of selectivity of action within the scope of anticholinergic manifestations. In the same manner, some ganglionic blocking agents also cause antimuscarinic effects and vice versa. Some antispasmodics can act as local anesthetics by blockade of nervous impulses at the nodes of Ranvier, but great caution should be exercised to attribute local anesthetic activity primarily to the

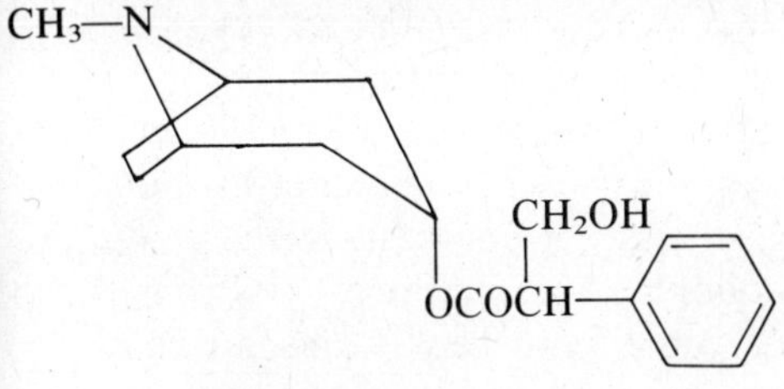

Atropine

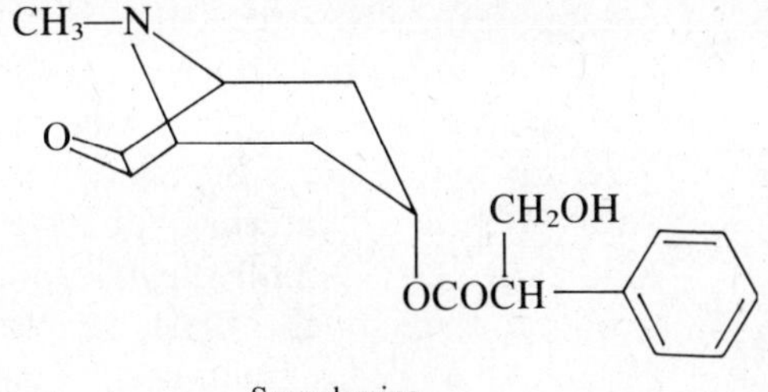

Scopolamine

anticholinergic effects of these drugs. Their principal mode of action is on the sodium channels of neuronal membranes.

Some mydriatics have CNS effects, especially in children, or can increase intraocular pressure in patients with simple glaucoma. The antihistaminic and CNS depressant side effects of anticholinergics have been used in the treatment of Parkinsonism. It should not be forgotten that meperidine was conceived as a "reversed" ester anticholinergic and its potent analgetic action was discovered accidentally in the pharmacological laboratory. Even psychotomimetic effects hang line Damocles' sword over the clinical development of anticholinergic drugs. All these activities are observed and developed by experimental biologists, and medicinal chemists often take a fatalistic attitude when their hopes are dashed by "side effects." However, after synthesizing a small series of congeners, trends veering away from disturbing secondary activities can be seen occasionally and can be enhanced by patient molecular modification, aided by quantitative evaluation of structure–activity relationships. The determination of trends in lipophilicity has been especially popular in these studies. Then, also, some old standby methods of altering the stability of a series of compounds can be applied, and thereby their resistance to metabolic removal from the sites of their action can be changed. The best-known examples of such modifications are exchanges of ester groups that are readily hydrolyzed by serum esterases for amide, ether, and other less-hydrolyzable moieties.

The assay methods for the various anticholinergic activities are fairly simple, and this has contributed to the popularity of screening compounds for such properties (403, 404). The relative chemical simplicity of many of the compounds that have been studied or predicted to be anticholinergics has further supported medicinal–chemical interest in this field.

The historical anticholinergic "lead" drugs were solanaceous alkaloids obtained from *Atropa belladonna* (deadly nightshade), *Hyoscyamus niger* (black henbane), and *Datura stramonium* (thorn apple, jimsonweed). The active principles of these plants are mainly (S)-(−)-hyoscyamine and its (±) racemate, atropine, as well as (−)-scopolamine. Atropine was isolated in 1831 (404, 405); its structure was established by Ladenburg (406) and its total synthesis was achieved by Willstätter (407). Scopolamine crystallizes from the mother liquors of (−)-hyoscyamine. Its epoxy ring structure offered greater difficulties than the structural elucidation of atropine but was achieved in the 1920s (408–411) and it was synthesized by Fodór (412).

Two routes of molecular modification offered themselves at first glance. One was to derivatize the alkaloid structures, for example by quaternization since ACh is a quaternary ammonium base ester. Second, since these alkaloids are esters of the dicyclic amino alcohol tropine and the tricyclic base scopine, respectively, these basic moieties could be dissected and imitated in many ways by simplifying the alkamine structures and altering the steric conditions of the ester components, which may be regarded as a blocking configuration. All these ideas have been worked over in innumerable variations. It was found that the tropine and scopine ring systems were unnecessary for producing anticholinergic effects and that simpler heterocyclic systems could be substituted without loss of overall pharmacological activity. Ultimately, it turned out that heterocyclic structures were not needed at all, and dialkylaminoalkyl groups took their place. Combined with changes in the blocking ester moieties, structures emerged that could have been conceived as antihistaminics (H_1 receptor antagonists) as well. There was additional opportunity for modification. The hydroxyl group of tropic acid could be replaced by hydrogen-bonding or other functions but it has remained optimal as far as potency goes. As in many other series, one of the enantiomorphs of suitable chiral isomers may be more active than the other; several physical properties such as dissociation constants and lipophilicity have been invoked in attempts to correlate structure and activity. A review of anticholinergics may be consulted for details (413).

2.6. DRUGS FOR RELIEF FROM PAIN

To the patient and to the physician, pain is a symptom of pathological disorder. Pain has physiological and psychological components and is feared by everybody, whether it consists of a local discomfort or a virtually intolerable agony inflicted by catastrophic injury or disease. Humans and animals alike seek relief from pain by every conceivable protective and corrective procedure and try to avoid it by precautions against injury, by submitting to local or general anesthesia, and by using chemicals that counteract or prevent painful pathogenic stimuli. Such chemicals are called analgetics (or analgesics). Compounds that block nervous transmission of noxious stimuli are called anesthetics; if consciousness is retained, they are classified as local anesthetics, if consciousness is lost, as general anesthetics. There may be overlaps depending on the modes of administration of such drugs.

Since pain is a warning signal of physiological disorder, the animal body has evolved a complicated system of producing and sending nociceptive messages to the CNS for decoding. This is achieved by biosynthesizing certain prostaglandins, which signal the pain, from precursor biogenetic lipids by the so-called arachidonic acid cascade. As a countermeasure, certain tissues can biosynthesize small peptides, which have the capability of obtusing pain in somewhat the same way as morphine alkaloids do; hence their name "endogenous morphines," or endorphins. Even smaller peptides with analgetic activity are the enkephalins, so called because they were first found in brain (Kephale = head). The pathogenic prostaglandins and the analgetic peptides have provided biochemical insight into the production of pain and into mechanisms of raising the pain threshold. Before these factors were recognized, the field of analgesia was largely empirical. Likewise, as details of the transmission of impulses across the nervous membrane became known, the significance of the sodium (ion) channel rose to prominence. This placed the understanding of the mechanism of action of local anesthetics on a clearer foundation but did not, in fact, provide new useful guidelines for their design. The medicinal chemist has to thank the natural-products chemist and biochemist for the fundamental knowledge derived from the study of batrachotoxin, tetrodotoxin, and saxitoxin but it will take as yet undiscovered ingenious molecular manipulations to harness these extremely toxic materials for their inherent potent local anesthetic activity. For a review of these toxins, which interfere with the normal mechanism of sodium ion migration through the neuronal membrane, and of the mechanism itself, see ref. 414.

These theoretical researches have stimulated the intellectual milieu of studies on pain-relieving agents but the question remains unanswered whether clinically useful agents will be elaborated on this basis in the foreseeable future. In the case of the toxins, will it be possible to separate toxicity

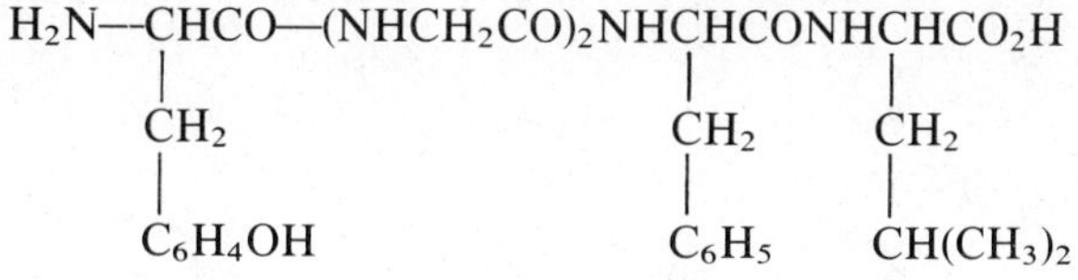

$$H_2N-CHCO-(NHCH_2CO)_2NHCHCONHCHCO_2H$$
$$| \qquad\qquad\qquad\qquad | \qquad | $$
$$CH_2 \qquad\qquad\qquad CH_2 \quad CH_2$$
$$| \qquad\qquad\qquad\qquad | \qquad | $$
$$C_6H_4OH \qquad\qquad\quad C_6H_5 \quad CH(CH_3)_2$$

Leu-Enkephalin

$$H_2N-CHCO(NHCH_2CO)_2NHCHCONHCHCO_2H$$
$$| \qquad\qquad\qquad\qquad | \qquad | $$
$$CH_2 \qquad\qquad\qquad CH_2 \quad (CH_2)_2$$
$$| \qquad\qquad\qquad\qquad | \qquad | $$
$$C_6H_4OH \qquad\qquad\quad C_6H_5 \quad SCH_3$$

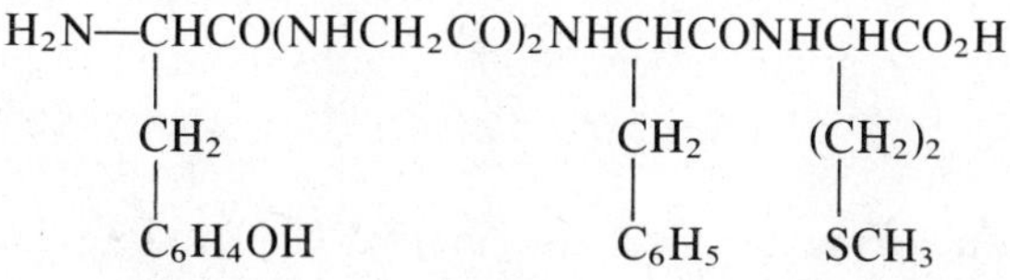

Met-Enkephalin

from properties affecting the sodium channel in neuronal membranes? In working on enkephalins (EK) and endorphins, the structural complexity encountered with the neurotoxins did not pose a problem. The structures of these peptide opioids were soon understood, and established methods of SAR studies in peptide series could be applied (415). At the 2-position, a D-amino acid is important for high potency, and this has been attributed to enhanced metabolic stability and improved conformation vis-à-vis the receptor. About 50 molecular modifications of β-endorphins have been prepared and tested, and more than a thousand analogs of the natural enkephalins have been synthesized, predominantly by the solid phase process (416), in the hope of finding an opioid-like analgetic without dependence liability (417). In animals, SAR emerged that even led to orally "superactive" analogs, for example, to [D-Ala^2, MePhe^4, Met(O)ol^5]enkephalin; this showed 30,000 times the activity of Met-EK and 1000 times that of morphine in the mouse tail-flick test (418). But clinical trials did not at all extend these data to patients; even the most potent analog did not behave like morphine (419). This lack of correlation of animal data (including monkeys) and clinical observations may be based on metabolic differences but remains unexplained at this time.

Since hopes for clinically *useful* analgetics among peptides have not yet been realized, the medicinal chemist has had to return to older models in searches for improved potent analgetic agents. Such work had its start before 1920 and is still going on. It has taught the medicinal chemist some important lessons.

The medical profession has retained morphine as a traditionally reliable potent analgetic ever since it was isolated in 1805 from the opium poppy, *Papaver somniferum,* by the German pharmacist Sertürner (19). Morphine 3-methyl ether (codeine) was purified in 1832 from the same source (20). The addictiveness, or dependence liability, of morphine was recognized early, particularly since opium smoking and the inevitable addiction to this ancient habit had worked its way from the Orient to Western countries. But it took the invention of the hypodermic syringe by Wood in 1853 and the widespread use and abuse of injectable morphine to raise morphine dependence to a major medical, psychiatric, and social problem. This was potentiated further when Wright acetylated both hydroxyl groups of morphine and thereby obtained heroin (420). Heroin was introduced as an analgetic a little more potent than morphine in 1898 by the same pharmacologist who introduced aspirin (421), and it was listed promptly in the pharmacopoieas. However, it did not take long before the euphoric action of heroin was recognized and its connection to dependence liability established. The outlawing of heroin in most countries did not remove the drug from the scene but only drove it underground.

Morphine contains an allylic alcoholic hydroxyl and a phenolic hydroxyl group; these are the easiest points of chemical attack and were altered in attempts to improve the properties of the parent alkaloid. Some of these derivatives were made at an early stage, some a little later (421–425). In 1929, the Committee on Drug Addiction of the U.S. National Research Council undertook a systematic research project designed to separate analgetic and dependence-liability properties in congeners and synthetic analogs of morphine alkaloids. The chemical work was done at the University of Virginia (L. F. Small, E. Mosettig, A. Burger, and coworkers), the pharma-

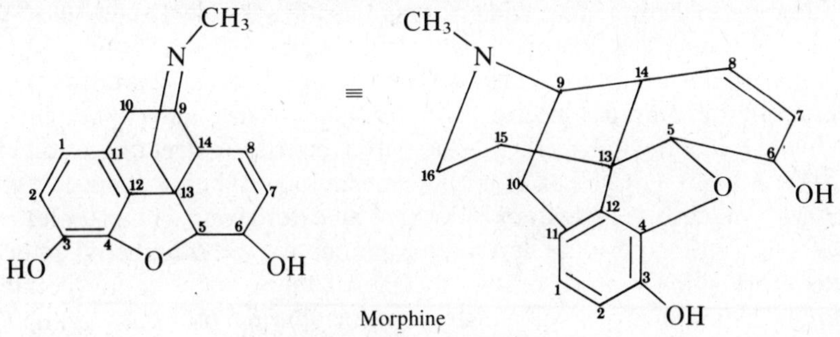

DRUGS FOR RELIEF FROM PAIN

Morphine

Heroin (420):   3,6-(OCOCH$_3$)$_2$
Hydromorphone (422, 424): 6-C=O, 7,8-dihydro
Oxymorphone (423):   6-C=O, 14-OH
Hydrocodone (424):   3-OCH$_3$, 6-C=O, 7,8-dihydro
Metopone (425):   5-CH$_3$, 7,8-dihydro, 6-C=O

cological study at the University of Michigan (N. B. Eddy), and the clinical evaluation at the Narcotic Prison Hospital of the U.S. Public Health Service in Lexington, Kentucky (C. K. Himmelsbach, L. Kolb et al.) (426). About 150 congeners and semisynthetic derivatives of morphine were studied. Only metopone (425) emerged as a useful clinical drug and demonstrated at last that analgetic and addictive activities did not have to parallel each other directly.

If morphine is depicted without much regard for stereochemistry, several ring systems appear prominent in its formula. One can spot hydrophenanthrene, furan, dibenzofuran, an octahydroisoquinoline, and more. There are three reactive functions: phenolic OH, alcoholic OH, and a tertiary amino group. When these functions are combined they appear as —CHOHCH$_2$NR$_2$, or —CHOH(CH$_2$)$_2$NR$_2$ attached to the ring systems discernible in the morphine formula. Over 300 such amino alcohols (and amino ketones) plus synthetic intermediates were prepared, but their pharmacological success was equivocal. If Hansch analysis had been known at that time, it would have recommended increased lipophilicity in order to increase analgetic potency in laboratory animals.

Looking back at these efforts years later, the German chemist Vongerichten wrote that a coincidence, that zinc-dust distillation of morphine yields phenanthrene, had led the American workers to ten years of slave labor making phenanthrene amino alcohols. He probably was right. We never considered the conformation of the piperidine ring in morphine or its relationship to other sections of the molecule. The major advance in preparing synthetic analogs of morphine came in 1937, not through thoughtful dissection of the morphine molecule but through a masterpiece of serendipity.

The 1930s saw the appearance of many new—though not very effective—antispasmodics of the general formula ArCO$_2$(CH$_2$)$_2$NR$_2$, Ar$_2$CHCO$_2$(CH$_2$)$_2$NR$_2$, and similar structures. They were conceived as potential blocking agents of acetylcholine [CH$_3$CO$_2$(CH$_2$)$_2$N$^+$Me$_3$], which also is an acylated amino alcohol. The rules of isosterism emerging in drug design at that time emphasized the need for similarities in electronic and steric character for similarly acting drugs but not necessarily the same sequence of atoms within their molecules. "Reversed" esters appeared as a good variation that should not interfere with anticholinergic activity and incidentally avoid questions of overlap in the patent courts. A structure of this type might be $ArCH(CO_2C_2H_5)(CH_2)_2NR_2$, the aminoethyl chain being linked to a benzyl-type carbon instead of by way of an ester group. As a synthetically convenient type, piperidine derivatives connecting to both phenyl and carbethoxy in position 4 were prepared. One of them, meperidine (pethidine, Dolantin, Demerol, etc.) was singled out for pharmacological workup (208, 209). It had been synthesized by Eisleb in 1930 (427).

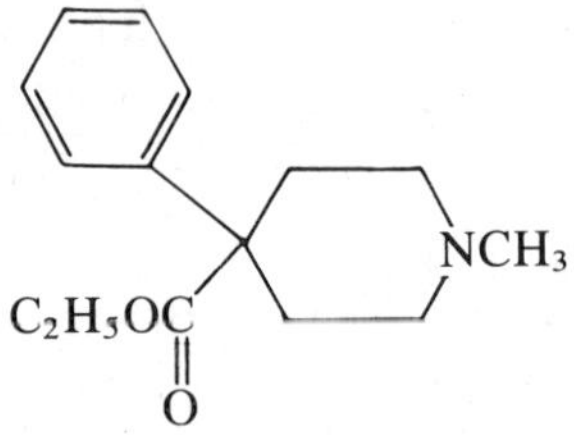

Meperidine

Pharmacologically, meperidine lived up to its expectations. It had moderate antispasmodic as well as sedative properties. When Schaumann tested it in the cat (208, 428), he was surprised by an exhibition of Straub-tail, a phenomenon associated with morphine in that species. Tests for analgetic potency revealed that meperidine had 10–12% of the overall activity of morphine. This was so unexpected that Schaumann took another look at the

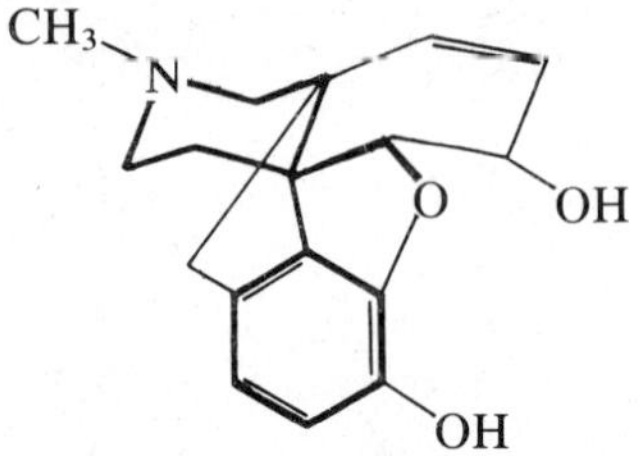

Morphine, with 1-methyl-4-phenylpiperidine-4-carbonyl emphasized

morphine formula and indeed spotted in it the segment necessary for meperidine. This deduction pointed out to medicinal chemists in general the overbearing importance of steric aspects of prototype molecules. The use of stereo-models and interpretations of "lead" structures on a conformational basis has become commonplace since that time.

The same deductions can be traced in the origin of methadone (6-dimethylamino-4,4-diphenyl-3-heptanone), a typical blocking-type structure with a bioisosteric ethyl keto group in lieu of the traditional methyl ester arrangement. This ketone and several cogeners were prepared by Eisleb in

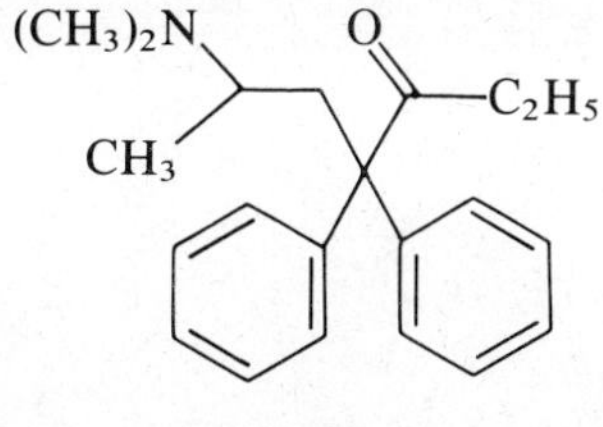

DRUGS FOR RELIEF FROM PAIN

Methadone

the course of his researches on blocking agents (429), but forewarned by the events with meperidine, he quickly discovered the potent analgetic action of methadone. If the cyclohexane ring of morphine is compared with an incipient aromatic ring, methadone constitutes a further simplification of the morphine structure. The methadone molecule has been modified in hundreds of ways, leading to a number of effective analgetics, such as dextropropoxyphene (430) and dextromoramide (431). All these compounds can be and have been used and abused, with resulting dependence liability.

Grewe synthesized morphinan, in which the furanoid ether linkage of morphine is missing (432). Its 3-hydroxy derivative, levorphanol (433) is

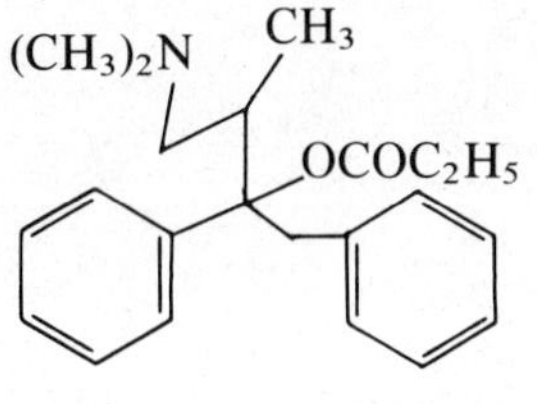

Dextropropoxyphene 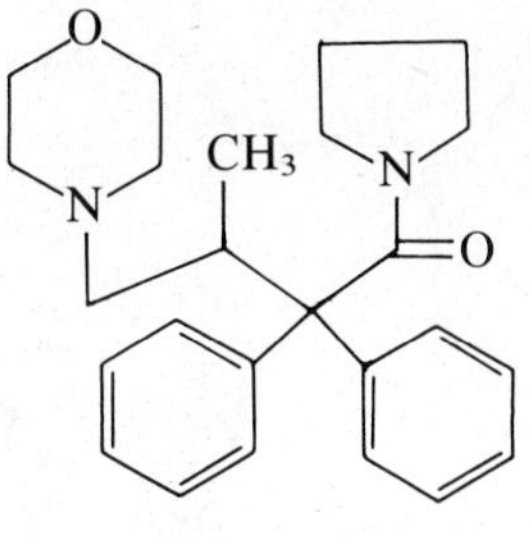   Dextromoramide 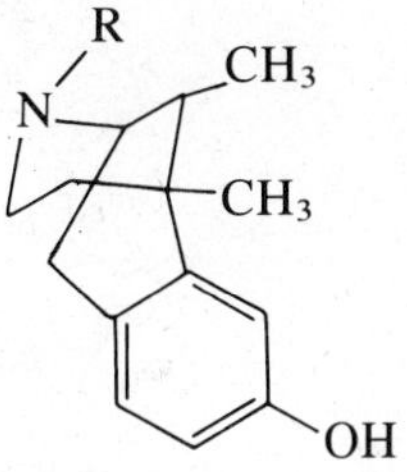   Phenazocine: R = $CH_2CH_2C_6H_5$
Cyclazocine: R = CH_2—◁
Pentazocine: R = $CH_2CH=C(CH_3)_2$

three to four times more potent than morphine but retains addictive properties. Further dissection of the morphinan skeleton led E. L. May and his coworkers at the NIH to leave only two methyl stumps where the cyclohexene ring had been (434). The resulting compounds, named benzomorphans, showed a definite though not complete separation of analgetic and dependence-producing effects. This became even more pronounced when the N-methyl group was modified as in phenazocine, which carries an N-phenethyl group (435). An even simpler structure, 5-(m-hydroxyphenyl)-2-methylmorphan, had analgetic activity equivalent to morphine in animal tests (434). This seems to be the simplest and least rigid structure still endowed with potent analgetic activity. In accord with this finding, the much more rigid oripavines, obtained from thebaine via Diels–Alder and Grignard reactions, exhibit exceptionally high potency (436). Etorphine is about 2000 times as potent as morphine, and buprenorphine has a very low level of

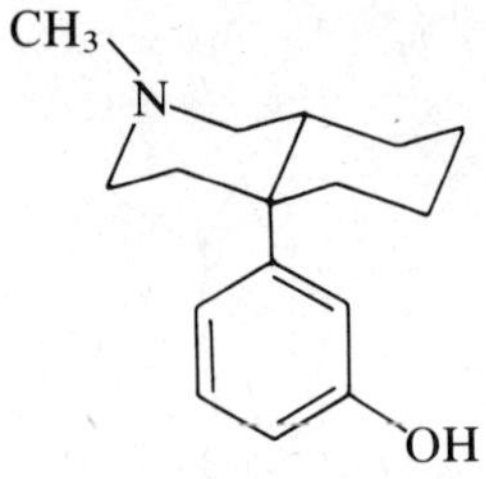

5-(*m*-Hydroxyphenyl)-2-methylmorphan

dependence liability and respiratory depressant side effects compared with those still seen in the benzomorphans. Buprenorphine also produces fewer psychotomimetic aftereffects, which are a serious problem with cyclazocine and *N*-allylnormorphine. The latter (123), patterned upon the *N*-allylnorcodeine described earlier (122), also causes psychotomimetic effects; it is a

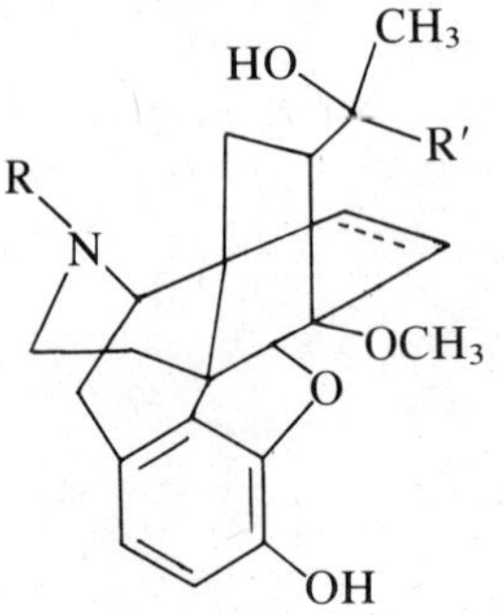

Etorphine: R = CH$_3$, R' = C$_3$H$_7$
Buprenorphine: R = CH$_2$ △, R' = *t*-Bu
(double bond reduced)

Oripavines

morphine antagonist as well as a clinical analgetic (437), although this could not be demonstrated by the hot-plate or tail-flick tests in animals. Such mixed agonist–antagonists have been encountered quite frequently (438). See (438a).

Two 14-hydroxynormorphine analogs are pure narcotic antagonists. They are naloxone (—NCH$_2$CH=CH$_2$) and naltrexone (—NCH$_2$Δ). They have helped to explain pharmacological puzzles in opioid study, and the orally active naltrexone has been used in the clinical treatment of opioid overdose.

2.6.1. Opiate Receptors

The abundance of SAR data of opioid analgetics, and expecially the stereospecificity of potent analgetics, has prompted much speculation about the nature of the opiate receptor (439, 440). The flat aromatic ring found in all potent morphine-like analgetics, the lipophilic portion, the basic nitrogen atom, and additional minor functions should have their counterparts on the receptor surface. Schematic drawings incorporating rectangles, circles, and so on, of dimensions calculated from the molecular areas of the groups in

DRUGS FOR RELIEF FROM PAIN 65

the drug molecules can be constructed but have no bearing on the actual nature of the receptor. It is generally believed that the receptor is composed of protein combined with membrane lipids, perhaps cerebroside sulfate and phosphatidyl serine since these lipids specifically bind morphine-like analgetics (441). These researches are fascinating avenues for explaining, *postfactum*, the localization of opiate analgetics and their antagonists but they have not contributed new ideas to the design of improved and more specific agents in this series.

2.6.2. Centrally Acting Antitussives

Like so many other drugs, the morphine type of analgetic possesses multiple activities. Besides those already mentioned, morphine, codeine, and other members of the group slow intestinal peristalsis and inhibit the cough reflex by a central mechanism. Codeine has been used as an antitussive for a long time in spite of its constipating action. As new potent analgetics were developed, they were tested as antitussives, after the effect of morphine and codeine had been established in animal tests (442, 443). It was soon discovered that the steric requirements for antitussive activity are not very rigid; (+)-morphine, which is devoid of analgetic activity, is an active antitussive agent although less so than (−)-morphine (444). The pharmacology and SAR of antitussives have been reviewed (445).

Cough can be suppressed by expectorants and mucolytics and by peripherally acting compounds (which may sometimes have a central component). In connection with potent analgetics, only agents that mainly act centrally will be considered here.

Modification of the structure of morphine has furnished a number of traditional antitussive analogs, led by codeine. Although codeine occurs only to the extent of 0.7–0.5% in opium, it is prepared commercially by *O*-methylation of morphine, or it can be made from thebaine by way of codeinone (446). The value of pharmacognosy may be seen in the fact that thebaine, a very minor constituent of opium (*Papaver somniferum* L. or *P. album* Mill.) is the major alkaloid of *P. bracteatum* and could become a good source of antitussives (see below). Codeine may owe its activity to its metabolic conversion to morphine, but other morphine ethers, especially pholcodine (3-morpholinoethylmorphine) are also useful antitussives, the latter suffering from fewer side effects than codeine (447). Among other structural analogs that are useful antitussives are 7,8-dihydrocodeine (448) and 7,8-dihydrocodeinone (449), in spite of their dependence liability. Apparently, the medicinal chemist can go through the same motions as in analgetic research and ask the pharmacologist to test the new compounds for antitussive activity. One way to decide what compound to look for next is to be interested in the chemical reactivity of available starting materials. For example, 6-halo derivatives exchange their halogen with azido ions ($-N^-\!\!-\!\!N^+\!\!\equiv\!\!N$); 6-deoxy-6-azidodihydromorphine ("azidomorphine") and its 14-hydroxy

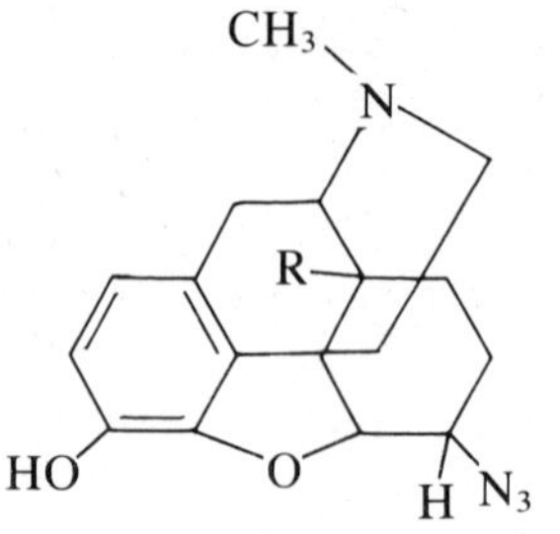

Azidomorphine: R = H
14-Hydroxyazidomorphine: R = OH

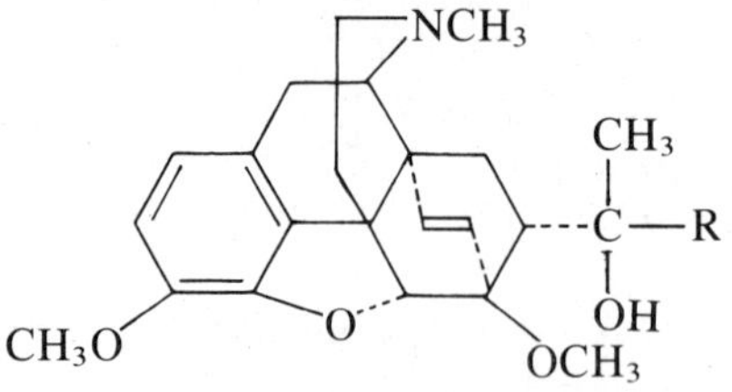

6,14-*endo*-Ethenotetrahydrothebaine
(7-substituted-16-methyl)

derivative are hundreds of times more active in antitussive tests than codeine (450, 451). Similarly, several 7-substituted 6,14-*endo*etheno-16-methyltetrahydrothebaine derivatives, synthesized from thebaine, have prominent antitussive activity (452). Such syntheses, while of great interest to organic or natural-products chemists, are not based on the logic one would hope for in drug design, except that hydrophobicity is heightened by the new moiety. When morphinans became accessible by total synthesis (432), 3-substituted derivatives were prepared and tested (433). These compounds are analogs of morphine but lack the furanoid oxygen ether atom, the 7,8-unsaturation and oxygen functions (OH, =O, etc.) in the alicyclic ring. The (−)-isomers correspond to morphine and, like the natural prototype, possess analgetic and antitussive properties and varying dependence liability. The (+)-isomers are essentially devoid of analgetic and addictive properties but are effective antitussive agents. Their clinical representatives are (+)-3-methoxy-*N*-methylmorphinan (dextromethorphan) and dimemorphan [(+)-3,*N*-dimethylmorphinan] (453). A compound synthesized as, and found to be, a narcotic antagonist (454) turned out to be a potent and long-lasting antitussive;

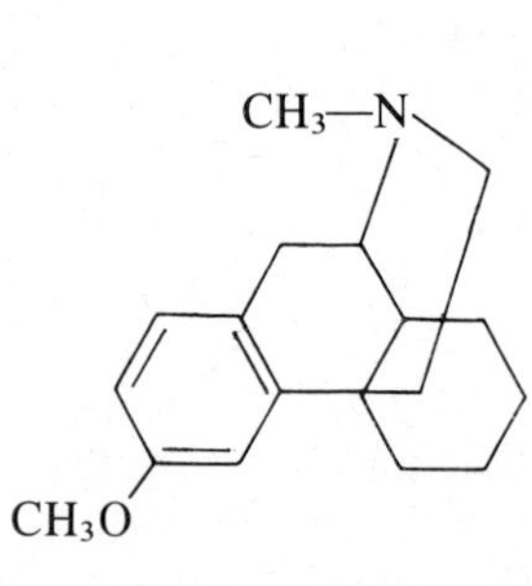

Dextromethorphan

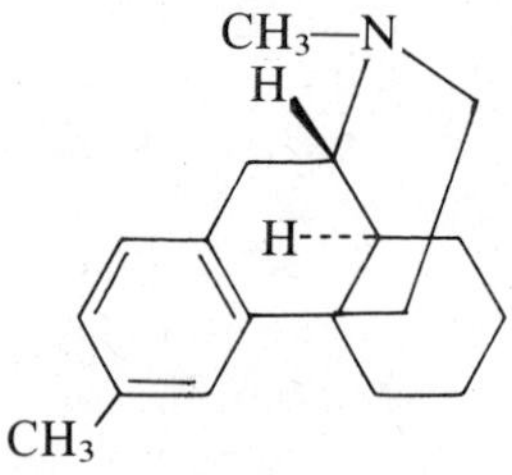

Dimemorphan

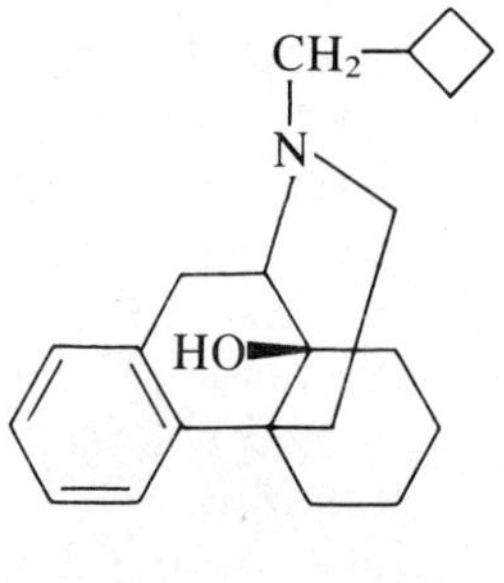

Butorphanol

its name is butorphanol. This line of discovery underscores the empirical restraints on these researches. For this reason, the subject will not be pursued further here. For other potent antitussives such as noscapine [(−)-α-narcotine], caramiphen, levopropoxyphene, and others found by screening, see reference 445.

2.7. ANTIHYPERGLYCEMIC AGENTS

About 22% of all cases of diabetes can be controlled by diet alone, about 33% by injections of insulin, and about 45% by oral hypoglycemic agents.

Insulin was isolated and purified in 1921 (455); it completely changed the prognosis of diabetic patients, especially those who suffered from the early-onset type of the deficiency. The manufacture of the hormone relies on the availability of animal pancreases from abatoirs, since the chemical synthesis (456–459) of insulin, even by the automated methods, has not furnished commercially useful quantities of the 30-amino acid peptide. Very recently, *E. coli* and yeasts have been induced by genetic engineering to yield adequate amounts of (human) insulin.

The medicinal chemist's task has included (i) the stabilization of the hormone, and (ii) experiments to vary its structure with the hope of improving duration of action and decreasing allergenicity. The addition of protamin increases the duration of action of insulin, and zinc ions prolong the activity further. Modifications of the two peptide chains of insulin have centered on co-oxidizing the reduced synthetic A chain with reduced natural B chain (460) and on converting porcin (B30 = Ala) to human insulin (B30 = Thr) (461–463). The insulin molecule contains a hydrophobic binding area at positions 24–26 that appears important for receptor binding. This is stabilized in native crystalline zinc complexes by their compact three-dimensional structures. These zinc insulins are the most active forms of the hormone, more so than any of the more than 150 synthetic analogs that have been studied (464). Insulin activity is still observed, even though to a lesser degree, if the B chain is shortened. The details of such modifications have been reviewed (465). Apart from the insulin-producing β-cells, other cells in the pancreatic islets also make glucagon, which is an insulin antagonist. It has been suggested that the ratio of glucagon to insulin is important in regulating glucose metabolism.

The D-cells of the pancreas, as well as the hypothalamus, are the source of another hormone, somatostatin, which contains 14 amino acids with two cystine groups joined in a disulfide ring. Somatostatin suppresses both glucagon and insulin secretion, its net effect being hypoglycemia. Early hopes that somatostatin represents a new "lead" for antihyperglycemic peptides have not been fulfilled. Molecular modifications of somatostatin have led to enhanced potency (466) in releasing insulin but not in releasing glucagon. Other peptide analogs suppress the release of growth hormone and glucagon but have less effect on insulin secretion. For a review, see reference 467.

Insulin and all related peptides are destroyed in the gastrointestinal tract and must therefore be administered parenterally. This is a drawback in the treatment of diabetes and has been an incentive in the search for orally active antihyperglycemic drugs.

In 1942, Janbon and colleagues (468) tested various *N*-heterocyclically substituted sulfanilamide derivatives for antibacterial properties. Among these was 5-isopropyl-2-sulfanilamido-1,3,4-thiadiazole; during clinical tests pronounced hypoglycemia was produced. This led to an extended study of other thiadiazoles (469), which revealed that their blood-glucose-lowering effect depended on the presence of a functioning pancreas gland. An inspection of the formula of the thiadiazoles shows that they contain the group $ArSO_2$—NH—C(=N—)—S, that is, an arylsulfonylthiourea type of arrangement. Therefore, Franke and Fuchs (470) tried out a series of sulfonylureas and introduced one of them, 1-butyl-3-sulfanilylurea (carbutamide), in Europe as an orally active antidiabetic agent. This drug was somewhat toxic,

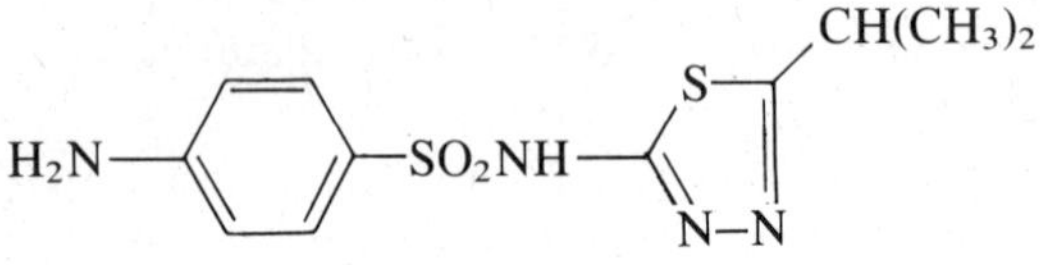

Janbon's thiadiazole "lead" (468)

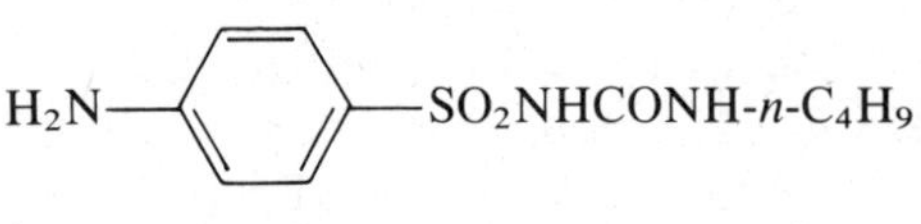

Carbutamide

but the simplicity of its structure became the signal for very extensive molecular modification. Since antibacterial activity was to be bred out of these compounds and the antihyperglycemic activity was to be emphasized, the 4-amino group that characterizes all bacteriostatic sulfanilamides was replaced by other aromatic substituents. This strategy worked and furnished dozens of candidate antidiabetics, of which only four are mentioned here.

p-$CH_3C_6H_4SO_2NHCONHC_4H_9$ Tolbutamide

p-$CH_3COC_6H_4SO_2NHCONH$—⬡ Acetohexamide

p-$ClC_6H_4SO_2NHCONH$-n-C_3H_7 Chlorpropamide

p-$CH_3C_6H_4SO_2NHCONH$-N⬡ Tolazamide

Tolazamide is actually a sulfonylsemicarbazide. All these compounds stimulate insulin release by the β-cells of the pancreatic Islets of Langerhans, but they also act by other possibly insulin-related, mechanisms.

An early observation that guanidine lowers blood sugar levels (471) was followed up by the synthesis of polymethylene diguanides $[H_2NC(=NH)NH]_2$ $(CH_2)_{10-12}$, called Synthalins. They have a fairly effective hypoglycemic action (472) but are toxic. Later (473) other biguanides were prepared and

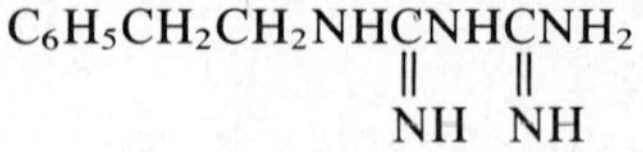

$C_6H_5CH_2CH_2NHCNHCNH_2$
$\|\|$
$NHNH$

Phenformin

GENERAL ASPECTS OF DRUG DESIGN

tested, one of which, phenethylbiguanide (phenformin, DBI), maintained itself as a clinical antidiabetic agent for several years. Dimethylbiguanide (metformin) (474), which had been synthesized earlier, was then also tested, as was the dibutyl homolog that has been used in Europe.

Many other structural types have been screened for hypoglycemic activity but there has been no chemical or biochemical rationale in these trials. The cyclopropane types of natural product, such as hypoglycins, as well as naphthaleneacetic acids, pyrazoles, hydantoins, and so on, form a long list of compounds without a thread of defendable thought. Metabolites of active drugs have been tested but no material has succeeded in combining potency with low toxicity.

2.8. GENERAL ASPECTS OF DRUG DESIGN

The preceding sections have presented short accounts of a few selected areas of medicinal research that occupied the mainstream of academic and industrial interest about 1960 and advanced the knowledge and methodology of medicinal chemistry, pharmacology, and all the medical sciences. Many other fields of therapeutic research running parallel to these developments have not been mentioned because they did not divulge novel ideas in drug design or the biochemical explanation of drug action. The early 1960's have been chosen arbitrarily, in order to review and consolidate the gains that had been made and to project trends of research which emerged at that time.

Several practical aspects of this consolidation have a bearing on this date of reflection. After the thalidomide episode, several members of the United States Congress running for reelection perceived the regulation of drug research as an issue that might have an exploitable appeal to the electorate. The Kefauver-Harris amendments to the Food, Drug, and Cosmetics Act of 1962 tightened up all aspects of supervision of drug development and gave the Food and Drug Administration broad powers to approve both the marketing of drugs and the methodology of chemical, biological, and clinical research on drug development. This tragically slowed almost all areas of drug research, but as a side effect it gave medicinal science time for unhurried thinking. Such thinking, necessary at all times, was particularly needed at that juncture because the overwhelming therapeutic development of the three preceding decades begged for a hiatus to look over calmly the state of the art. There was no slowing of new ideas, although the marked slowing of the commercial introduction of new drugs may belie this statement. But "new" drugs did not include novel drugs that kept on appearing, albeit less frequently. It should be kept in mind, however, that all human activities, including science, proceed in cycles, and that slower periods follow and precede more feverish periods of progress.

In medicinal chemistry, the early 1960's marked the surfacing of well-defined methods of quantitative studies of structure–activity relationships

(QSAR) with the publication of the linear multiple regression model by Hansch and Fujita (129), the additive model by Free and Wilson (475), and the similar interaction model by Boček and Kopecky (476, 477). Although these ideas have not affected drug research profoundly, they have added a new practical dimension to restricting the onus of systematic molecular modification.

The succeeding sections draw a general picture of some of the basic procedures that embody the characteristics of medicinal chemistry as they emerged in the past two decades.

2.8.1. Screening

All drugs are still discovered or developed by screening. Random screening of collections of chemicals is practiced less and less, although soil samples are still processed randomly in search of antibiotics, and crash programs suddenly ordered by political decisions (as in the case of cancer chemotherapy in the early 1970's) still have to be initiated by this method. The intellectual input by chemists in random screening is near zero. After the initial rush has abated and a few "leads" have emerged, preselection of analogs of such "leads," however far-fetched structurally, begins to limit the number of candidate compounds in a given test system. This is done by educated guesses, intuition, and experience in medicinal research. It is often a committee decision, especially if limits of similarity to or structural differences from available "lead" compounds have to be weighed against the cost of tests of additional substances. Statistical approaches to preselection have also been tried (478).

2.8.2. Molecular Modification

It has always been the desire of medicinal chemists to curtail the number of molecular variants to those which convenience, synthetic expediency, and structural analogy would suggest. Until recently one had to start modifying a "lead" structure on intuition and prepare and test as many modified analogs as possible. Then one could attempt to rationalize structure-activity relationships and hopefully find as few exceptions as possible to the relationships emerging from these experiments. The rules thus established usually permit cautious extrapolations from which predictions about additional analogs could be made optimistically. The proof of the pudding rests, of course, with the test results of each compound in the series.

With the advent of Hansch analysis (QSAR, 129), the initial number of modified structures became smaller than in systematic modifications based on educated guesses. Even at best, however, four, six, or more—usually 10 to 15 modifications—have to be prepared, their distribution coefficients have to be determined, and other physical properties have to be called upon before predictions can be made about SAR. Again, the biological tests of

GENERAL ASPECTS OF DRUG DESIGN

all analogs made provide the verification of any predictions one had ventured to make.

Two new approaches to such studies have been proposed. One is based on the metabolism of the "lead" compound in two species of laboratory animals. The other relies entirely on the predictable chemical reactivity of the "lead" compound, from which a chemist can deduce its transition state and often the molecular species that will react with a critical enzyme.

2.8.3. Drug Metabolism

The biotransformation of experimental and clinically used drugs has been studied extensively for almost all agents that have been candidates for clinical trials. Investigations of the metabolic fate of a drug have become part of the regulatory requirements for approval of a drug for clinical use. In most, if not all, cases the biotransformation of drugs depends on the animal species and the animal organ studied. Drugs may be bio-deactivated (detoxified) or activated under such conditions. Usually not one but numerous metabolic pathways have been observed for a given drug, some occurring simultaneously, others depending on the experimental conditions. Human drug metabolism may take the same direction as a particular animal species for one drug and quite another species for another drug. This makes the extrapolation from laboratory to clinical experiences almost impossible for unrelated structural types. These data have been reviewed in many monographs, of which only a few recent ones are cited (479–482).

The many facets of drug metabolism are of biochemical, pharmacological, and clinical interest. It should be realized that metabolism of a drug to multiple metabolites means that treatment of a pathology with this drug results in the presence of multiple foreign compounds in the body. If these compounds are biologically inactive and are excreted in a reasonable time, they offer no toxicological problems. If any one of the metabolites is biologically active or persists in the organism, complicated problems of toxic reactions may ensue.

The medicinal chemist deals with several problems of drug metabolism. If a drug is deactivated too rapidly, it may become necessary to block metabolic deactivation by substituting at or sterically hindering the molecular positions at which deactivation occurs. Of course, this creates a new synthetic analog of the drug, with all the vicissitudes of reinvestigation that are required for any new test compound. A typical early example was quinine, which is hydroxylated metabolically at position 2' to a carbostyril. Blocking of the 2'-position by aryl, CF_3, or halogens gave some antimalarially active compounds, but each with its own set of toxicological properties. By the same token, a drug whose pharmacokinetics slow down an adequate rate of biotransformation might be equipped synthetically with structural features that would promote metabolism. Again, this will furnish new compounds that have to run the gamut of complete pharmacological workup.

One example for planned activation is lucanthone, whose methyl group is bio-oxidized to methylol (CH_2OH). Synthesis of the alcoholic derivative (hycanthone) confirmed that the oxidation product was the metabolite that exerted the activity (483).

With such experiences to rely on, the medicinal chemist can venture to make certain predictions concerning the probable metabolic fate of a given drug. Apart from rare exotic biotransformations, one can anticipate stepwise oxidations, reductions, saturations or desaturations, alkylations, and dealkylations among common primary biotransformations. Occasionally a more radical oxidation may break C—C bonds or cleave rings. The primary metabolic products are frequently conjugated with sulfuric acid, glucuronic acid, amino acids, and so on, as a preamble to solubilization and excretion. Many medicinal chemists are engaged in synthesizing these metabolites. Occasionally, they will take the metabolic pathways as guides to novel "lead" structures. For example, chlorguanide, an antimalarial, cyclizes metabolically to triazine derivatives, and these have served as new "leads" to cyclic analogs, including aminopyrimidines, that ultimately led to novel antimalarials containing these structural moieties.

2.8.4. Transition-State Analogs

It is logical to expect that enzymes react most effectively with substrates elevated from their ground state to transition states, that is, with activated structures of fleeting existence. Analogs of transition-state "intermediates" might be uniquely suited to inhibit the particular enzymic reaction because they would be tailored to fit the active site of the target enzyme (484). The most plausible examples of transition-state analogs have been found in the pyrimidine series, for example, in the cytidine deaminase-catalyzed conversion of cytidine to uridine. The intermediate formed by the addition of water to the 4-position of cytidine contains the arrangement —HN—C(NH_2)(OH)—, and this could be imitated in tetrahydrouridine, which indeed binds to cytidine deaminase some thousandfold more tightly than the substrate cytidine and ten-thousandfold more tightly than the product, uridine. Tetrahydrouridine is thus an effective inhibitor of cytidine de-

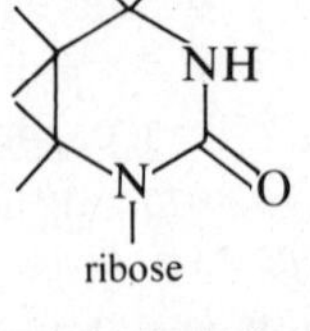

Tetrahydrouridine

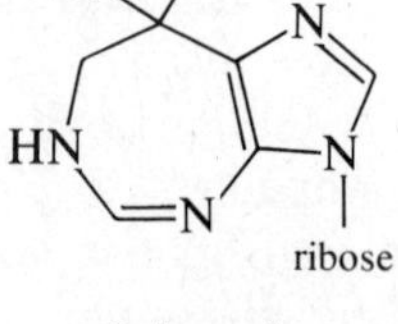

Coformycin

aminase. A similar situation exists for adenosine deaminase, which is inhibited by the antibiotic coformycin (485, 486) and its 2'-deoxy analog.

GENERAL ASPECTS OF DRUG DESIGN

Transition-state analogs have also been constructed in other structural series (487). The application of these principles to clinically useful drugs is still awaited.

2.8.5. Suicide Enzyme Inhibitors

A variation on the same theme of presenting to the enzyme a specific structure for its inhibition is seen in the so-called suicide enzyme inhibitors (488) or K_{cat} inhibitors. They are compounds that possess latent reactive functional groups which are unmasked by the enzyme, the enzyme becoming inactivated by its own mechanism of action. Such inhibitors should bind tightly (usually covalently) to the enzyme and cannot be washed out readily. The enzyme is thereby inhibited permanently, and it can be regenerated only slowly by biosynthesis (489–491). A number of examples of the amino acid decarboxylases, transaminases, racemases, dehydrases, hydroxylases, and several other enzymes have been reviewed (492).

A typical example for the suicide inhibition of aldehyde dehydrogenase *in vivo* has been presented by Wiseman and Abeles (493). This enzyme is inhibited by coprine. Coprine is hydrolyzed to cyclopropanone hydrate, which is dehydrated to cyclopropanone; the latter adds to the thiol group of the enzyme, forming a stable thiohemiketal. This product is an analog of the thiohemiacetal, thought to be the key intermediate in the enzyme-catalyzed oxidation of aldehydes. An analogous inactivation mechanism has been postulated for the irreversible inhibition of monoamine oxidase (MAO)

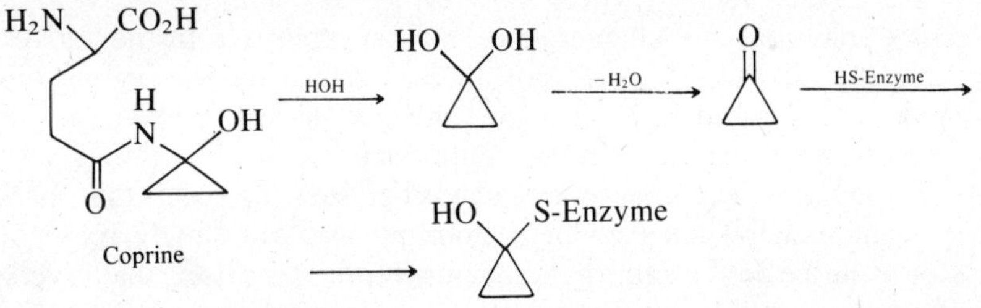

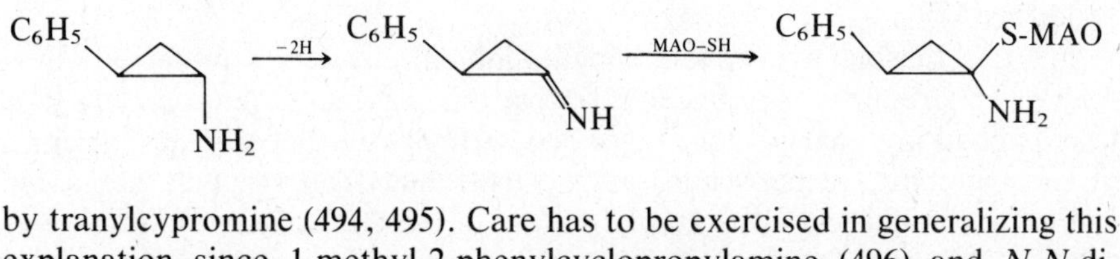

by tranylcypromine (494, 495). Care has to be exercised in generalizing this explanation since 1-methyl-2-phenylcyclopropylamine (496) and *N,N*-dimethyl-2-phenylcyclopropylamine (497) are also potent inhibitors of MAO but cannot furnish intermediate cyclopropanonimines.

As a means to improve the rational design of biologically active substances, suicide enzyme inhibition will have to be counted on with some reservations. Compounds that could become terminally unsaturated or could be converted to α,β-unsaturated carbonyl compounds by their reactions with enzymes are primary candidates for suicide enzyme inhibitors. Their selectivity may be enhanced over that of similar pre-prepared unsaturated compounds because they become additionally unsaturated while reacting with a specific biocatalyst. Most of the mechanisms postulated have been proposed to explain apparent specificity but need further corroboration (492).

Another example of a suicide enzyme inhibitor is norethisterone (17α-ethynyl-19-nortestosterone), which effectively and irreversibly inhibits estrogen synthetase (aromatase) at 2×10^{-6} M concentration. By contrast,

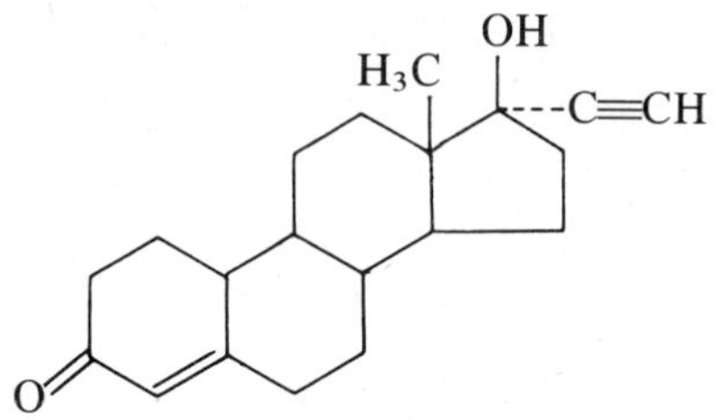

19-Norethisterone

ethisterone (17α-ethynyltestosterone), which contains an angular 10β-methyl substituent, no longer causes suicide inhibition, apparently because of the added bulk perpendicular to the steroid ring structure (498).

2.8.6. Active-Site-Directed Irreversible Inhibitors

Reversible inhibitors are characterized by their ability to fit the active site of an enzyme and deny the substrate access to this catalytic region. This is achieved by a snug fit of the analog, snugger than the substrate's to be effective. Nevertheless, the concentrations and affinity constants of substrate and inhibitor will counteract each other with the result that the substrate, resupplied by the enzyme's environment, will slowly remove the inhibitor from the active site. In therapeutic terms this means that reversible inhibition may not last long enough to allow recovery from a pathological condition.

For this reason, irreversible enzyme inhibition offers a preferred alternative. This reduces the concentration of the enzyme either completely or to a level so low that pathogenic effects virtually disappear. The "design" of such inhibitors usually aims at two molecular features. One has some similarity to either a substrate or cofactor of the enzyme or to a section of the peptide chain that has been recognized as a critically reactive part of the enzyme's active site (histidine or serine groups in trypsin, chymotrypsin, acetylcholinesterase, etc.). Such moieties direct the proposed inhibitor to the active site, presumably because of the site's recognition of structures which are analogous to the accustomed substrate. The second feature, that makes the active-site-directed inhibitor irreversible, is some reactive group

GENERAL ASPECTS OF DRUG DESIGN

that is supposed to react at a point near the active site with the formation of a firm, usually covalent, bond to the enzyme's peptide backbone or side chains. The most common reactive moieties of this kind are alkylating groups whose leaving groups interact with electrophilic sites of the enzyme by way of carbonium ions, aziridinium ions, and similar nucleophiles. The art of designing an effective inhibitor of this kind consists of placing a nucleophile at a place in the molecule where it will be most reactive and least encumbered by the bulk of the rest of the molecule, which should resemble recognition features at the active site. Since most active sites are found in steric cavities provided by three-dimensional convolutions of the peptide chain of the enzyme, the inhibitor molecule should fit sterically into such cavities, interact loosely with critical groups along the walls of the peptide chain cavity, and extend its nucleophilic group toward an anchoring electrophile of the peptide chain or a cofactor. The resulting firm attachment of the inhibiting molecule will prevent the inhibitor from being swept away by higher concentrations of substrate.

The design of such inhibitors could theoretically be done by chemically probing which of a limited number of nucleophilically equipped analogs of substrates or of proteinogenous amino acids are particularly reactive and willing to attach themselves to a simple electrophile. The proof of this approach would rest with a test reaction *in vitro* involving the enzyme in question.

This process was elaborated by B. R. Baker in a monumental series of papers and has been reviewed (499). In a few cases, where the enzyme has been sequenced and the amino acids in various locations of the active site are believed to be known, these aminoacyl or analogous aminoalkyl groups can be replaced by those of substrate analogs. More commonly, substrate analogs are attached to alkylating moieties. Among the latter we find nitrogen mustards, α-chloro ketones, sulfonyl fluorides, and similar groups. To be sure, such a selection directs our choice to a few potential candidate structures with a better-than-average chance of antienzyme activity. But the prolific output of hundreds of papers by Baker, each listing many dozens of candidate compounds, should be a warning signal that accurate drug "design" even in such a well planned series still relies heavily on screening. For a review of the present state of the art, after 15 years of valid attempts to sharpen up rationalization, see reference 487.

The specific *ab initio* design of a drug to interact effectively with a receptor site is still in its infancy, but it is at present the only experimental way of creating a "lead" without a random approach. It calls on the biochemical knowledge of a putative region of catalytically active biomacromolecules that could serve as a description of the target of drug action. One reason why more accurate and predictable drug structures cannot yet be designed on this basis is that we do not know the primary, let alone tertiary, structure of nonrigid hollow areas that enclose active enzyme sites. At the present rate of progress, this problem should become amenable to more accurate analysis within two or three decades. Concomitantly, it will be the task of organic–medicinal chemists to correlate more closely the interdependence of more specific and selectively reactive groups with substrate or amino acid analog structures. Perhaps the nucleophilicity of such alkylating agents will have to be exchanged with electrophilicity or free radical attack on positions near the active site. Such studies can be carried out in model reactions and tried out on enzymes after most of the physical–organic problems of reactivity versus bulk and steric encumberances have been ironed out.

For an account of inhibitors of dihydrofolate reductase using x-ray diffraction structures of drug-enzyme complexes, see reference 42a.

2.9. MOLECULAR MODIFICATION

Molecular modification is the chemical part of comparisons of structure–activity relationships that depend on pharmacological studies of the compounds supplied by the chemists in collaborative research programs. Indeed, molecular modification is the central scientific activity of medicinal chemistry. It is therefore of interest to explore briefly the philosophy that serves as a background for molecular modification.

As in other scientific disciplines, medicinal chemists specialize in one or several of the almost 100 areas of drug investigation. Some medicinal chemists become experts in antituberculous agents, others in antidepressant drugs, and so forth. In the pharmaceutical industry a medicinal chemist will be shifted occasionally from one major field to another, and academic medicinal chemists may enter new and perhaps unrelated research areas prompted by biochemical key discoveries or the more mundane availability of funds from research grant agencies. As part of their work, medicinal chemists build up an interest in and an intimate speaking acquaintance with the biochemical, pharmacologic, and therapeutic ramifications of their current specialty. Acquiring such background knowledge is important to medicinal chemists provided some clear biochemical contribution to the etiology of a disease can be discerned. They can seize upon such biochemical disclosures, be they on the substrate or enzyme level, and design potentially inhibitory molecules based on this knowledge. To illustrate the potential and the limitations of molecular modification, a hypothetical case of a drug with two major activities is presented in Section 2.9.1.

2.9.1. Separation of Biological Properties

Isoniazid had been found to be a highly effective antituberculous drug (187–189); it was derivatized in the normal course of events to minimize minor CNS side effects and to protect patent positions. Among its N-alkyl derivatives, N-isopropylisoniazid (iproniazid) (500) was tested extensively and

MOLECULAR MODIFICATION 77

$$\text{Py}-\text{CONHNHCH(CH}_3)_2$$
Iproniazid

introduced clinically. Unfortunately, its use was accompanied by a high incidence of psychotomimetic and hepatotoxic side effects.

The antituberculous activity of iproniazid was valuable and the drug might have become a permanent addition to the medicinal armamentarium had it not been for those side effects. The CNS effects were soon recognized to be due to inhibition of monoamine oxidase (MAO) and the consequent high concentrations of the biogenic amines that had survived degradative enzymatic deamination (501). Until the drug had to be withdrawn because of its hepatotoxicity, it was used clinically as an antidepressant.

If medicinal chemists would wish to use iproniazid as a "lead," they would have the option to rid it of either its MAO-inhibitory, its antituberculous, or its hepatotoxic effects in order to attain a drug without (or with a minimum of) side effects. If these medicinal chemists would work in the field of antimycobacterial chemotherapy, removal of MAO-inhibitory and hepatotoxic activities would be demanded. If they were members of a team of psychopharmacologists, they would want to restrict antimycobacterial and hepatotoxic effects. In either case, molecular modification would have to be undertaken according to some thoughtful scheme. But what would be the differences? What can one change in iproniazid anyway? Assuming that metal chelation plays a role in its antibacterial action, substitution of the hydrazide hydrogens is bound to introduce unwanted bulk. Should the pyridine ring be reduced? Probably not, because flat aromatic areas are needed in so many instances. Substitution of the pyridine ring offers innumerable options, but which of these might be conducive to enhance one of the two useful activities and decrease the other one? Should the pyridine ring be replaced by other aromatic rings? Perhaps yes if MAO inhibition is to be increased and antituberculous activity decreased. This regularity was elaborated by systematic molecular modification of aralkyl hydrazines and aroyl hydrazides without initial thought of iproniazid as a "lead," that is, it would be an afterthought rather than planned strategy. Should the hydrazide nitrogens, one or both of them, be replaced isosterically by O or CH_2? Are there predictable analogies for structures such as $PyCOCH_2NHR$ or $Py\text{-}CONHCH_2R$ to enhance one of the two properties in question? Again, a few such analogies can be culled from the literature and might serve as modest guidelines for providing antimycobacterial results. Or if N-isopropylhydrazine is lost by drug metabolism, should this be inhibited by sterically hindering substituents in the 3 and 5 positions of the pyridine ring?

Additional questions of this kind may be posed for a molecule as simple as iproniazid and answered without a firm decision. After a number of examples of a given type of modification have been produced, measurements of partition coefficients and one or two other physical constants might show some worthwhile trends but there will be disappointing exceptions. In other words, any type of molecular modification will give rise to some SAR data, but *a priori* predictions for the separation of two unrelated biological properties remain unreliable.

In more complex molecules, where one modification could be applied in many structural places, the situation will become too complicated to be followed logically. What these examples try to show is that molecular modification for the enhancement of only one of two biological properties at the expense of the other will follow more or less the same pathways until a trend is established. This takes about a dozen compounds in a series, perhaps 8–10 if Hansch analysis can be applied at that point. This means that the medicinal chemist is restricted to the intelligent *methodology* of molecular modification but cannot yet aspire to lay out a reliable course of action in a new series of compounds. Experience and conclusions based on analogs in similar series will be the only aids in steering the chemist in an acceptable direction.

For these reasons, the methodology of molecular modification will be examined as the one area in which medicinal chemists have rationalized an originally unruly and random choice of compounds to be tried. For the sake of simplicity, one can hope to achieve either enhancement of potency or decreases in toxicity—seldom, if ever, both.

2.9.2. Methodology of Molecular Modification

The oldest starting points for molecular modification were naturally occurring biologically active substances of plant and animal origin. The products we succeed in isolating were not meant to be therapeutic agents in medicine. If they happen to have therapeutic activity, this is just our good luck, but it would be too much to expect that they be free from side effects. Therefore, molecular modification will improve selectivity of action in many cases and will simplify the structural units necessary for biological activity if we remove systematically those groups or moieties that serve no therapeutic purpose.

Such trimming of the structure of natural products has been done in very many instances. One of the earliest examples was the alkaloid cocaine, whose carbomethoxy group and tropine skeleton were cut down stepwise until the dialkylaminoalkyl benzoate structure emerged as the segment responsible for local anesthetic activity. Similarly, the morphine structure could be simplified, as shown in the benzomorphans, morphinans, and related analgetically active fragments. Many hormonally active polypeptides have been pared down to tri-, tetra-, or pentapeptide segments and similar small units that retain all or most of the original biological activity.

There are no general ground rules for such degradative studies. Obvious functional groups and substituents are removed first or occasionally modified (an amide for an ester group, a carboxyl being reduced to CH_2OH or CH_3, etc.) in order to probe the significance of the naturally present function for the biological activity. This must be done step by step so that one can evaluate each structural change. However, this should not discourage chemical boldness; a case in point is the deletion of the "alcoholic" ring of morphine, leaving two methyl groups as stumps after this amputation, as it was done by May (434) with the benzomorphans. In saturated condensed ring systems, it may be necessary to probe the effect of stereochemistry on the biological activity. If one of the rings is tilted with respect to the other one, as in decahydronaphthalenes, one of the geometric isomers may fit into the receptor space while the other cannot be accomodated. As Pauling put it (502),

> The specificity and the physiological activity of substances is determined by the size and shape of molecules, rather than primarily by their chemical properties, and the size and shape find expression by determining the extent to which certain surface regions of the molecules can be brought into juxtaposition.

If direct isomerization cannot be achieved, it may be necessary to synthesize the unnatural stereoisomer and test it. The first experiments in such series will be empirical; additional cases can probably be reasoned out by analogy.

The most appealing naturally occuring "leads" are found among the products of intermediary metabolism. Here is a vast reservoir of prototype substances that wait to be imitated by molecular modification. The altered and modified analogs have a better-than-average chance to play a role as metabolite antagonists and thus get enmeshed in the metabolic network. How far these alterations should be pressed is a matter of judgment and luck. It would be ill-advised to think that only closely related analogs can replace natural metabolites effectively. In most cases one wants to obtain antagonists rather than agonists of the natural metabolites. There will be a steady transition from agonist to antagonist and usually a region in which the synthetic analog will exhibit both activities depending on the conditions of the test. Since the antagonist, unless equipped with an alkylating group, must have functional groups that anchor it more firmly at the receptor than the agonist, any aromatic rings, unsaturated groups, and other features that increase local electron densities will increase the likelihood of electron transfer reactions to the target macrobiochemical. The soundness of these thoughts has been documented in very many cases and will be summarized in the discussion of bioisosteric replacements where electron distribution and analogy of steric conditions are intertwined.

It is fortunate if one can pinpoint any disease-related metabolite that might lend itself to molecular analogization. The advances in the biochemistry of diseases have singled out ever new biochemicals to which one should turn one's attention. Until a short time ago, for example, fatty acids and their derivatives had been given relatively little heed as "lead" compounds although such substances as the tuberculostatic chaulmoogric and hydnocarpic acids had long been known, and tuberculostearic and related acids begged for modification. With the revival of arachidonic acid and the realization of its chemical conversion to prostaglandins, thromboxanes, and so on, the incentive for molecular modification of such structures was increased greatly. Other structural areas have not yet been tapped and offer welcome alternatives to the traditional choices of biochemical prototypes such as biogenic amines, amino acids, purines, pyrimidines, and other structures related to peptides and nucleotides. The distinction between designed antimetabolites and synthetic compounds selected from screening procedures becomes blurred if the synthetic substances are later found to be involved in biochemical inhibitions. Even if they have been discovered by random screening, structural or functional analogies (redox possibilities, chelating ability, etc.) can sometimes be read into their formulas *post factum*. An experienced medicinal chemist might select them from a random collection of chemicals on the basis of such considerations. In any case, synthetics with interesting biological activities or analogies have offered more "lead" possibilities than any other choice of selection.

2.9.2a. Biological Guidance. Considerable imagination is required for the molecular modification of existing experimental or clinical drugs whose side effects, discovered during pharmacological study, suggest a use in a disease unrelated to the anticipated utility of the agent. The best known examples are the bacteriostatic sulfanilamides whose minor antihyperglycemic and diuretic activities have been raised to practical utility by molecular modification.

One such property of sulfanilamides is the production of alkaline urine and metabolic acidosis (503). This is a consequence of the inhibition of carbonic anhydrase (504) in the kidney (505) by sulfanilamide and leads to increases in Na^+ and HCO_3^- excretion, a diuretic potential of the drug. The diuretic potency of sulfanilamide itself is too low, and many N^1-substituted derivatives, whose bacteriostatic action had already been studied, were therefore tested. *In vitro* carbonic anhydrase inhibition was observed best with N^1-heterocyclically substituted derivatives (506, 507). Continued molecular modification (over 2000 analogs) ultimately abandoned the sulfanilamide skeleton and placed the sulfonamide group into heterocyclic nuclei, for example into 1,3,4-thiadiazole, in which the traditionally untouchable primary amino group of the bacteriostatic sulfanilamides was acetylated. The result was acetazolamide, a clinically useful carbonic anhydrase inhibitor that causes diuresis and also, by inhibiting the enzyme in the eye, relieves intraocular pressure in glaucoma.

In an effort to improve carbonic anhydrase inhibition further, benzenesulfonamides that no longer contained a *p*-amino group were studied just as they had been encountered in the antihyperglycemic series. A large number

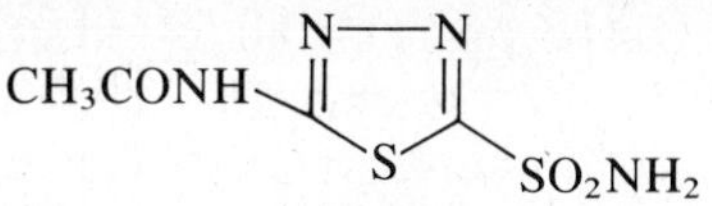

MOLECULAR MODIFICATION 81

Acetazolamide

of benzenesulfonamides with electron-withdrawing substituents were tested, and the best results were obtained when a second sulfamoyl group was introduced *meta* to the first one (508). Such disulfonamides markedly increased the excretion of Na^+, K^+, Cl^-, and $^-HCO_3$ and increased the Cl^-/Na^+ ratio. Derivatives carrying a carboxyl group were particularly effective. Since *m*-disulfonamides undergo cyclization readily to 2*H*-1,2,4-benzothiadiazine 1,1-dioxides (292), such compounds were included in the testing program and were found to be potent orally active diuretics (509). Chlorothiazide can be reduced at the N=C(R)— bond to give hydrochlorothiazide, which has become a preferred diuretic agent. The thiazide diuretics also effectively lower elevated blood pressure.

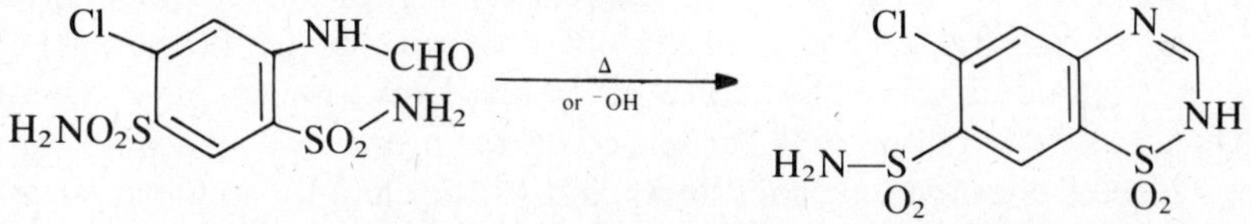

Chlorothiazide

The sequence of events that led from sulfanilamide to (hydro)chlorothiazide teaches a number of lessons in molecular modification. First, the inhibition of carbonic anhydrase was much more pronounced in N^1-heterocyclically substituted sulfanilamides than in the original unsubstituted prototype, and this suggested expansion of the "lead" to heterocyclic sulfonamides. For modifications of other drugs in other series the lesson is not to restrain oneself to immediate and obvious derivatives and analogs of the "lead" but to look at analogs with different structural "backbones" equipped with similar functionalities, as long as the overall shape of the new molecules is similar to that of the prototype. Careful stepwise molecular changes can reveal better than anything else potential mechanisms of action and the properties required for potency in a given biological test. However, such restrained modification is liable to furnish compounds with a profile of action very similar to that of the "lead." If one is satisfied with that activity profile, there is no need for modification. The common scientific reason for undertaking molecular modification, let alone legal and commercial considerations, is to improve the therapeutic profile of the "lead," that is, to increase potency and specificity. Therefore it becomes reasonable to probe among a small number of not too closely related analogs at an early stage of molecular modification. Why is this not done routinely? Because of inertia: the prototype compound and a few closely related derivatives must have been obtained by chemical methods well studied for the necessary reactions. Obviously, it is easier to adhere to these methods than to work out new ones that may lead to more complex structures.

The second lesson of the chlorothiazide story is the need to study the chemical reactions of the primary or secondary "leads", in this case *o*-amide-substituted *m*-benzenedisulfonamides. Some medicinal chemists may have shied away from the transition of the well-established monocyclic *m*-benzenedisulfonamides to bicyclic analogs. Such hesitation is less likely to happen after this and other experiences in various structural series but it will persist to some extent in the literature.

2.9.2b. Homologation and Chain Branching. Any compound containing hydrogen linked to carbon, nitrogen, oxygen, sulfur, and so on, can be homologized by substituting for the hydrogen any alkyl group, saturated, unsaturated, unbranched or branched. Longer alkyls may also contain carbon chains interspersed by hetero atoms, such as ether oxygen, imino nitrogen, sulfur, and so forth. The biological properties of homologous compounds in series of this broad definition show regularities of increase and decrease depending on the structure of the alkyl substituent. On the whole, lengthening of the (saturated) carbon chain from one to 5–9 atoms causes an increase in pharmacological effects, but further lengthening brings about a rather sudden decrease. This had been observed as early as 1869 for the cell-depressant ("hypnotic") activity of alcohols (34). A maximum is reached for 1-hexanol to 1-octanol, then activity declines with a further lengthening of the alkyl chain and disappears in hexadecanol. Branching raises activity, and so does transition from primary through secondary to tertiary alcohols in an isomeric series. Similar increases paralleling lengthening of the chain up to a maximum, followed by a parabolic decline if the chain is increased further, have been documented in many instances (510). These observations parallel the lipophilicity of the respective homologs, and for secondary and tertiary alkyl compounds, their ability to participate in S_N2 reactions. One of the earliest quantitations of relationships between lipophilicity and (local anesthetic) potency was expressed by Overton in 1901 (128). In more recent times, mathematical relationships between chain length, boiling and melting points (criteria of intermolecular cohesiveness), and molecular volume have been calculated for unbranched aliphatic hydrocarbons and simple derivatives (511).

As in any reaction between two chemicals, physical adsorption is the first step in the reaction between drugs and cell chemicals. This is the reason why the fit between these reagents is of prime importance; it permits the two reagents to establish loose bonds whose sum total amounts to sufficient strength to hold them in position for chemical interaction by whatever mechanism is involved. The adsorption process is reversible as seen, for example, in the termination of local anesthesia as soon as the anesthetic has been swept away or metabolized. This is expressed in the Michaelis–Menten

$$S + E \underset{k_2}{\overset{k_1}{\rightleftharpoons}} ES \overset{k_3}{\longrightarrow} E + P$$

equations (512) for enzyme-catalyzed reactions of a substrate S, where E is the enzyme, ES is the complex formed by adsorption of S to E, P represents the reaction products, and k values are the rate constants associated with each step. Good steric fit will assure the stability of ES and prevent its reversal to S + E by direction of k_2. This will give the k_3 reaction a chance to yield product or products P. The irreversible effects leading to cell death must be due to deep changes of solubilities and absorption rates that disorganize the normal chemical reactions within the cell. For this purpose, penetration of the cell membranes is a prerequisite, and this is favored by increased lipophilicity, up to a point. When the distribution coefficient no longer reflects solubility in water, no further biological activity can be expected. The chemist's ability to predict this range by chemical model reactions will make possible an *a priori* choice of a cutoff point for biological testing. As one goes up a homologous series, toxicity to cell processes as measured by the equitoxic reactions in the external cell medium decreases by a factor of three with each additional carbon atom (513). For attempts to predict the position of the highest active member of a homogous series from the effective concentration (thermodynamic activity) of equiactive homologs, see references 514–516.

Branching of alkyl chains introduces steric bulk protruding from the parent backbone. This can result in impeding the chemical reactivity of functional groups in the vicinity of the branch. An example is phenethylamine, which is deaminated rapidly by monoamine oxidase whereas its methyl homolog, amphetamine $[C_6H_5CH_2CH(CH_3)NH_2]$, is a slow substrate and a relatively poor inhibitor of the enzyme. Primary amines (RNH_2) are often more (selectively) toxic and more active than secondary amines such as $RNHCH_3$, and the latter are followed by tertiary amines, for example, $RN(CH_3)_2$. Examples are seen in the series of 8-aminoquinoline antimalarials, where primaquine is much more effective than its secondary and tertiary amine homologs, and in desipramine, which is more active as an antidepressant than its parent drug, imipramine, from which it is derived as a metabolite.

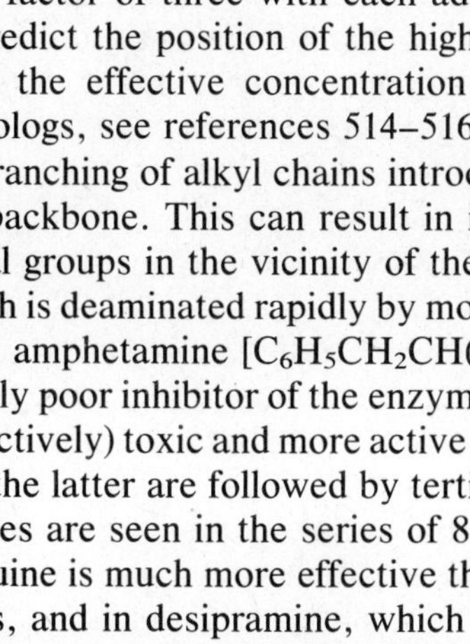

Primaquine

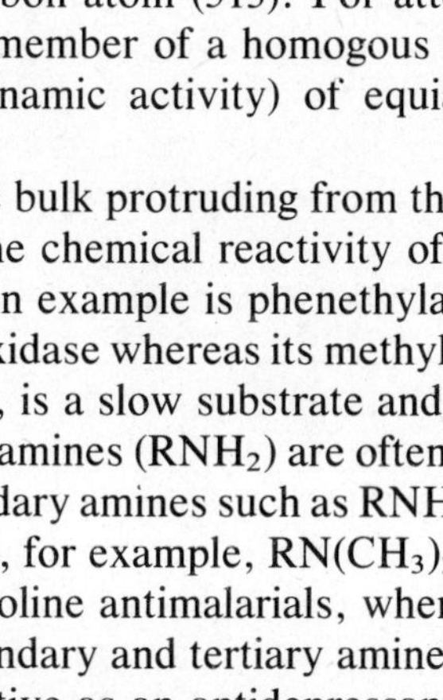

Imipramine: R = CH_3
Desipramine: R = H

If an unsaturated alkyl such as allyl or propargyl is introduced as a replacement for a small alkyl group, reversals of activity are sometimes experienced. Such instances are seen in *N*-allylnormorphine, *N*-allylnormeperidine, and similar compounds that go through a transition from potent analgetics through mixed agonist-antagonists to pure analgetic antagonists (naloxone). It might be interesting to contemplate whether the unsaturated analogs may act as suicide enzyme inhibitors in the biosynthesis of the endorphins.

Homologation has played an important role in the development of neuroleptic and tricyclic antidepressant agents from antihistaminics. The antihistaminics contain, in general, the chain —C—C—NR_2, and although some of them have a neuroleptic component, antipsychotic activity is not unveiled in most cases until the carbon chain is lengthened to C_3—NR_2. The slight increases in lipophilicity achieved by this homologation cannot explain this tilting of activity profiles. Neither can homologation of the traditional —$N(CH_3)_2$ group, which usually damages pharmacological utility, nor the advantage gained by incorporating the tertiary amino group of the side chain in a ring be attributed to changes in one or two physical properties of these compounds. Of these properties, lipophilicity has been emphasized by many investigators. It is of great importance for the crossing of channels in the cell membranes and can be measured readily by determining the distribution coefficient of the compound between water and a water-immiscible solvent, usually 1-octanol (517), but many other physicochemical parameters have to be considered (518).

2.9.3. Bioisosterism

In SAR studies and drug design it is always necessary to compare the formal and three-dimensional structures with the substituent and functional groups of compounds that show a similar spectrum of biological activities. In most instances one will find similarities in molecular shape and overall chemical functions and will base one's explanation of biological similarities on these resemblances. This total complex of analogies that comprises steric, electronic, and molecular orbital comparisons is called bioisosterism.

The virtual equivalence of many physical properties of benzene and thiophene, and to a lesser extent of pyridine, were noted by Hinsberg (3). Obviously, this equivalence must be attributed to the mutual interchangeability of divalent sulfur and vinylene (—CH=CH—), and of trivalent nitrogen ((—N=) and carbon (—CH=), if one disregards the hydrogens on carbon. Hückel (519) extended the concept of equivalents to other compounds, comparing (not very accurately) methyl to fluorine, methylene to nitrogen, and methyne (CH≡) to oxygen. When quantum theory took over, thiophene could be pictured as a hybrid of six resonance states (520), and molecular orbital treatment provided evidence that *d*-orbital participation should explain the structural resemblance of thiophene and benzene (521, 522).

A property to which much attention was paid in the early stages of studies on isosterism was isoelectric distribution. As long as only physical properties of pairs of compounds were compared, this offered an attractive point of departure. The physicist Grimm (523) accordingly arranged a few atoms and groups according to their total number of electrons (we would now think of orbitals). Some examples with a bearing on organic compounds in which such atoms and groups may occur are listed below (523):

Grimm's Hydride Displacements (Number of Electrons)

6	7	8	9
=C=	—N=	—O—	—F
	—CH=	—NH—	—OH
		—CH$_2$—	—NH$_2$
			—CH$_3$

Molecules differing only by hydride substitution, as in the vertical columns of this table, may be chemically quite different but similar in some physical properties. Of course, the size of a compound increases as one descends the vertical columns, and in addition, the properties of the resulting compounds must not be influenced strongly by dipole moments. Thus, even in such simple cases the role of several molecular properties has to be considered, and a compromise between these several effects has to be reached in formulating explanations of physical similarities. These overlapping and mutually influencing effects defy complete accuracy of pinpointing structure–property comparisons. In medicinal chemistry, the added overbearing dimension is the reaction of each compound at macromolecular catalytic sites of unknown structure. Even if one shuts one's mind to this undefeatable reality, the concept of bioisosterism, as useful as it is, cannot be equated with undeniable evidence. It has, however, remained the best possible approximation for explaining and predicting chemical and biological similarities and analogies for 50 years and has not been replaced by more defendable assumptions. Even the various schemes of mathematical evaluations of SAR have not eliminated exceptions although, in structurally limited series of compounds, they have narrowed down the approximations estimated by educated guesses within the framework of bioisosterism.

It is a question of semantics how bioisosterism is defined. The original definition (183) described the phenomenon that compounds of related chemical structure have similar or antagonistic biological properties; that is, they exert some activity at the same biologically active site. Little comfort can be gained from designing a structure to give a certain biological result only to find that the hoped-for action has been reversed. At least the new compound will not be totally inactive in a given test. It is more advantageous to regard bioisosteres as groups or molecules that have chemical and physical similarities producing *broadly* similar biological properties (196). Readers interested in the shifting concepts of this evolving field and striving for greater reproducibility in structurally diverse series should consult reviews in which various similarities of physical properties have been stressed (129, 192, 194, 195, 524).

Bioisosteric replacement is the principal guide followed by medicinal chemists in developing analogs of a "lead" compound, whether as agonists or antagonists of biological effects. The parameters being changed are molecular size, steric shape (bond angles, hybridization), electron distribution, lipid solubility (= hydrophobicity), water solubility, the pK_a, the chemical reactivity to cell components and metabolizing enzymes, and the capacity to undergo hydrogen bonding (receptor interactions). Even if the bioisosteric replacement is relatively minor (Cl for CH_3 or vice versa), many of these parameters will be disturbed. In the case of Cl vs. CH_3, size, similar shape, and electronic effects will predominate; if these are important, the overall properties of the two compounds may be adequately similar. But Cl may block metabolic hydroxylation whereas CH_3 may be bio-oxidized and the compound may thus have a shorter half-life.

The purpose of molecular modification is usually to seek subtle changes in the compound that should not alter some properties but change others in order to improve potency, selectivity, duration of action, and reduce toxicity. Bioisosterism makes it possible to limit some of these changes. All things considered, retention of overall molecular shape is the overriding condition (502) for analogy of action. Changes in lipophilicity can be read into the formula of a compound qualitatively by an experienced chemist without actually checking on the distribution coefficient. Any reduction of lipophilicity will also reduce the ability of the compound to penetrate cell walls, and the drug will be absorbed and transported less effectively. This can be counter balanced by substituting the molecule with lipophilic groups at sterically undemanding positions. The "bioisosteric" replacement compound may thus look quite a bit different from its prototype. Increasing selectivity relies upon retaining desirable properties and varying unimportant parameters. However, an unimportant parameter for the established biological activity may be a decisive factor for a side effect.

In the design of bioisosteres, an appreciation of the biochemical mode of action may play an important role. For example, aspirin acetylates prostaglandin synthetase and thereby deactivates this enzyme, which ordinarily catalyzes the biosynthesis of nociceptive prostaglandins. Isosteres of aspirin [o-$CH_3COXC_6H_4COOH$] (aspirin: X = O) in which the phenolic oxygen atom has been replaced by "classical" (523) isosteric groups or atoms are inactive because they cannot release the acetyl group at all (X = CH_2) or at an adequate rate (X = S, NH). Another aspect of biochemical activity is hydrogen bonding of a compound to the carbonyl groups of the peptide backbone or to various functional side chains of enzymes and probably receptor sites. The bonding groups do not have to be located in an identical manner to attain hydrogen bonding. This means that isomers and reversed

QUANTITATIVE STRUCTURE–ACTIVITY ANALYSIS

functions (—COOCH$_3$ and —OCOCH$_3$) as well as functionally similar structures (—COOR and —COCH$_2$R) can be substituted for each other bioisosterically. Such changes permit a much wider choice of bioisosteres than the original physically narrowly defined examples (1). An excellent list of successful bioisosteric replacements in many structurally and biologically different fields has been published by Thornber (196, Table 4). The same article also lists examples (196, Table 3) of groups that can be used as candidates for bioisosteric replacements regardless of the biological activity desired. In this way, chemists can make SAR predictions on their own with a fair degree of hope that biological testing will confirm the chemical predictions. Typical and widely acclaimed examples of such bioisosteric groups are

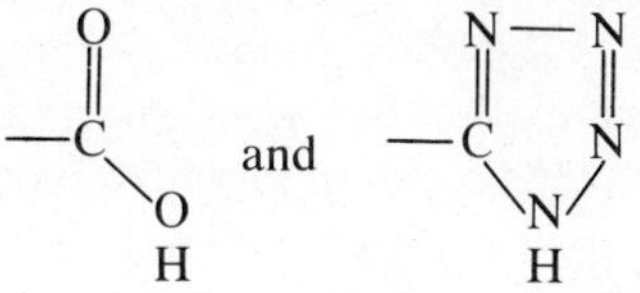

$$-\underset{\underset{H}{O}}{\overset{O}{\underset{\|}{C}}} \quad \text{and} \quad -\underset{\underset{H}{N}}{\overset{N=N}{\underset{\|}{C}}}\underset{N}{\overset{}{\diagdown}} $$

in which both steric and ionizing analogies are considered, and

$$\text{Ar—OH}^+, \text{ArNH}^+\text{SO}_2\text{CH}_3, \text{ArNH}^+\text{CN}$$

where steric similarities have given way to the more striking analogies of similar ionization. The examples of replacing thiourea groups by *N*-nitro- or *N*-cyanoguanidine groups have been mentioned previously (399, 401) (Section 2.4).

2.10. QUANTITATIVE STRUCTURE–ACTIVITY ANALYSIS

In the past two decades, mathematical and statistical analysis of SAR has been used increasingly to supplement, verify, and sharpen up SAR analysis based on intuition and experience. The quantitative methods (QSAR) have remained equivocal; they have not been able to replace the intuitive approach although they have been of aid in reducing the number of educated guesses in molecular modification. Nevertheless, they have contributed directly to the philosophy and practice of drug design and medicinal research.

After a "lead" has been discovered by any of the empirical methods discussed in Chapter 2, QSAR can help in recalling similar structures or biological activity profiles by computer analysis. With only the traditional intuitive experience, the medicinal chemist must determine, by a process not unlike gambling, where to turn at the end of a small-sized or medium-sized series of analogs synthesized to evaluate a "lead." QSAR may calculate the significance of the analogs at hand and thereby suggest additional analogs to be synthesized. If the mathematical analysis is incisive enough to detect conformational needs in a given test series, expansion of purely chemical substitution may be augmented by stereochemical studies. Finally, if a series of analogs shows no discernible trend of increased potency or specificity, QSAR may indicate that the best compound obtainable in this series has already been prepared and that work on these compounds should be terminated. Since it is always a painful decision to give up a research project, the moral support by one's mathematical—and therefore supposedly more authoritative—colleagues is appreciated by the "noncomputerized" chemists in the laboratory. If a competitor later finds an improved trend in the abandoned series, one can blame the mathematical statisticians for the decision to abandon the work.

Description of the molecular structure, electronic orbital distribution, reactivity, reaction rates, and the role of structural and steric components and substituents of chemical—mostly organic—compounds has been the subject of mathematical formulation by physical–organic chemistry for half a century. Its conclusions were based on physical measurements, and quite naturally those measurements were preferred that could be performed most easily and conveniently with the available instrumentation. Among these are the determinations of pK_a and lipophilicity. Lipophilicity (hydrophobicity) can be measured readily by distribution of the compound between an aqueous and nonaqueous, water-immiscible solvent. The optimal nonaqueous solvent chosen is 1-octanol (129, 525). The octanol–water partition coefficient is designated P, and the Hansch value π is the effect of a given substituent on log P. The importance of log P for drug potency had been appreciated earlier by Fieser (526) and by Brodie and Hogben (527). The term hydrophobicity was used in their studies instead of the equivalent term, lipophilicity, but did not become a central feature of drug design until the seminal work of Hansch (129, 525).

The other physical property most widely measured at an early date was the pK_a. An optimum was observed for antibacterial sulfanilamide derivatives (528), and dissociation constants were measured for adrenergic amines (529). The contribution of QSAR after 1964 was to quantitate and evaluate relationships between such physical properties and biological activities and to improve the methodology of the measurements. For this purpose, the chemical structure of a compound has to be transformed into a set of numerical descriptors, much as methods of abridged nomenclature have been used to list structures of even great complexity in abstracts journals or for computer memories.

The most widely acclaimed QSAR method is the linear free energy or extra-thermodynamic method of Hansch (129, 525). It assumes an additive effect of various substituents in electrostatic, steric (repulsion), hydrophobic (lipophilic), and dispersion data in the noncovalent interaction of a drug and biomacromolecules. These are principally the following: (i) the Hammett σ values, which are the logarithm of the effect of the substituent on the acid dissociation constant of benzoic acid; (ii) the Taft E_s value, which can be calculated from the rates of ester hydrolysis and is proportional to the radius of spherically symmetric substituents; (iii) the Hansch π value as defined above; (iv) molar refractivity (MR), which is derived from the effect of a substituent on the refractive index of a compound and is a parameter of

$$\log \frac{1}{c} = A(\pi) + B(\sigma) + C(E_s) + D(MR) + a$$

dispersion interactions. The Hansch equation expresses the proportionality of these factors where log 1/c is the relative potency of the analog and c the concentration required to produce some standard biological response in a test procedure. The coefficients of the constants A, B, C, and D are fit by least-square multiple regression analysis. π is the partition coefficient value for lipophilicity, σ is Hammett's value for the electronic property, and E_s is Taft's steric parameter describing the size of the substituent group. This equation has been elaborated best for simple aromatic substituents which, if desired, can be varied bioisosterically. If B and C were zero, then the potency would be a function A of π only. In case of predictive drug design, substituent groups would be bioisosteric if they have similar π values and would be independent of their σ and E_s values (530).

If one does not wish to measure one or several physical properties of a compound, one can avail oneself of the method by Free and Wilson (475). This statistical method makes the assumption that the introduction of a particular substituent at a particular molecular position always leads to a quantitatively similar effect on biological potency of the whole molecule, as expressed by the equation

$$\log \frac{1}{c} = x + \sum^{mn} a_{ij} G_{ij}$$

In this equation c is the same as in the Hansch equation, i is the number of the position at which substitution occurs, j is the number of the substituents at that position, m is the total number of the substituent positions, and n the total number of substituents. The term a_{ij} refers to the presence (1.0) or absence (0.0) of the substituent or substituents j at position i. The group contribution values of G_{ij} are obtained by multiple regression analysis.

Best results with the Free–Wilson calculations are obtained in series with several positions available for substitution, and only if each substituent at any location is present in at least two compounds of the series.

A number of hybrid Hansch/Free–Wilson equations have been proposed. They are summarized in a recent review chapter (531).

QSAR have been used to forecast biological activities with varying degrees of reliability. A number of success stories are listed in an excellent review by Martin (532). This review also catalogs the lack of universal success of predicting potency and mentions the following preventable circumstances that had led to failure. "(1) The prediction was based on a poorly designed series or an invalid or ambiguous regression equation; (2) it was based on an extrapolation outside the range of the physical properties represented by the original substituents, . . . and (3) the conditions of the biological tests were different." One can expect that such experimental errors will occur less frequently as more medicinal chemists become expert in QSAR and that the reliability of the method will improve.

On the other hand, statistical data can only furnish average values, and individual variations are disregarded by statistical methods. In addition, a biological effect is barely ever the result of a single stimulus or inhibition. In almost all cases of drug action, a cascade of stimuli or inhibitions contributes to the total profile of the biological effect. Computer memory helps to balance the influence of several physical properties and metabolic rates on the action of a drug, but it is not yet comprehensive enough to assure the predictability of biological effects on the basis of calculated statistical data. For details of difficulties confronting fundamental chemical questions that need to be considered (a lack of absolute ways to describe molecules, descriptors of their three-dimensional features, and difficulties in correlation of hydrophobicity and electronic effects), see Martin's review (532)).

2.11. CONFORMATIONAL ANALYSIS

It has been pointed out in preceding sections that suitable steric shape must be a very important characteristic of a molecule if it is to fit (make contact) at a bioreceptor and be bonded to it long enough to allow time for action. As Ehrlich put it in 1913, "*Corpora non agunt nisi fixata.*" The greatest advances in determining the shape (conformation) of a compound occurred when x-ray crystallographic spectrophotometry became widely available and the interpretation of the spectra by computer analysis reduced the time needed for these calculations to a very few weeks as compared with the many months required without computers. Still, these measurements were restricted to the crystalline state of a drug until recently, and a debate is still going on regarding what bearing such a conformation has on conformation in solution where physiological reactions take place. It has been argued that the conformations having the lowest energy level will predominate in the solid as well as the liquid state, even for flexible structures. However, a crystal structure determination only yields a "snapshot" of a particular conformation and does not reveal other equienergetic shapes.

X-ray diffraction spectra can be supplemented or confirmed by quantum chemical calculations that provide charge distributions and other physical data derivable from wave functions. The Extended Hückel Theory method, formerly used widely, is no longer regarded as reliable. Instead, the most commonly applied computer calculations are the Complete Neglect Differential Overlap or the Intermediate Neglect Differential Overlap methods [CNDO/2, INDO], Modified INDO [MINDO], and Perturbative Configuration Interaction using Localized Orbitals [PCILO] procedures. For the liquid state, NMR analysis furnishes data for geometric isomers and their conformations. From such data, support of solid-state conformations for liquid-state conditions has been obtained in some cases but differences have also appeared. For example, dopamine in solution exists about equally as the *gauche* and *trans* conformers, whereas in the crystalline state it is present only in the *trans* conformation (533, 534). An atlas of the three-dimensional structure of drugs, especially neuroleptics, is available (535).

Since all these measurements yield only an inventory of all energetically possible and impossible conformations *in the abscence of receptor* (536), the question remains open as to the actual stereochemistry of a drug at biologically active sites during and after establishing bonding. Both flexible drug structures and receptor sites on protein molecule or nucleic acid surfaces can change their conformations in adaptation to the mutual effect of receptor and drug, respectively. That means that substrate-active (or drug-active) site complexes must be isolated and measured conformationally. Such information is still in its infancy (43).

One way to simplify the abundance of conformations in a flexible molecule is to incorporate the most flexible portion of the structure in a rigid ring. For example, the ubiquitous flexible ethylamine chain of biogenic amines can be fixed sterically by incorporation in a ring. However, at least one or

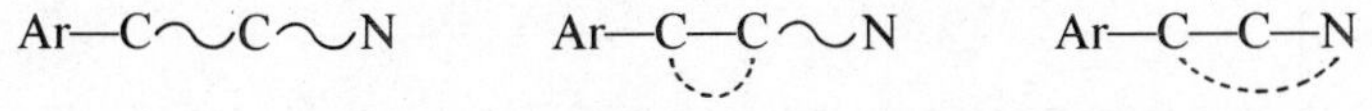

$$\text{Ar—C}\frown\text{C}\frown\text{N} \qquad \text{Ar—C—C}\frown\text{N} \qquad \text{Ar—C—C—N}$$

two atoms must be added, or several more, to make up the rigid ring structure. This, of course, disturbs lipophilicity, affects the pK_a of the amine, and creates totally new conditions for the resulting compound, which may be far from optimal for the formation of a drug-active-site complex. Moreover, the new compound will be metabolized differently, and this in turn may interfere with its travel to the active site and thus affect potency. With all these alterations in mind, changes in conformation induced by artificial procedures as outlined above may render valuable theoretical information but introduce new uncertainties into practical drug design.

3

SELECTED EXAMPLES OF DRUG DESIGN

In this part of the volume a more detailed account of a few selected topics in medicinal chemistry will be found, but only those topics are discussed that offer instructive insights into medicinal–chemical thinking and planning. The sections of this chapter are arranged not in chronological sequence but according to the pharmacological effects of the various classes of drugs.

3.1. THYROID HORMONES AND THYROMIMETICS

The relatively simple structure of the iodinated thyronines found in the thyroid glands, their accessibility to model building and conformational analysis, and their reactions with specific proteins have combined to make the thyroid hormones one of the most gratifying topics of medicinal investigation. Their biosynthesis can be followed with great reliability and in considerable detail. Mechanisms of action can be deduced from the measurable adsorption of the hormones on certain proteins whose function in the cell nucleus is fairly well established. The stereochemistry imposed on the combined aromatic rings by the large substituents, and the requirement for lipophilicity for these substituents, further define the shape and character of the thyroid hormones and their analogs. From these data, a general shape of the receptor of these compounds could be derived, and a receptor protein was extracted and studied (537). Thus, these hormones present almost an ideal picture of all the important data a medicinal chemist could hope to obtain in a given structural area. For a detailed account, see the excellent survey by Jorgensen (538).

The most important thyroid hormones are thyroxine (T_4) and 3,3′,5-triiodothyronine (T_3). Mono- and diiodinated thyronines are present in minute amounts. T_4 is metabolically 5′-deiodinated to T_3 to some extent, and T_3 is somewhat more active than T_4 depending on the test method. 3,3′,5′-Triiodothyronine (reverse T_3), an isomer of T_3, has also been detected in thyroid glands, and in the serum of human fetuses and newborns. It is hormonally inactive. Apparently, then, optimum potency is attained in T_3 and T_4, and a study of the conformation and other molecular properties of these two hormones is decisive in outlining SAR. However, it would be inadvisable to interfere with the oxidative iodination of T_3 to stop the biosynthesis at the most potent hormonal stage (T_3) since the two final iodinated hormones have been assigned special potencies in the different physiological processes that they catalyze.

The biosynthesis of the thyroid hormones depends on the availability of dietary iodide ion. In addition, compounds containing organically bound iodine are deiodinated in the liver with the formation of iodide. Iodide from all sources is concentrated in the thyroid gland by an active transport system. Interestingly, other ions that resemble iodide in shape (Br^-, ClO_4^-, BF_4^-) and some linear pseudohalides (SCN^-) are also concentrated or at least act as competitive inhibitors of iodide uptake. A membrane-bound hemoprotein enzyme, thyroid peroxidase, oxidizes iodide to iodine by an as yet uncertain mechanism, and the iodine in whatever form present iodinates tyrosine radicals from thyroglobulin *ortho* to their phenolic OH. In the course of this reaction, histidine is also iodinated but only little; most of the iodination furnishes 3,5-diiodotyrosine and about half the percentage of 3-iodotyrosine. These products are then "coupled" under the influence of thyroid peroxidase. First their reactive resonance forms couple to form an unstable quinol ether of T_4 or T_3, respectively, one iodinated monomer being converted to a dehydroalanine group that can tautomerize to an imine and be hydrolyzed to (isolatable) pyruvic acid. The peptide chain falls apart simultaneously at points adjacent to the iodination sites. About 10–50 atoms of iodine (corresponding to 3–12 hormone molecules) are involved out of about 120 tyrosine residues in human thyroglobulin. The iodinated thyroglobulin is digested by proteases and releases the iodinated thyronines. T_3 and T_4 resist intracellular deiodinases although they are reduced when in the general circulation. The slightly soluble hormones are bound rapidly by serum carrier proteins, mostly α-globulins, of which TGG (thyroxine-carrying globulin), an acidic glycoprotein of molecular weight ca. 63,000, has been identified. A second carrier of T_4 is TBPA (thyroxine-binding prealbumin) whose amino acid sequence (539) and tertiary struture are known (540). This oblong protein contains a channel through its long axis in which the T_4 is embedded and held by ion pair association of its carboxyl group and the lysyl ϵ-amino groups of TBPA. Moreover, TBPA contains two folds on its surface which, in size and shape, complement a region of the α-helical background of DNA and which contain many ionic side chains as points of interaction with the nucleic acid. One could interpret TBPA as a model for chromatin serving as a nuclear receptor protein (541).

It is not surprising that these data, rarely matched in diversity and detail in other areas of medicinal chemistry, have given rise to extensive biochemical speculations concerning the mechanism or mechanisms of action of the thyroid hormones. Perhaps the most impressive support for these conjectures has come from studies of the actual molecular shape of the iodothyronines. Because of the steric effect of the large iodine atoms in positions 3 and 5, the diphenyl ether structure is skewed, the planes of the two aromatic rings being perpendicular (542). The amino acid side chain, in the energetically most favored conformation, has the carboxyl group most distant from the aromatic ring bearing the side chain. The phenolic 4'—OH group is coplanar with its aromatic ring. (For references, see 538, 543.) Figure 1 shows the 3'-iodine atom in L-T_3 in the distal conformation; this is apparently the hormonally active shape of the molecule.

Of the many molecular modifications of the thyroid hormones (538), only a few are mentioned here.

The ether oxygen can be replaced isosterically by sulfur or methylene without qualitatively impairing thyromimetic activity. Any of these bridges (O, CH_2, S) provide an interesting bond angle of ca. 120°. A phenolic hydroxyl in position 4' is important for hydrogen bonding to transport proteins; it can be replaced by isosteric groups such as 4'—NH_2 or by derivative groups that can be metabolized to 4'—OH (4'—OCH_3, 4'—OAc) but at the cost of reduced hormonal activity. The 3'-position *ortho* to the phenolic OH and distal to the other aromatic ring must be occupied by a lipophilic substituent. This can be a halogen, preferably iodine (or bromine) or an alkyl group approximately the same size as iodine, such as isopropyl. 3,5-Disubstitution is essential to the lipophilicity and geometry of the molecule. It does not have to be 3,5-diiodo; the iodine atoms can be replaced by nonpolar substituents, for example methyl, as long as they hold the aromatic rings perpendicular to each other. The acidic [alanine] side chain can also be varied, to acetic or propionic acid dimensions, but should preferably be *para* to the linkage connecting the aromatic rings.

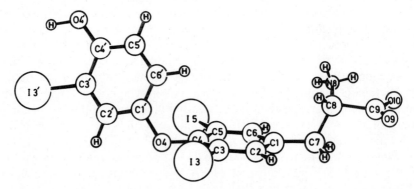

FIG. 1. L-3.5.3'-triiodothyronine (L-T_3) showing the numbering of the atoms. L-T_3 is shown with the aromatic rings in the "skewed" conformation, $\phi = 90°$, $\phi^1 = 0°$, and with the alanine side chain in the cisoid conformation, $\chi^2 = 90°$. The carboxylate residue is most distant from the aromatic ring, $\chi^1 = -60°$, and coplanar with the NH atoms, $\psi^1 = 0°$.

This condensed summary of structural requirements for thyromimetic activity does not do justice to the large number of molecular modifications that have been studied (see 538). They have permitted proposing fairly detailed suggestions concerning the role of individual functional and substituent groups in receptor binding and ligating to storage and transport macromolecules. In this field, perhaps more than in any other, a responsible picture of the hormone–receptor interaction has evolved. According to Jorgensen (538), a thyroid hormone appears as a three-dimensional matrix (diphenyl ether, diphenylmethane, etc.) with a spatially defined pattern of substituents that contain a coded message to the receptor. This message is carried by plasma transport proteins of the hormone to what appears to be an acidic protein associated with DNA. The interaction of hormone and nuclear protein disturbs the conformation of the receptor and this leads to transmission of a second message to the DNA. It is amplified by the genetic creation of specific messenger RNAs which initiate the synthesis of proteins and enzymes that participate in the growth, development, and metabolism characteristically catalyzed by thyroid hormones.

3.2. THE D VITAMINS

Most vitamins are biocatalysts that are virtually unique in their action. In a few cases, multiple forms of the same vitamin (e.g., vitamin B_6) can be interconverted to each other, but peak activity is observed in only one of these derivatives (in the case of vitamin B_6, in pyridoxal phosphate).

Vitamins classified D are concerned with calcium metabolism and deposition. They share these biological tasks with the parathyroid hormone. In many classes of vitamins, natural-product chemistry and synthetic organic chemistry have solved the major chemical problems. The role of the vitamins in biological and biochemical mechanisms has been assigned to biochemical

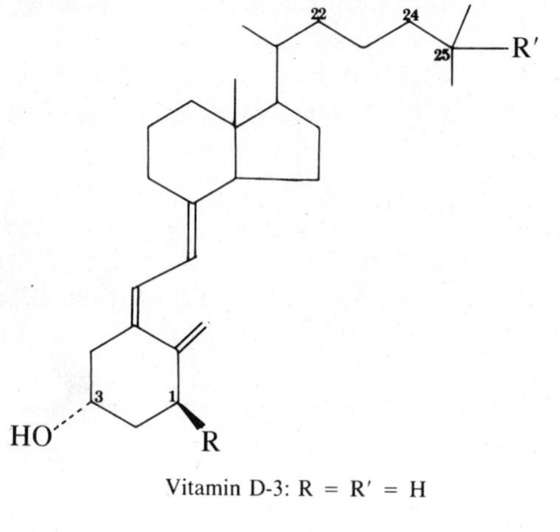

Vitamin D-3: R = R' = H

research. In the case of the D vitamins, the discovery of activity in products from several unrelated sources has given medicinal chemistry the opportunity to elaborate structure–activity relationships.

Vitamin D-3 (544) is metabolized to OH derivatives at positions 1, 24 and 25. The enzymes concerned with these hydroxylations are, for example, "25-OH-D_3-1α-hydroxylase," "1α,25-$(OH)_2$-D_3-hydroxylase," and so on. Depending on the assay method in the chick or rat, conclusions can be drawn about the rate of importance of the three OH groups in regulating blood calcium homeostasis. In addition, several *in vitro* tests including competitive binding to chick intestinal cytosol proteins and stimulation of bone calcium in cultured fetal rat bone can be used to evaluate the relative order of activity and special effects of vitamins D and their analogs. From these tests one can conclude that all three hydroxyls of 1,25-$(OH)_2D_3$ are necessary for optimal response in these tests. However, the 3- and 25-hydroxyls are not essential for minimal activity. This is based on lower rates of activity found in 3-deoxy and especially 1α-OH-25-fluoro-vitamin D_3 (545) and demonstrates that metabolism of 1α-OH-D_3 to 1,25-$(OH)_2D_3$ is not needed for activity in some of the test systems. 24-Hydroxylation has been studied with both 24(S) and 24(R) isomers but neither epimer is significantly more active (546). The conformation and unsaturation of the side chain is even less restrictive, but the stereochemistry at C-24 cannot be altered if the compound is to be a substrate of 1α-hydroxylase. Shortening of the side chain abolishes *in vivo* activity.

3.3. PEPTIDES AND PROTEINS

In this section, small and large peptides and biologically active proteins are taken up from a medicinal chemists's point of view, regardless of their type of biological activity. Thus, both peptide hormones and peptide antibiotics are discussed because the chemical techniques dealing with these compounds are the same. So are molecular modification, degradative and synthetic procedures, and spectroscopic and x-ray studies of the conformation of biologically active peptides. The mechanisms of their action are quite different, depending on their biological regulatory or chemotherapeutic mission. As more and more medicinal chemists become involved in the unraveling of biochemical mechanisms and metabolic processes, such data might be allocated to the scope of this section. They are of greatest interest to biochemists, to our search for the understanding of the chemistry underlying life processes, and for the classification of peptides and the emergence of unifying fundamental concepts in explaining their mode of action. However, the biochemical insight into these mechanisms has helped predictive analog design only to a limited extent.

The first important facets of peptide chemistry are isolation and purification techniques; chromatographic methods including TLC and HPLC combined with rapid and highly sensitive bioassays have made possible unprecedented progress in this field. Elucidation of the amino acid sequence by instrumental standardization (amino acid analyzer) is followed by rapid techniques to gain information about the three-dimensional structure by x-ray crystal structure analysis, which has now been reduced to a few days for smaller peptides. In addition, conformation can be determined in solution by NMR, ORD, or CD spectroscopy, which are of special value for cyclized peptide structures. All these studies belong in the province of modern natural-products chemistry with medicinal motivation. Medicinal chemistry enters the picture with the design of analogs. About 500 analogs of oxytocin and vasopressin have been synthesized, both by the historic stepwise method of protection and deprotection as each new amino acid is added and by the solid-phase ((Merrifield)) automated synthesis method (547). Even with such a complicated polypeptide as insulin, 150 or so analogs have been prepared. Instead of amino acids, other acids have been introduced at various positions to establish the significance and specificity of the natural sequences. Encouragement for such molecular modifications has been gained from evolutionarily selected variations of some amino acids in the peptide hormones of several vertebrate species (548).

In many cases, commercial synthesis has been required to furnish the peptide hormones to the clinician because the natural glands were inadequate as their sources. Peptide antibiotics, on the other hand, are manufactured universally by fermentation of the source organisms. One would have anticipated that synthetic analogs, which may have a better selectivity or toxicity ratio, would be welcomed with open arms by the medical profession, but that has not been so. Of the hundreds of analogs of oxytocin and vasopressin, only two have been accepted for clinical use. One is [1-deamino,D-Arg^8]-vasopressin, an antidiuretic useful in diabetes insipidus (549), the other the related [1-deamino,Val^4,D-Arg^8]-vasopressin, which has an antidiuretic/pressor ratio of >125,000:1 compared with 1:1 for Arg-vasopressin (550).

Almost all peptide hormones must be administered by the intravenous route, a procedure that severely limits their use for ambulatory patients. Peptide antibiotics suffer from the same drawback except for those with D-amino acids and similar features that render them less acceptable as substrates of proteolytic enzymes. However, thyrotropin-releasing hormone (TRH), a tripeptide of the formula pGlu-His-Pro-NH$_2$ is orally active although its plasma half-life in rats is only 2 to 4 minutes.

The present state of peptide hormones has been reviewed authoritatively (551). For this reason, we will deal here only with cases of structural studies and modifications that demonstrate general principles and suggestions of procedures in this field.

3.3.1. Calcitonin

Calcitonin, a 32-amino acid peptide from the parathyroid gland, has hypocalcemic effects. The sequence of amino acids varies somewhat depending on the animal species. Large differences in the sequence are tolerated without abolishing the immunological-biological response, but on drastic alteration only a small percentage of the effect remains unless the new amino acids have the same polarity. Two cysteine groups at positions 1 and 7 are joined in a cyclic disulfide bridge; this bridging cannot occur if the cysteines are replaced by methionine, and indeed activity is lost in such an analog (552). However, the —S—S— bridge can be replaced by —CH_2—CH_2— when two aminosuberic acid groups are substituted for the cysteines, and the resulting analog retains 80% of the biological activity of natural eel calcitonin (553). This means that the electronic contributions of the sulfur atoms of the cysteine bridge are far less important to activity than the conformation of the molecule brought about by the bridge. In addition, the —CH_2—CH_2— analog cannot be reduced and therefore the amino acid chain cannot be opened.

3.3.2. Peptide Analogs

For peptide analog synthesis, efficient and stereospecific syntheses have to be worked out, giving the protein specialist ample opportunity to develop the best of his art. For example, a method involving minimal protection of functional groups (554, 555) has proved well suited to prepare analogs of GnRH (gonadoliberin, LHRH) with the least amount of final purification. Structure–activity relationships in such series depend on sensitive and sometimes contradictory assay procedures, which have to be standardized carefully. Agonists as well as antagonists have been sought in this series, the latter for their potential contraceptive utility. Antagonistic activity has been achieved by replacing amino acids No. 6 and/or No. 10 of the decapeptide structure of gonadoliberin by other amino acids (556) or even by synthetic bioisosteres such as α-azaglycine (557, 558), but it would be premature to draw definite conclusions from these experiments.

    The reader will find the origin of the many peptides and the properties that led to the recognition of their far-reaching hormonal activities in a review by Meienhofer (551). It is not enough to be a good organic or natural-products chemist to work with these substances. Nor does biochemistry provide the experience needed to deal with biologically active peptides, especially if they are available only in minute quantities and recognizable only as maxima in spectral or chromatographic curves. Peculiarities in solubility, stability towards hydrolysis and oxidation, adsorption characteristics on chromatographic supports, avoidance of racemization, and other properties encountered with these materials demand a measure of experience before a chemical research program in this field can be undertaken. In this regard the peptides, small or large, possess more peculiarities than the steroids, prostaglandins, and other natural products that have become accessible to organic chemical techniques in general. That means that medicinal chemists wishing to prepare analogs and study SAR will have to train to become peptide and protein chemists before assuming synthetic assignments.

The usual operations leading to analogs include first of all derivatization. N-Acylation of the N-terminal amino group or of other amino groups along the chain may or may not destroy biological activity. This holds also for the conversion of terminal carboxyl groups to ester, amide, or alcohol functions. Phenolic hydroxyls in tyrosine residues and other functional groups can be covered up by gentle etherification or esterification. The activity of such functionally altered derivatives can shed light on the significance of the reactive groups affected.

Most other structural modifications demand total or partial synthesis. Each amino acid present in the natural peptide hormone can theoretically be replaced by other aminoacyl groups, although this is not always readily feasible. As candidates for such substitutions one can choose other proteinogenous amino acids, their stereoisomers (these resist proteolysis), N-methyl derivatives, or purely synthetic materials that fit into the peptide chain and affect its polarity, stereochemistry, hydrogen bonding ability, and other properties that control molecular shape. The peptide chain may be lengthened—or more commonly shortened—in order to determine the minimum features necessary for some degree of biological activity. Fragments of the original peptide often serve as starting points for modified analogs, and this may cut down the synthetic work involved.

The increasing disclosure of three-dimensional structures of peptides and proteins makes it possible to plan stereochemical modifications with more confidence. However, the ultimate choice of therapeutically active peptides, such as that of captopril starting from cobra venom peptides, largely remains a screening operation.

In this brief discussion, we review some of the important approaches to the chemical modification of insulin that demonstrate the methodology in current use.

Of the 30 amino acid groups of mammalian insulin, four are tyrosine (Nos. 14 and 19 of the A chain, 16 and 26 of the B chain). These tyrosine groups can be iodinated for radioimmunoassays. The six carboxyl groups can be esterified (HCl—MeOH, CH_2N_2, Et_3OBF_4, BF_3—MeOH) (559). Three amino groups including the ϵ-Lys at B-29 can be acylated selectively, for example, by reversible protection with Boc, trifluoroacetyl, or methanesulfonyl-ethyloxycarbonyl groups (560). More than 150 synthetic analogs and derivatives have been synthesized (560); although no more potent nor metabolically more stable analog has been found, SAR have revealed interesting regularities. One of them is the dispensibility of seven of the amino acid groups of the B chain for activity, and these deletions have been correlated with the apparent importance of the remaining residues for receptor binding

(560). Acylation or aminoacylation of the terminal amino group of A1 glycine does not diminish activity substantially; for details see (561). For additional SAR data, see (560–563).

3.4. ANTICOAGULANTS

3.4.1. Polysaccharides

Three types of medicinally interesting polysaccharides are found. One is in the capsules of various bacteria, making them type-specific in the causation of certain infectious diseases. The second type is the blood-group-determining saccharides which depend on their terminal glycoside group for their specificity. The third are heparin and heparin-like substances, which occur in blood and animal tissues and possess anticoagulant properties. Here only an example of the chemistry of heparin is given, to illustrate what contribution a medicinal biochemist could make to this field.

3.4.1a. Heparin. Heparin was discovered by a medical student at Johns Hopkins University (564), and its anticoagulant properties were described by his faculty mentor (565, 566). A review of its chemistry and actions may be found in reference 567. Heparin is isolated commercially from such animal tissues as bovine lungs, and the intestinal mucosas of pig and cattle. It is a polydisperse, anionic, sulfated mucopolysaccharide with molecular weights estimated at 3000–57,000. Some authors believe only fractions above molecular weight of 7000 are active, although an enzymatic degradation product of molecular weight of ~5000, called *lambda*-heparin, still has anticoagulant activity in mice.

The structure of heparin (fractions) is represented by the partial formula.

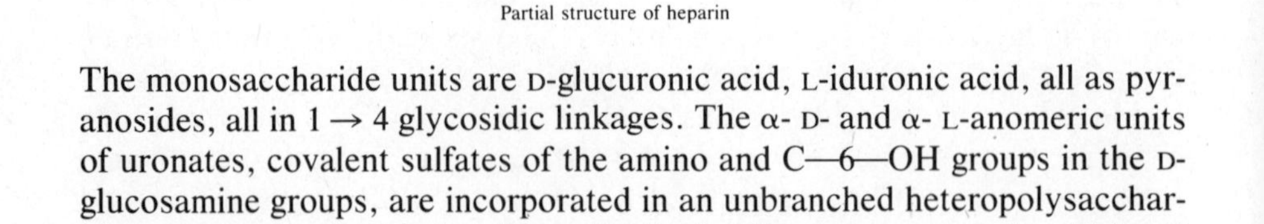

$CH_2OSO_3^-$ ... CO_2^- ... $CH_2OSO_3^-$ ... $CH_2OSO_3^-$ ... CO_2^- ... $CH_2OSO_3^-$

^-O_3SNH ^-O_3SO ^-O_3SNH ^-O_3SNH $AcNH$

Partial structure of heparin

The monosaccharide units are D-glucuronic acid, L-iduronic acid, all as pyranosides, all in 1 → 4 glycosidic linkages. The α- D- and α- L-anomeric units of uronates, covalent sulfates of the amino and C—6—OH groups in the D-glucosamine groups, are incorporated in an unbranched heteropolysaccharide macromolecular chain with polyanionic character, possessing high pK values.

The medicinal chemist can attempt to shorten or simplify these mixtures of macromolecules by enzymic degradation or by synthesis. For example, an α-(1 → 4) linked amylose chain containing 2-sulfonamido and 5-carboxyl groups has been prepared and found to have some anticoagulant properties.

ANTICOAGULANTS

Since the effect of different heparins on blood coagulation factors (antithrombin III, factors IXa, Xa, XIa, thrombin, etc.) depends on the molecular size of the fraction used, separation of as many fractions as possible would be a worthwhile problem. Heparin binds to the lysyl residues of antithrombin III and thereby speeds up the inhibitory effect of this material on the other "factors," that is, coagulation enzymes.

3.4.1b. Heparinoids. Materials that have a structural analogy to heparins are called heparinoids. They are sulfuric acid esters of various polysaccharides, prepared from these materials by treatment with chlorosulfonic acid. Some or all of the secondary alcohol groups in the most important of them, the pentosans, are esterified in this manner (568). These heparinoids are in some respects more active than heparin in animals (569). Another example is dextran sulfate; dextrans are linear glucose polymers produced by *Leuconostoc mesenteroids* or *L. dextranicum;* they have molecular weights of 40,000–70,000. Dextrans of this range of molecular weights have been used as antithrombotic agents.

That many diverse simple structural types are useful as oral anticoagulants indicates that a search for other compounds incorporating some of the properties of heparin and heparinoids may yet uncover simplified candidates for clinical use.

3.4.2. Oral Anticoagulants

A disease of cattle characterized by internal bleeding was described in 1922 (570) and traced to the ingestion of improperly cured Melitotus hay (sweet clover hay). The causative agent was identified as 3,3-methylenebis(4-hydroxycoumarin) (dicumarol) (571), a substance that had been synthesized earlier (572). It was suggested that this compound should be tried as an oral anticoagulant (573) to attain an apparent progress over the previously available parenteral heparinoids. Another coumarin derivative, called warfarin,

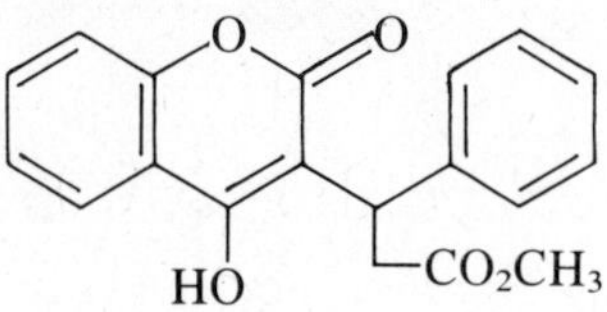

Warfarin

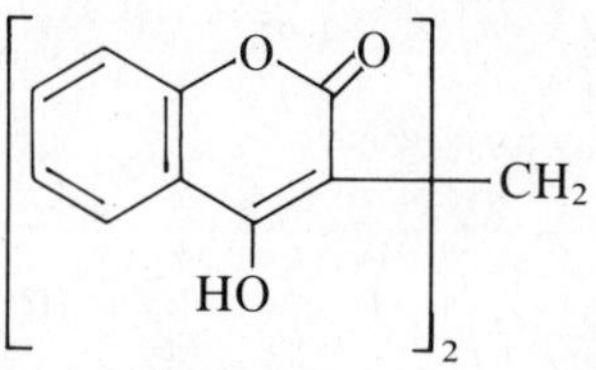

Dicumarol

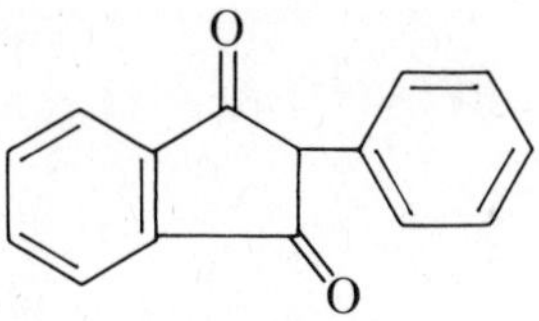

Phenylindanedione

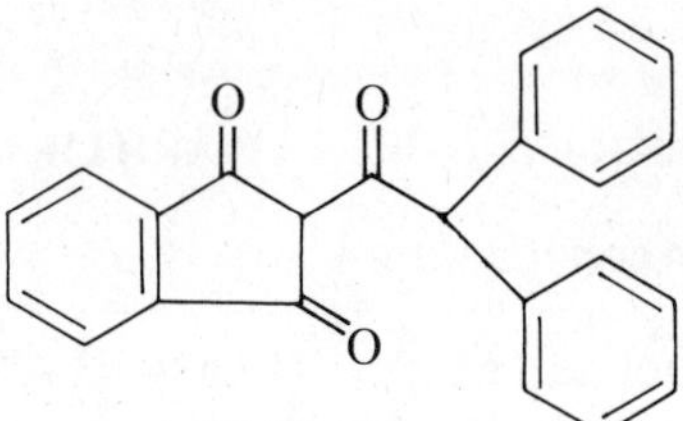

Diphenadione

had been used as a hemorrhagic rat poison (574, 575) before being introduced as a clinical anticoagulant.

With these "leads" and several simple synthetic approaches available, a large number of molecular modifications have been undertaken. The traditional modifications concerned changes in the phenolic hydroxyl, the ketone groups, and substitutions in the aromatic rings. No unifying guidelines could be elaborated; comparisons of physical properties such as ionizability, lipid solubility (576), and overall size as estimated by molecular models (577) remained indecisive. Ring contraction, another traditional method, furnished indanedione derivatives such as phenylindanedione (phenindione) and diphenadione. Like the coumarins, they inhibit prothrombin biosynthesis as well as the synthesis of factors VII, IX, and X. A chemical working hypothesis trying to explain the activity of both types of ring compounds points to the possibility that they can form cyclic ketals (see structures).

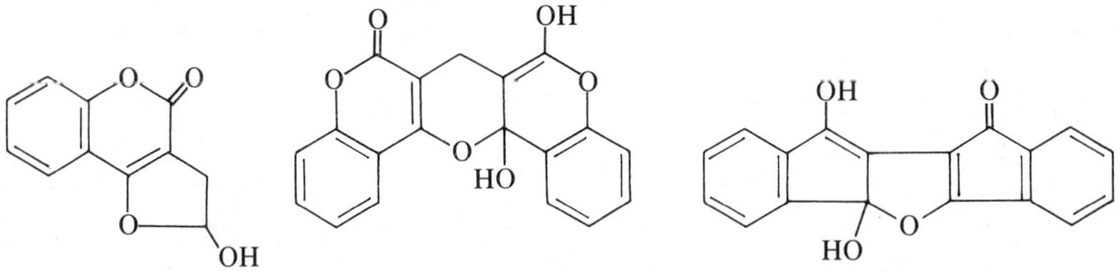

Potential cyclic ketal structures of oral anticoagulants

These cyclic ketals (578) can arise only if the side chain of the cyclic ketone has a carbonyl or a potential carbonyl group in an appropriate position (2' or 3' in the coumarins), or if two molecules are involved in the formation of the activated species. Perhaps vitamin K furnishes a cyclic form of analogous structure (579), and the oral anticoagulants could conceivably compete with vitamin K for active sites.

Postulated cyclization of vitamins K

3.5. CHOLINERGIC, CURARIFORM, AND RELATED AGENTS

3.5.1. Acetylcholine

Acetylcholine (ACh), $CH_3CO_2CH_2CH_2N^+(CH_3)_3$, is a structurally simple neurohormone. It contains an ester and a quaternary ammonium group connected by two carbon atoms. It is biosynthesized from choline and acetate, with acetyl coenzyme A as a source of acetate, under the catalytic influence of choline acetylase (580). Acetylcholine is hydrolyzed rapidly by the action of acetylcholinesterase (AChE) which thus prevents the accumulation of unwanted spasmogenic concentrations of ACh at nerve endings or the motor endplate.

Pharmacologically, two alkaloids, muscarine and nicotine, mimic ACh. To account for these actions, two ACh receptors are postulated, a muscarinic and a nicotinic receptor. These receptors have been studied in many different ways. A classical approach is to define the distance of characteristic functions and the conformation of the alkaloids (the N–N distance in nicotine,

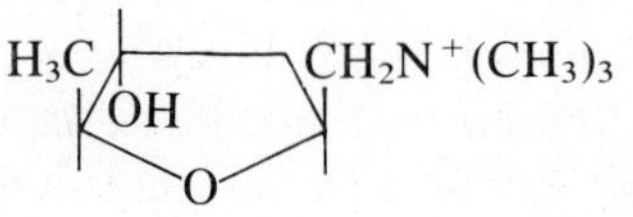

Muscarine

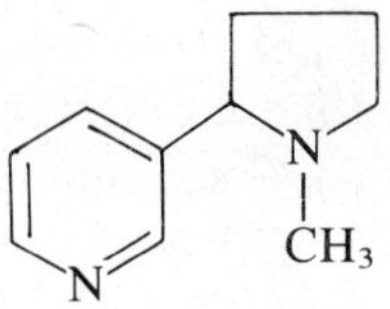

Nicotine

or the CH_3C—O . . . N distance in muscarine) and to deduce regions and sizes of the receptors corresponding to those distances. SAR studies of this kind are complicated by the possibility that nicotinic agonists, including nicotine itself, act indirectly through the release of endogenous ACh (581).

A second way to investigate ACh receptors has taken advantage of the direct isolation of receptor proteins from sources rich in these materials. They are the electric tissues of the giant South American freshwater eel, *Electrophorus electricus,* and that of the giant electric ray, *Torpedo marmorata.* From these tissues, Triton X 100 solubilization and further fractionation has rendered purified proteins of molecular weights 230,000 that dissociate into subunits of molecular weights 45,000 and 54,000, respectively (582). Similar, although not identical, fractions have been obtained from other marine sources. For an up-to-date review, see ref. 583.

The similarity of the amino acid sequence of AChE and receptor subunits has given rise to the speculation that the two types of protein are similar. ACh can exist in skewed or extended conformations, and the relatively small differences in the spacing of its ester and ammonium groups are believed to account for the behavior of ACh at the enzyme and the receptors. The anionic subsite of the receptor protein (to which the quaternary ammonium head becomes attached) is close to a reducible disulfide bridge (584); reduction of this bridge to two SH groups must lead to conformational changes and thereby to changes in responses to ACh and other related agents. In addition, neuromuscular blocking agents of the decamethonium type affect receptor-subunit dissociation and thereby inhibit the formation of ion channels (582).

So far, it has not been possible to construct satisfactory models of ACh receptors or AChE to which models of agonists or antagonists could be fitted.

Therefore, studies of analogs of ACh, and especially of competitive and blocking drugs, have remained the topic of wide-ranging molecular modification.

Acetylcholine has such a simple structure that it offers only few opportunities for the creation of congeners. One can alter the quarternary ammonium head, replace the acetyl group by other acyl groups, and expand or branch the —CH_2CH_2— chain. The ester group can be replaced by other moieties; for example, choline ethers are often active depending on the methods of biochemical or biological screening. AChE inhibitors that prevent ACh concentrations from decreasing can be screened against the enzyme *in vitro* or *in vivo*. Neuromuscular blocking agents are assayed advantageously *in vivo*.

A mix of justifiable molecular modifications has arisen by basing them on both ACh and muscarine. This cholinergic alkaloid also contains a quaternary ammonium group, and its furanoid ring system is a variation on the connecting alkylene chain. Its alcoholic hydroxyl can be oxidized to C=O or reduced to CH_2, the furan ring can be aromatized, the furan oxygen exchanged for isosteric atoms (S), and so on. The stereochemistry of muscarine can be altered or reversed, with three asymmetric carbon atoms providing considerable amplitude for these changes. An exchange of the hydrofuran ring for pyrrolidone is seen in tremorine and its metabolite, oxotremorine, whose potent, primarily muscarinic activity was spotted by pharmacological screening (585). They produce parasympathetic activation.

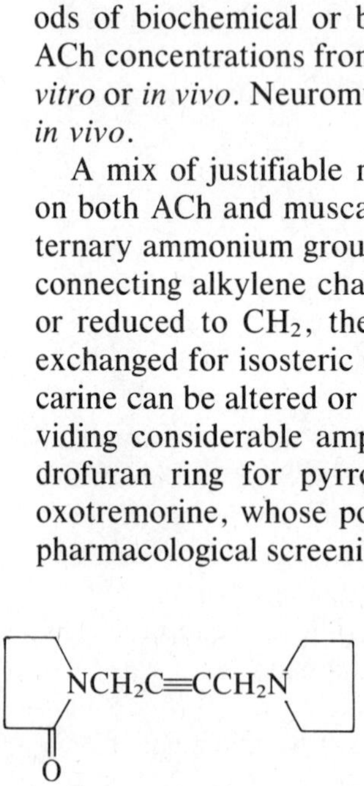

Oxotremorine 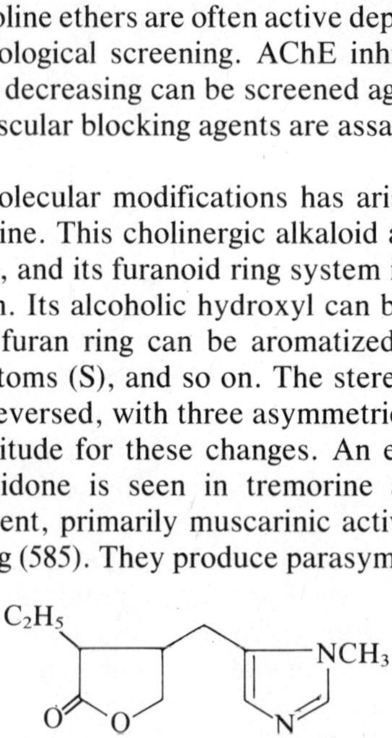   Pilocarpine 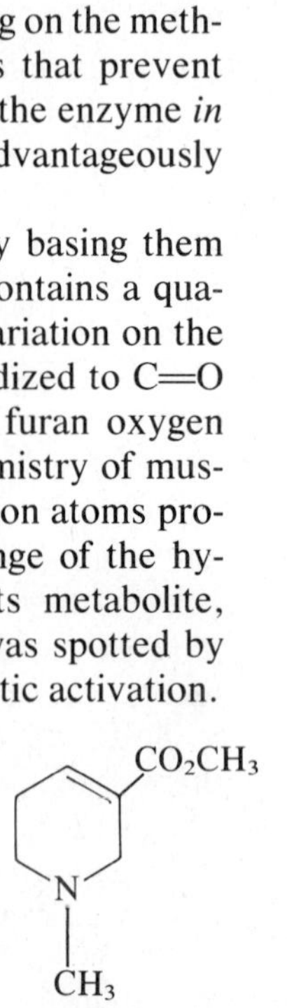   Arecoline

The naturally occurring alkaloids arecoline and pilocarpine also behave as stimulants of muscarinic receptors (586). Arecoline shares with ACh the two-carbon distance between the ester group and the amine position, but like pilocarpine it is not a quaternary ion except by protonation. Pilocarpine contains an oxotetrahydrofuran ring but its tertiary nitrogen is four carbons away in an imidazole structure. These combinations have all been explored in many modifications (587) aimed at increasing agonist selectivity or creating antagonistic anticholinergic agents. The greatest emphasis in reconciling these structural differences has been placed on molecular rigidity to overcome conformational changes of flexible molecules. This has been achieved by incorporating flexible portions in rings, for example, —CH—CH— / CH_2 for —CH_2—$CH(CH_3)$— (588, 589). Interatomic distances and angles in such systems necessary for muscarinic activity have been defined (581, 590).

3.5.2. Neuromuscular Blocking Agents

Neuromuscular blocking agents are used primarily as muscle relaxants in surgery, intubation, electroshock therapy of depressive disorders, and certain spastic conditions. They interfere with various manifestations of ACh either by depolarization of the motor endplate, by competition with ACh for a place at receptors, or by interfering with the biosynthesis of ACh. Their pharmacology and uses have been reviewed in an excellent article by Stenlake (583).

The classical prototype of neuromuscular blocking agents is curare, a poisonous mixture of many alkaloids and impurities from the bark of tree vines of the *Strychnos* and *Chondodendron* species native to South America. Of the many active principles, only two types will be mentioned. Their botanical sources are distinguished by the containers in which they were originally packed. Typical tube curare alkaloids are tubocurarine and its congeners, while the calabash alkaloids are represented by the structurally more complex toxiferines. In 1851, Claude Bernard (31, 591) recognized that the action of curare left both nerve and muscle separately excitable and therefore must affect the neuromuscular junction. When it was found that curare extracts contain quaternary ammonium salts, Crum-Brown and Fraser (32) tested other quaternized alkaloids and simple synthetic tetraalkylammonium ions for the same paralytic effects. Their observations were the earliest studies of structure-activity relationships of organic compounds.

Almost 70 years later the structure of (+)-tubocurarine was established, as a bis-quaternary ion to begin with (592) but then corrected to a monoquaternary bis-isoquinoline alkaloid (593). The corresponding bis-quaternary compound is identical with the alkaloid chondocurarine, and the so-called "dimethyltubocurarine" is in fact the fully O-methylated bisquaternary compound, O,O-dimethylchondocurarine.

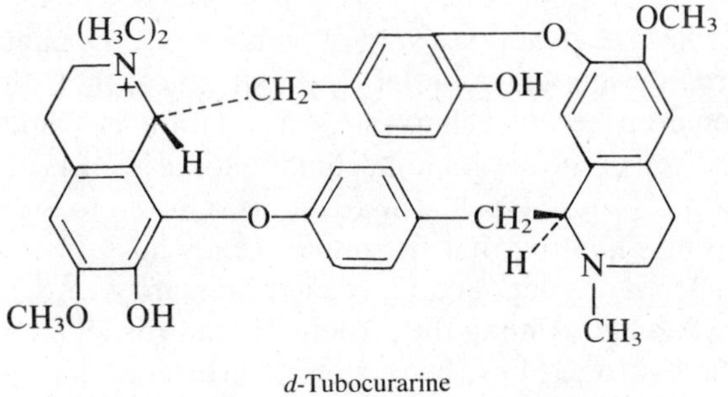

d-Tubocurarine

The temporary confusion about the mono- or bis-quaternary nature of these alkaloids paid off, nevertheless, in an unexpected manner. Models were built of the mistakenly bisquaternary *d*-tubocurarine, and armed with centimeter rulers and conversion tables, several pharmacologists measured the airline distance between the two onium nitrogens. This turned out to be about 14.5Å. Reasoning that this distance was significant for curarimimetic action, a polymethylene chain was arranged between two trimethylammonium heads, and thus the bismethonium ions were designed (594, 595). Neuromuscular blockade is optimal for $n=10$ and minimal for $n=5$ or 6, but these two homologs block cholinergic transmission at autonomic ganglia instead. Hexamethonium has been used to lower elevated blood pressure. Pharmacologically, the neuromuscular blockage produced by decamethonium differs from that of d-tubocurarine in that it is accompanied by long-lasting depolarization of the post-junctional membrane. This is a lesson for drug design: Whereas the overall effect of two drugs whose functions are placed similarly in space may be similar, their mechanisms of action may or may not differ.

A 10-carbon conformationally flexible alkylene chain between the onium heads appears to be optimal for neuromuscular blockade based on depolarization. The composition of the chain can be varied widely so long as no sterically hindering side chains or interspersions by rigid bonds (C=C, C≡C, alicyclic or aromatic rings, etc.) are brought into play. On the other hand, replacement of CH_2CH_2- by $C(O)O-$ does not interfere with potency; suxamethonium is a potent short-acting depolarizing agent (596). Bulkier analogs appear to be less tightly bound to the receptors and thereby cannot induce the conformational changes that appear to cause depolarization. The question whether such compounds can deny acetylcholine access to the receptors and thus act as competitive inhibitors of the neurohormone has not yet been answered fully.

$$(CH_3)_3N^+-(CH_2)_n-N^+(CH_3)_3 \qquad (CH_3)_3N^+(CH_2)_2OC(CH_2)_2CO(CH_2)_2N^+(CH_3)_3$$
$$\underset{O}{\|} \quad \underset{O}{\|}$$

Hexamethonium: $n = 6$
Decamethonium: $n = 10$ .... Suxamethonium

This volume has the deliberate goal of pointing out the chemical thoughts that go into drug design and some of the biochemical facts that explain the mechanisms of drug action. Therefore, no attempt is made to detail the hundreds of analogs of the methonium compounds as well as of tubocurarine, and combinations of these types. The reader is referred to Stenlake's review (583). Let me only mention that the onium heads have been varied widely to give the respective compounds a greater similarity to the bulky groups of the curare alkaloids. Among the structural features employed to this end were polycyclic hydrocarbons, benzyltetrahydroisoquinoline alkaloids, and steroids. Depending on the environment in which these many analogs have been prepared, studies of drug metabolism and differential modes of action have played a major role. Not all these compounds act only as depolarizing agents, but some of them inhibit AChE and thereby introduce a new set of properties that are not necessarily conducive to therapeutic uses. However, these are empirically determined biological properties that medicinal chemists might have considered but not anticipated.

3.6. ANTICHOLINERGICS AND HISTAMINE-1 RECEPTOR ANTAGONISTS

3.6.1. Anticholinergics

Anticholinergics are drugs that mimic the effects of cutting the parasympathetic nerves that supply ACh to various organs. ACh appears at the neuromuscular junction during the spontaneous functioning of the postganglionic fibers of these nerves and causes involuntary spasms of the muscle. The muscular contraction is normally ended by hydrolysis of ACh under the influence of acetylcholinesterase (AChE). Since ACh is the chemical neurotransmitter at other locations as well—at autonomic ganglia and the junctions of parasympathetic nerves with somatic (voluntary) muscles—different types of anticholinergic drugs counteract its actions at these three types of synapses. Those that block neuromuscular transmission at somatic muscles are curariform drugs and their variants and congeners. At autonomic (both sympathetic and parasympathetic) ganglia, the ganglionic blocking agents may be found useful. Drugs that inhibit AChE provide conditions virtually indistinguishable from spasms caused by excess unhydrolyzed ACh. There is structural overlapping between all these types of drugs, and it is not surprising that one or the other of these activities predominates in the profile of overall anti-ACh agents. Again, some of the "spasms" are not attributable to ACh but may be caused by serotonin, histamine, and experimentally by ACh-congeners or barium ions.

Anticholinergics counteract gastrointestinal spasm, gastric secretion, and peptic ulcer, and they dilate the pupil of the eye and paralyze accommodation. Most anticholinergics have all these properties to varying degrees. The extensive research programs in this field have had the aim of separating these properties as far as possible but few definite guidelines for achieving this purpose have emerged. Many a program undertaken to develop an antiulcer agent—the most profitable anticholinergic activity—has ended up triumphantly announcing the discovery of a "specific" mydriatic.

As will be seen, the structures characteristic of antispasmodic anticholinergics overlap with those of antihistaminics (H_1 receptor antagonists). Pharmacologists involved in a testing program for anticholinergic activities will therefore inevitably test the same compound for antihistaminic properties. *In vitro* the same (Magnus) test measuring the relaxation of an intestinal strip that had been stimulated to contract by ACh or by histamine can be used for this purpose. The *in vivo* tests are adapted to the more specific localized effects of the two agents or their congeners.

The classical natural products that became the prototype anticholinergics are the solanaceous alkaloids, (±)-atropine [and its enantiomer, (−)-hyoscyamine], and (−)-scopolamine (hyoscine).

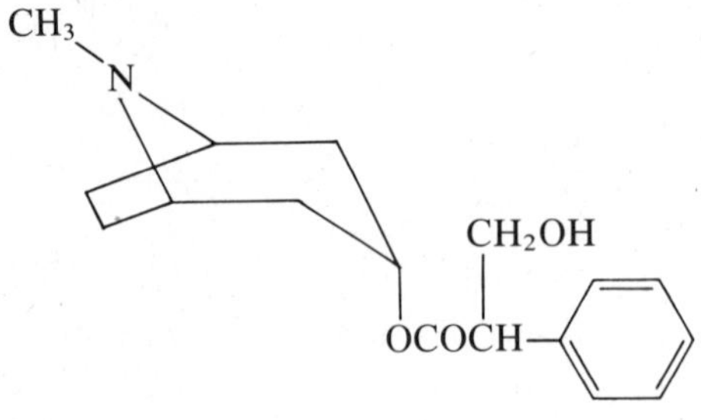

Atropine

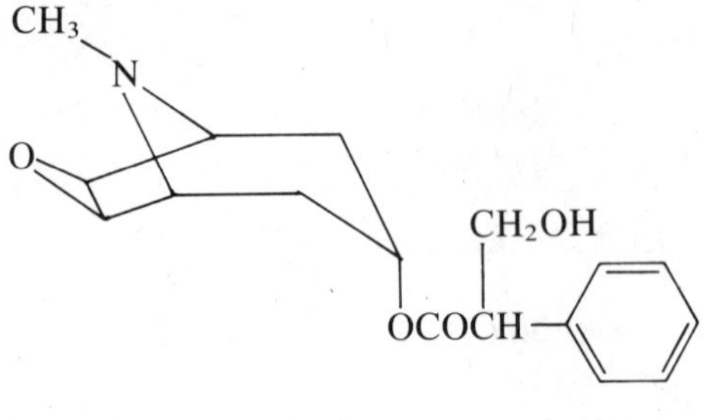

Scopolamine

Medicinal chemists have studied atropine and scopolamine in a manner very similar to that of cocaine. The SAR's of the *Datura* alkaloids follow a similar pattern. Metabolism abolishes all activity by cleaving the ester group, by hydroxylating the aromatic nucleus, and by glucuronidation of the alcoholic fragments. Quaternization of the alkaloids or *N*-oxidation leads to derivatives a few of which can be used as therapeutic agents like their parent tertiary bases. Among these are atropine *N*-oxide (genatropine) and scopolamine *N*-oxide (genoscopolamine). The tropic acid moiety has been replaced by numerous other carboxylic and hydroxy acids, for example, mandelic acid [$C_6H_5CHOHCOOH$] as in the tropine ester homatropine. As could be expected, quaternization decreases the ability of the parent alkaloids to penetrate membranes, especially the blood–brain barrier, and this minimizes the CNS effects of the quaternary derivatives. Other advantages and disadvantages of the quaternary compounds as well as of the tertiary alkaloids and their semisynthetic relatives have been described in a review (597). On the whole, traditional empirical and restrained molecular modification of the solanaceous alkaloids has yielded a few analogs with a narrowed and therefore more useful activity profile.

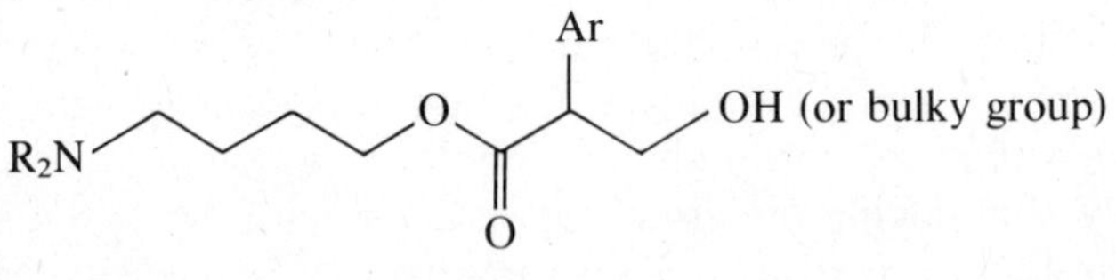

Atropine and scopolamine can be viewed as esters of amino alcohols with a branched and somewhat bulky carboxylic acid. Such an interpretation leaves the imagination a free rein for molecular modification. If analogs are to be constructed, any amino alcohol might be tried and esterified with any carboxylic acid from a chemical catalog. Or else, the amino alcohol might be quaternized or replaced by a diamine, which will then give an amide instead of an ester on combination with an acyl moiety. If the (ether) oxygen is replaced isosterically by CH_2, a ketone will result. There are other possibilities of retaining molecular shape—apparently the overriding requirement—while changing functional groups, and all this has been done in thousands of cases. A short list of analogs with a claim to potency or specificity of action or both may be found in reference 597, pp. 381–386.

It is reasonable to speculate that such molecules are attracted to a negatively charged anionic area of the muscarinic receptor. The compound should then be held in place by weaker but cumulatively not negligible forces such as dipole–dipole, hydrophobic, and van der Waals interactions. The most striking feature of the drug's cationic head in regard to antispasmodic potency is that the positively charged group may not exceed a certain size. This means the cationic head must fit into a definite space on the receptor surface. N-Methyl substituents are usually optimal; potency decreases for N-ethyl compounds and declines greatly for further homologs, such as n-propyl or n-butyl. However, N-isopropyl derivatives often are as active as N-ethyl homologs; apparently the branching detracts less from fit than additional length of the alkyl chain. This effect may also be due to the greater electron-repelling power of isopropyl, which would lend more basicity to the charged atom.

It has often been postulated that two phenyl groups in the bulky acyl moiety produce more potency than one phenyl plus some other bulky or cyclic group. This would imply that two flat-surfaced benzene rings are better than one and that their combined van der Waals forces are important for increased biological potency. There is not enough evidence to support such preferential actions of aromatic groups. Perhaps any cyclic group can create a protective area that sterically hinders the approach of the neurohormones to the active site and its environment.

The alcoholic hydroxyl group of the tropic acid section is not a requirement for activity, although it contributes to hydrogen bonding at the receptor. Quite commonly, activity is bolstered by the presence of hydroxy acid moieties. Other features that increase potency (but seldom specificity of action) are the configuration of the alcoholic hydroxyl group of tropine or scopine and the configuration of hydroxy acid moieties. All this points to the importance of steric and spatial conditions in the molecules of choline antagonists.

3.6.2. Antihistaminics (H_1 Receptor Antagonists)

A medicinal chemist, asked to make some general predictions about structural requirements for antihistaminic drugs, would find it difficult to define significant differences between those for anticholinergic and antihistaminic characteristics. This overlap is not restricted to structural features; very many drugs used for the treatment of seasonal rhinitis, hives, asthma, and other allergic conditions attributed in part to histamine also produce anticholinergic symptoms such as dry mouth, dizziness, blurred vision, and fatigue and other CNS depression. In the long run, a pharmacologist will test every compound expected to be anticholinergic for antihistaminic properties and vice versa. This does not mean, however, that no guidelines have been developed at all to separate these two trends of action. Nevertheless, structural predictions have remained questionable and require stepwise verification by experimental biologists.

This situation is aggravated by the fact that ACh and histamine are not the only mediators of the ultimate causes of the respective symptoms associated with bronchospasm and other conditions in this overlapping area. Serotonin, bradykinin, prostaglandins E_2 (PGE$_2$) and $F_{2\alpha}$ ((PGF$_{2\alpha}$), slow-reacting substance of anaphylaxis (SRS-A) and probably other biochemicals appear to participate as causative agents in a variety of allergic conditions. From a purely chemical point of view, the classical antihistaminics will be discussed here as a sequence to anticholinergic agents. They were discovered by pharmacological testing (prevention of bronchospasm caused by histamine aerosols or histamine administration) in rodents.

Taking a "lead" from choline ethers, Fourneau and Bovet (385) tried out various tertiary-amine analogs, some of them (piperoxan) disguised as benzodioxane derivatives, in antihistaminic tests and found them active. Iso-

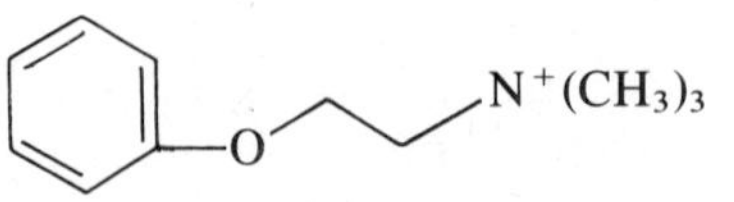

Aryl choline ether

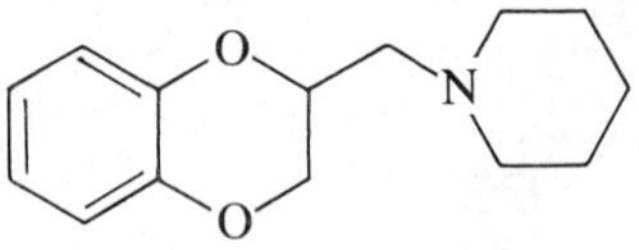

2-(N-Piperidinomethyl)-1,4-benzodioxane

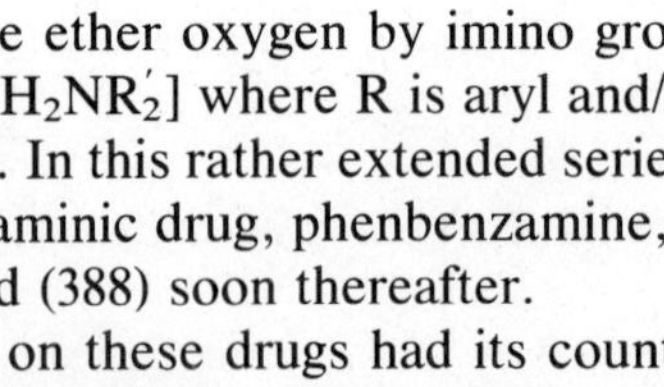

β-(5-Isopropyl-2-methylphenoxyethyl)dialkylamines

steric replacement of the ether oxygen by imino groups led to ethylenediamines (387) [R$_2$NCH$_2$CH$_2$NR$_2'$] where R is aryl and/or alkyl or aralkyl and R' is a small alkyl group. In this rather extended series of diamines, the first clinically useful antihistaminic drug, phenbenzamine, was prepared (392) in the 1930s and introduced (388) soon thereafter.

The French research on these drugs had its counterpart in very similar developments in the United States. Here tripelennamine was introduced at about the same time (598). It bears a striking relationship to pyrilamine, which was developed in Paris (599, 600). As one might expect, the two compounds and several of their close congeners and isosteres (e.g., 2-thenyl in place of benzyl) resemble each other pharmacologically except for minor details. Since both research units, although separated geographically and by

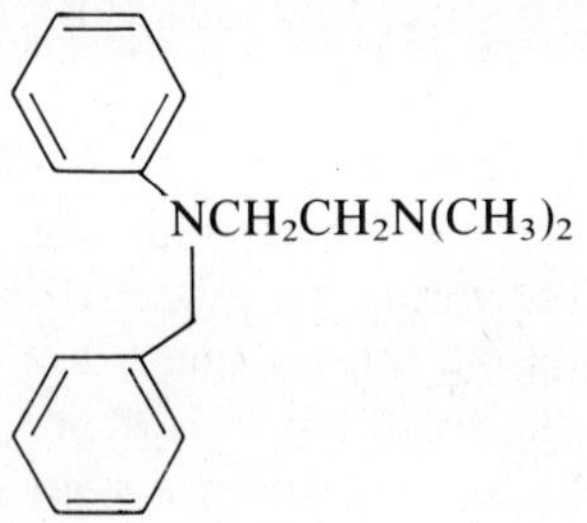

ANTICHOLINERGICS, AND HISTAMINE-1 RECEPTOR ANTAGONISTS

Phenbenzamine

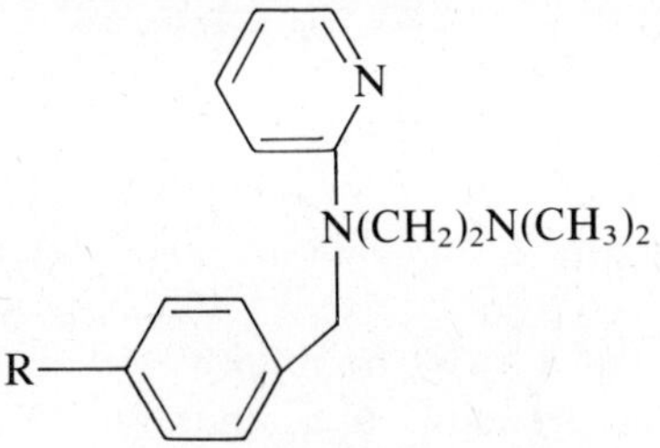

Pyrilamine: R = OCH₃
Tripelennamine: R = H

the events of a world war, derived their "lead" from the work of Staub (389), it is a compliment to the reliability of drug design in this series that such similar and useful drugs resulted from these efforts.

Although these derivatives of ethylenediamines began to look most promising and were modified successfully in extensive structural variations, aminoalkyl ethers were also pursued further. The earliest of these compounds was β-dimethylaminoethyl benzhydryl ether (diphenhydramine) (200, 391);

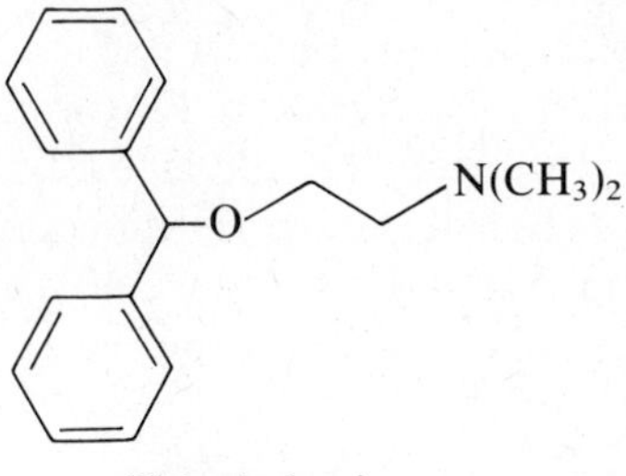

Diphenhydramine
(also base of dimenhydrinate)

it is both anticholinergic and antihistaminic, as could be expected from its blocking structure. An additional clinical benefit of diphenhydramine, perhaps somehow related to its anticholinergic CNS depressing properties, is its activity against motion sickness (dimenhydrinate). This was discovered when a patient, taking the drug for allergies, was relieved of motion sickness while traveling in a lurching bus to a hospital for desensitization (394). The success of these early antihistaminics led to extensive molecular modification, with hopes of separating the discrete biological properties, increasing specificity, and eliminating the bothersome CNS side effects. Every classical art of molecular modification was called into play. The ethoxy ether chain was lengthened, branched, and the ether oxygen replaced isosterically by CH_2. The aromatic rings were replaced by others, especially hetero- rings, and substituted by every imaginable substituent. If only one ring carries a substitutent, the benzyhydryl carbon atom becomes asymmetric, and the *dextro* enantiomorph in some cases was more potent than the *levo* isomer (601). Replacement of the NMe_2 group by azetidine or aziridine groups results in more potent, irreversible short-acting histamine antagonists (602). Quaternization of the NR_2 group increases parasympatholytic properties. Introduction of a pyrrolidino group instead of dimethylamino steps up antihistaminic activity, especially if the alkylene chain is made rigid by unsaturation as in triprolidene (603, 604). On the whole, H_1 receptor antagonist activity requires two bulky blocking groups, an intermediate thin chain of atoms, and a basic nonbulky tertiary amine head capable of protonation. These requirements are similar to those for anticholinergic agents, and it takes careful empirical juggling of functional effects to increase one of these activities at the expense of the other. For details of such trials, see a recent review (605).

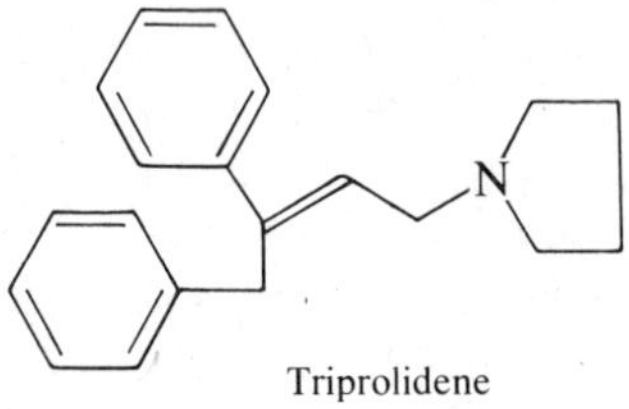

Triprolidene

The two blocking groups (usually but not necessarily aromatic or heteroaromatic rings) can also be joined covalently, forming a more or less rigid tricyclic system converging upon the alkyleneamino chain. Examples are phenothiazine derivatives and the isosteric thioxanthenes. The phenothiazines were first studied in 1945 in France (606, 607) and formed the research background to the antipsychotic drugs a few years later. Among the anti-

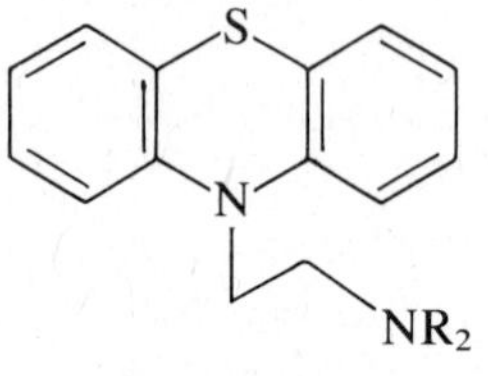

Phenothiazine 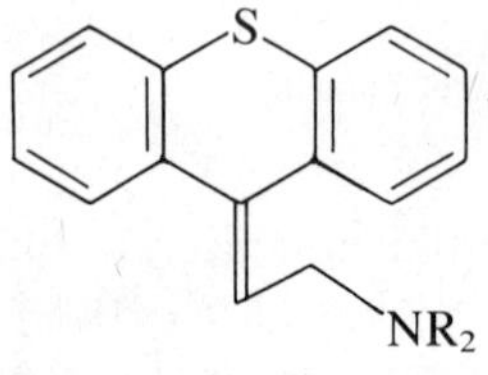   Thioxanthene 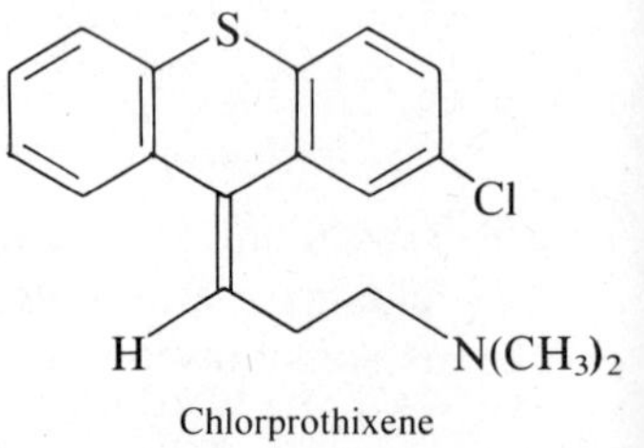   Chlorprothixene
              Antihistaminics

histaminics thus discovered are fenethazine [—$CH_2CH_2N(CH_3)_2$] and promethazine [—$CH_2CH(CH_3)N(CH_3)_2$]. Extension of the dialkylaminoethylene chain to dialkylaminopropylene is not conducive to antihistaminic activity (608) but promotes neuroleptic properties. An example for the thioxanthenes is chlorprothixene. The "*trans*" isomer, in which the aminoalkyl group lies on the side of the unsubstituted aromatic ring, is more antihistaminic (608) than the *cis* isomer, whereas the opposite is true for neuroleptic activity (609).

The same molecular modifications mentioned for compounds with separated aromatic blocking moieties have been applied to the condensed-ring systems. Notable are isosteric replacements of the sulfur atom of phenothiazine by —CH=CH— and —CH$_2$CH$_2$—. The resulting azepine derivatives have their aromatic rings tilted at considerable angles. From accumulated experiences, it looks as though greater tilting favors antihistaminic potency. Care must be taken in interpreting this conclusion. It holds best for a strip of intestinal tissue suspended in a bath to which the drug is added. In this type of test the transport across membranes, adsorption on serum proteins, and drug metabolism can be avoided. The composite *in vivo* activity resulting from all these factors does not project a picture of the true antihistaminic activity of an agent.

The central ring of tricyclic antihistaminics need not contain hetero atoms. A number of compounds with a central cycloheptadiene or -triene ring have antihistaminic activity (610, 611). For additional examples, see (605).

3.7. LOCAL ANESTHETICS

It is generally stated that local anesthetics are agents that penetrate nervous membranes at the Nodes of Ranvier and, once there, antagonize ACh-transmitted impulses. This would stamp local anesthetics as special anticholinergics, but at the concentrations at which local anesthetics are used, their anticholinergic—as well as antihistaminic and antibacterial—properties are of little importance. The mechanism of action of local anesthetics involves the closing of the channels of the nervous membrane through which sodium ions migrate. Two marine toxins, saxitoxin and tetrodotoxin, have served as tools in establishing this mechanism. For details of their pharmacological behavior, see ref. 612. They are structurally complex polycyclic hydrophilic guanidines that block the constricted portion of the sodium channel (the "selective filter") by binding to a biotoxin receptor and thus antagonizing the flux of partly dehydrated sodium ions (612, refs. 16, 56, 269–274). Similar but not necessarily the same type of blocking mechanisms have been postulated for the conventional aminoester and aminoamide local anesthetics and their analogs.

The discovery of the useful local anesthetic properties of cocaine and the development of procaine and other aminoalkyl ester drugs (613, 614) have been described briefly in Section 1.3.4. Procaine and amylocaine emerged from these experiments as the "leads" for hundreds of other aminoalkyl esters to be tested as local anesthetics.

The *p*-amino group of procaine is not a prerequisite for local anesthetic activity. Many other aminoalkyl esters of aromatic acids are not substituted by functional groups. However, the reason for the trial that led to procaine was the local anesthetic activity of benzocaine (ethyl *p*-aminobenzoate) (79), originally prepared for other reasons (78). Benzocaine is used widely as a topical anesthetic.

One would not expect local anesthetics to be abused, but that is what has happened to procaine. It is used in Rumania as a rejuvenating agent for

p—H$_2$NC$_6$H$_4$CO$_2$CH$_2$CH$_2$N(C$_2$H$_5$)$_2$

$$C_6H_5CO_2-\underset{\underset{C_2H_5}{|}}{\overset{\overset{CH_3}{|}}{C}}-CH_2N(CH_3)_2$$

Procaine

Amylocaine (Stovaine)

gullible persons who pay fancy prices for ineffective treatment. The only excuse for this abuse could be that the drug is hydrolyzed to p-aminobenzoic acid, which is known as a biosynthetic component of folic acid. There are less expensive ways to enrich one's diet with this vitamin.

The changes made in the molecules of the early ester anesthetics were largely predicated on the ease of preparation of the amino alcohol moieties. Considerations of lipophilicity, pK_a, and so on, barely entered the picture. Elongating the alkyl chain, branching it, changing the NR$_2$ group in every conceivable way has been combined with substituting the aromatic ring of the acyl portion, using many different aromatic groups, condensed and uncondensed, and other manipulations. The relative ease of coping with the synthetic problems in the ester series has made work on this type of local anesthetics very popular throughout the world. Physicians have learned to prefer special ester anesthetics for special clinical maneuvers, but that is outside the expertise of the medicinal chemist. Provided that a given compound in this series has an adequately low toxicity, its basicity is one of the few properties that can be controlled chemically. This is important because, after a half century of debating this point, the weight of current evidence favors the cationic form for the major expression of local anesthetic activity (615–617). It should be noted, however, that local anesthetics penetrate neuronal membranes largely in the unprotonated form. Another weighty consideration that can be influenced chemically is the hydrolytic stability of the ester linkage. Almost all the ester anesthetics are hydrolyzed by serum esterases and therefore suffer from a relatively short half-life. Add to this that few of them are vasoconstrictors (cocaine has this property), and they are therefore liable to be diluted and removed by the circulation. Since vasoconstriction is a pharmacological property, little can be done by chemists to increase or decrease it except to recommend the use of an admixture of an adrenergic vasoconstrictor, or to use a constant-infusion technique to maintain anesthesia.

The enzymatic hydrolysis of the ester group motivated Swedish investigators to search for less readily hydrolyzable anesthetics. This problem had already been addressed by Fourneau's sterically hindered ester, stovaine (amylocaine) (84). Isosteric replacement of the alkoxy oxygen of the —COOR group by NH [—CONHR] or CH$_2$ [—COCH$_2$R] suggested itself. One amide had been tried in 1900 (86), but it was too irritating. 2-Dimethylamino-2'-acetotoluidide, made as an intermediate during an unrelated study (87), produced numbness of the tongue when tasted, and as a follow-up similar amides were prepared. Apart from the known greater hydrolytic stability of amides compared with that of esters, additional steric hindrance of the amide linkage should make the NHCO group even less susceptible to

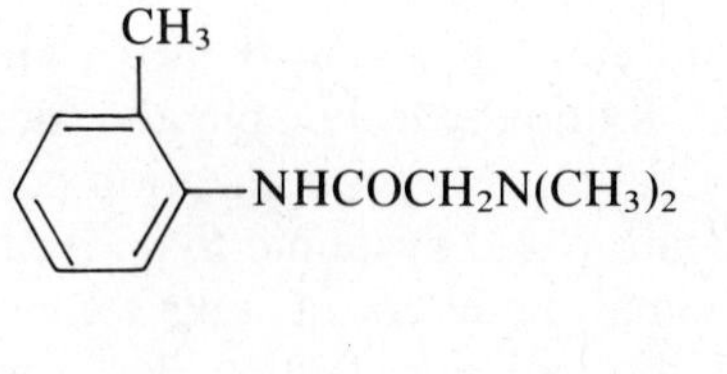

2-Dimethylamino-2'-acetotoluidide: 2-methylphenyl—NHCOCH$_2$N(CH$_3$)$_2$

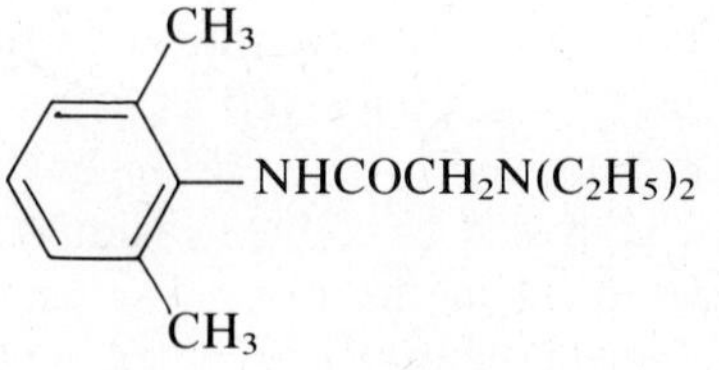

Lidocaine: 2,6-dimethylphenyl—NHCOCH$_2$N(C$_2$H$_5$)$_2$

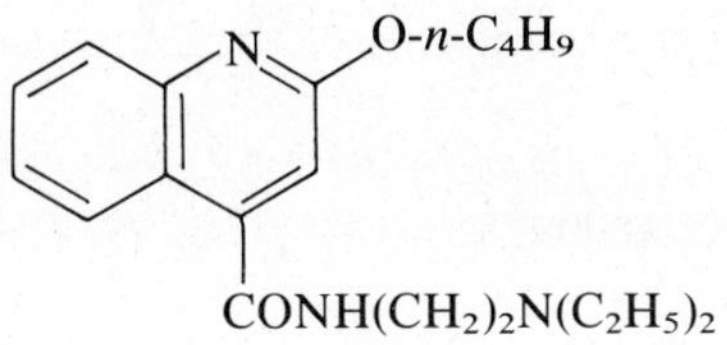

Dibucaine: quinoline with O-n-C$_4$H$_9$ at 2-position and CONH(CH$_2$)$_2$N(C$_2$H$_5$)$_2$ at 4-position

hydrolysis. The result of this plan was lidocaine (88, 618), which is indeed 112 times more stable than the o, o'-unmethylated lower homolog (619).

Again, extensive molecular modification was undertaken in order to improve further on the already very acceptable lidocaine. All the successful local anesthetics of the anilide type contain at least one or usually two o-substituents in the anilide ring. If the amide function is reversed from anilide to benzamide types, steric hindrance seems to be less strictly required. In dibucaine (620) the (reversed) amide group causes the compound to be a very potent local anesthetic but its toxicity is considerable. This drug had been conceived as a potential antipyretic and its local anesthetic activity was discovered by pharmacological screening.

Similar serendipitous events surround the development of some other compounds that have found use as local anesthetics. One is oxethazaine (621), which was a synthetic by-product in reactions designed to prepare the monomeric amine, C$_6$H$_5$CH$_2$C(CH$_3$)$_2$N(CH$_3$)COCH$_2$NH(CH$_2$)$_2$OH. The

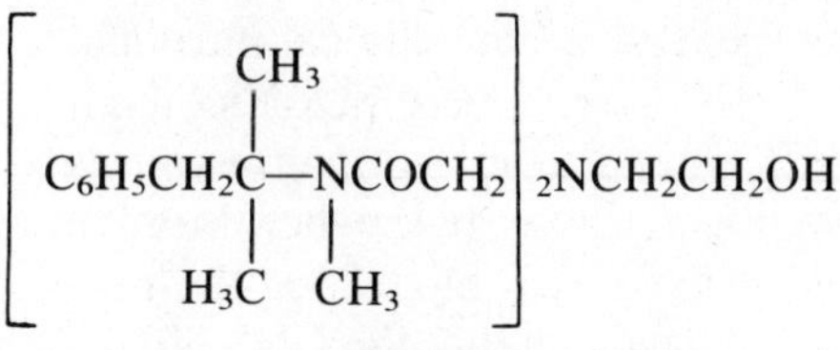

Oxethazaine: [C$_6$H$_5$CH$_2$C(CH$_3$)(H$_3$C)(CH$_3$)—NCOCH$_2$]$_2$NCH$_2$CH$_2$OH

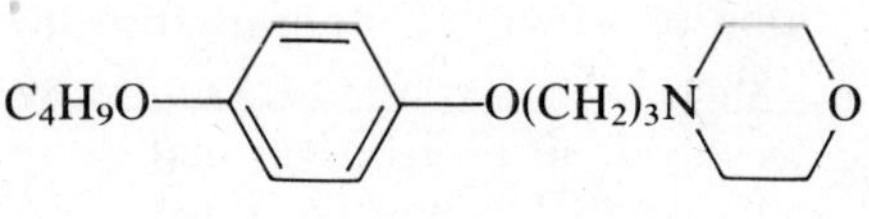

Pramoxine: C$_4$H$_9$O—C$_6$H$_4$—O(CH$_2$)$_3$N(morpholine)

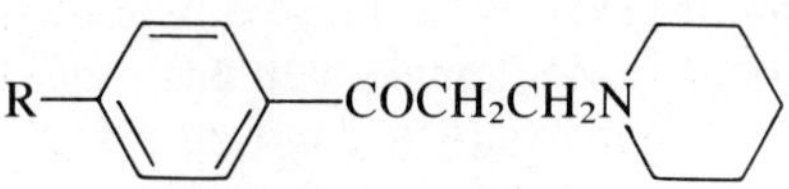

Falicaine: R = OC$_3$H$_7$
Dyclonine: R = OC$_4$H$_9$

R—C$_6$H$_4$—COCH$_2$CH$_2$N(piperidine)

others were dimethisoquin, a basic isoquinoline ether prepared for other purposes (93), and pramoxine (622), an aromatic ether taking its "lead" from dibucaine. Both dimethisoquin and pramoxine are used topically only.

These are but examples of structurally different local anesthetics which, at best, were conceived as being enzymatically nonhydrolyzable and therefore promised to have a longer duration of action. No generalization could be elaborated to reduce residual tissue irritation and systemic toxicity; the drugs used in the clinic represent the optimal members of long series of analogs and can hardly be expected to be improved upon further. Examples of Mannich-type aminoketones [—COCH_2(CH_2)$_x$NR_2 instead of —COO(CH_2)$_x$NR_2] are the homologous drugs, dyclonine and falicaine (623) which share most of the properties of the amino ether anesthetics.

3.8. NEUROLEPTIC AGENTS

Neuroleptic drugs have been given a number of names in the course of their clinical applications. In the language of the lay public, the term "tranquilizers" is probably known most widely. This is not a fortunate designation since it also covers antianxiety agents, frequently with a muscle relaxant component and often spoken of as "minor tranquilizers". The principal clinical use of neuroleptics is in the treatment of psychoses, and clinicians as well as medicinal chemists often speak of them as antipsychotic drugs. Terminology is of importance in this case because the designation "neuroleptic" or "antipsychotic" evokes certain structural correlations and sets aside other chemical structures associated with antidepressant, antianxiety and hypnotic-sedative drugs. No drugs have yet been discovered that cure mental disorders; all the agents available so far only suppress the manifestations of neurotic, psychotic, and depressive diseases, and of anxiety states. Nevertheless, medicinal chemists trying to give direction to their research and cooperating with a team of psychopharmacologists will find it convenient to base their synthetic planning on structures suggestive of a given psychopharmacological activity.

The desperate plight of psychotic, endogenously depressed, and manic-depressive patients was mitigated to a small extent by the clinical introduction of convulsive therapy (insulin shock, 1933; pentylenetetrazole convulsions, 1934; electroshock convulsions, 1937). Chemical aberrations as causes of psychiatric disorders had been suggested as early as 1845 when Moreau proposed that hashish intoxication might be used as a model for the psychoses seen in insanity. Sigmund Freud later suggested that the CNS effects of cocaine might be applied to the pharmacotherapy of psychoses. The psychological effects of some drugs such as the psychedelic (mind-manifesting) agents have been described by medicinal scientists (624, 625), psychopharmacologists (626, 627), psychiatrists (628), and others. From these studies, a spate of hypotheses for the biochemical explanation of several types of mental diseases has arisen, based on measurements of catecholamines, 5-hydroxytryptamine, acetylcholine, γ-aminobutyric acid, and other established or putative neurotransmitters in CNS tissues. The greatest obstacle to successful studies of mental disorders is the technique of animal models of such diseases. Although the effect of neuroleptic drugs on unlearned as well as conditioned behavioral responses has been perfected to a remarkable degree, the explanation of such experiments and the carry-over to clinical CNS disorders leaves much to be desired. The medicinal chemist feels frustrated by the limitations of psychopharmacological test methods because they do not provide adequate guidance in the design of more potent and more specific drugs. Behavior, both normal and abnormal, results from many varied components that cannot be affected readily by a unified pharmacotherapy. Some progress has been made in basing the design of neuroleptics, antidepressants, and antianxiety agents on structural analogs of neurotransmitters, especially 5-HT and GABA. However, none of those experimentally active drugs has advanced to clinical use. Perhaps this is an indictment of the validity of psychopharmacological test methods or of the complexity of overlapping facets of clinical mental disorders.

All psychopharmacological drugs encountered so far are multiactive agents. This overlap extends not only to the various types of behavioral disorders but also to many peripheral autonomic manifestations. The choice between equipotent agents is usually dictated by the range or absence of such side actions. Among these are antihistaminic, gastric secretion stimulatory, anticholinergic and local anesthetic properties and the ability to lower the body temperature. In addition, reserpine and tetrabenazine deplete biogenic amines in the brain and peripheral tissues. The phenothiazines and butyrophenones exhibit effects on amine transport at various cell membranes and affect the metabolism of biogenic amines. Few if any of these measurements have had an impact on drug design and the improvement of drug specificity but they have rounded out the understanding of possible mechanisms of the biochemical action of neuroleptics. In few other areas of medicinal science has the biochemical study of these drugs contributed so much to our desire to appreciate the mode of action of drugs as for agents that modify behavior.

3.8.1. Reserpine

The discovery, or rather, the rediscovery of the antipsychotic action of reserpine may serve as a good example of the involvement of natural therapeutic compounds in the development of modern drugs. Reserpine is an alkaloid occurring in a complex mixture of nitrogenous bases in the shrub *Rauwolfia serpentina*, which grows in India and in countries surrounding the Pacific basin. The powdered roots and extracts of *Rauwolfia* had been used for centuries in Hindu medicine and were mentioned in the *Veda* as a means of treating all kinds of diseases, including insanity and strokes. The accuracy of such ancient diagnoses did not need to be too high to endow the plant with a credible therapeutic folklore. In the West, the medicinal use of *Rauwolfia* was mentioned as early as 1755 (629). A more reliable description of the tranquilizing and antihypertensive properties of rauwolfia extracts was reported in 1931 by Sen and Bose in an Indian journal (630) but difficulties in intercontinental communication hid this publication for another 18 years until Vakil rendered a clinical report in a more widely read British journal (278). Reserpine was isolated from rauwolfia extracts in 1952 (280), and its structure was established (631) and confirmed by synthesis (632). The pharmacology of reserpine indicated that it was tranquilizing and antihypertensive (633), the principal activities previously noted for crude rauwolfia extracts (278, 630). The neuroleptic properties of reserpine were corroborated clinically soon thereafter (634).

Reserpine

Reserpine is used less as an antipsychotic drug today than it was 25 years ago. Its hypotensive properties have proved more durable clinically, but the use of reserpine on the whole has declined. Many attempts have been made to separate sedative and antihypertensive activities but these experiments have remained only partly successful. Molecular modifications of reserpine and its natural alkaloidal congeners have centered on alterations of the ester groups at C_{16} and C_{18}, the ether group at C_{17}, and on substitution in the aromatic portion of the indole ring system at C_{10} and C_{11}. Some of the saturated rings, especially ring E, which carries the ether and ester functions, have been aromatized by semisynthetic procedures. None of these standard methods of molecular modification has resulted in valuable drugs.

3.8.2. Phenothiazines

The events that led to the discovery of the first effective synthetic neuroleptic agent, chlorpromazine, are examples of the guidance that clinical pharmacology can give medicinal chemists. The phenothiazine derivative promethazine had been prepared as a potent antihistaminic agent (608) based on the observation (607, 608) that the blocking morieties (Ar—X—Ar) of anticholinergics and antihistaminics could be joined in the *ortho* positions to

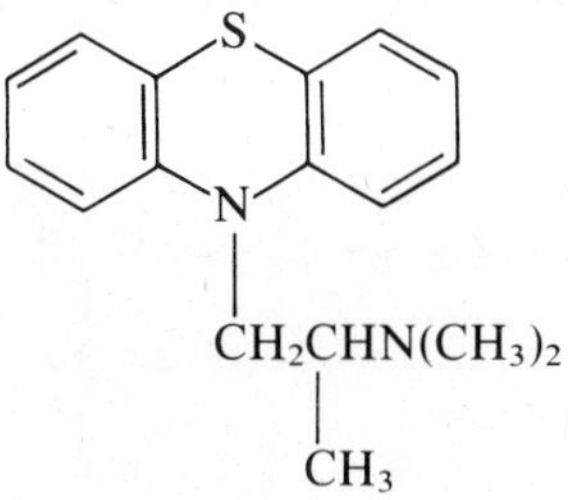

Pyrathiazine: R = CH₂CH₂N⟨pyrrolidine⟩

Methdilazine: R = CH₂—⟨pyrrolidine-NCH₃⟩

Trimeprazine: R = CH₂CHCH₂N(CH₃)₂
                          |
                          CH₃

Promethazine

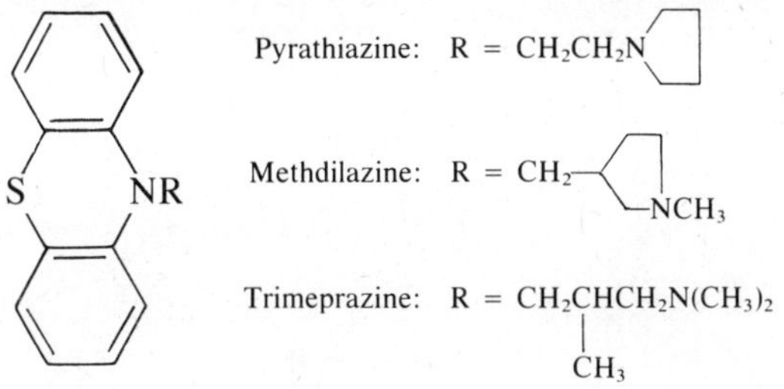

give 5- or 6- membered tricyclic compounds, for example, fluorene, dihydroanthracene, carbazole, or phenothiazine derivatives (Ar—X—Ar) with (Y) bridge. The dibenzazepines with 7-membered central rings were to join such "tricyclics" years later. Promethazine not only contained a suitably near-flat tricyclic ring system but also a traditional side chain in which the basic tertiary amino group and the phenothiazine ring nitrogen were separated by two carbon atoms. Other representatives of such structures were pyrathiazine, trimeprazine, and methdilazine (635, 636). In trimeprazine, the N—N distance had been lengthened to three carbon atoms. The most potent antihistaminic of these analogs was promethazine; in addition, it caused pronounced sedation and prolonged barbiturate-induced sleeping time in laboratory animals. A similar lengthened chain with concomitant CNS-depressant properties had been encountered in the antihistaminic–anticholinergic drug diphenhydramine [$(C_6H_5)_2CH$—$OCH_2CH_2N(CH_3)_2$], in which the ether oxygen atom isosterically takes the place of one methylene group. Such chain lengthening provided an incentive for Charpentier (254) to attempt to separate CNS-depressant and antihistaminic activities by further modification. The sedative–depressant properties increased at the expense of antihistaminic properties in derivatives in which the two nitrogens were separated by three carbon atoms. The resulting compound (promazine) was further modified by ring substitution, especially by the traditional chlorine, which is supposed to exert both an activating and "dampening" effect. More likely, chlorine stabilizes the aromatic phenothiazine system against premature oxidative detoxification. In any event, chlorpromazine turned out to be a pharmacologically multiactive agent (637) and had unusual actions on the CNS. These effects were evaluated by Delay (253), who recognized the clinical potential of chlorpromazine as an antipsychotic agent. The drug also performs well as an antiemetic.

Chlorpromazine has become the standard drug by which the activity and pharmacological profile of other congeners are usually measured. The bewildering multiplicity of its biochemical and pharmacological manifestations is also expressed in the many clinical side effects of the drug. The two most worrisome are unwanted sedation and, even more so, the frequent incidence

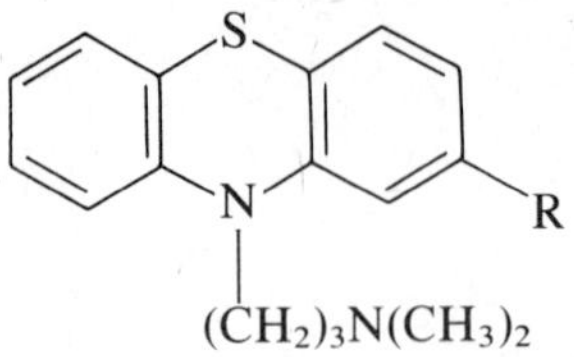

Promazine: R = H
Chlorpromazine: R = Cl

of extrapyramidal side effects on prolonged administration. These are observed as parkinsonism-like tremors and other symptoms of this disease syndrome. An explanation of this occurrence is seen in the decrease of the concentrations of dopamine at its receptors caused by the phenothiazine neuroleptics. Dopamine, not utilized in receptor reactions, is degraded to homovanillic acid in the brain; a drop in dopamine concentrations in certain brain tissues (caudate nucleus, substantia nigra) is known to be associated with the etiology of parkinsonism.

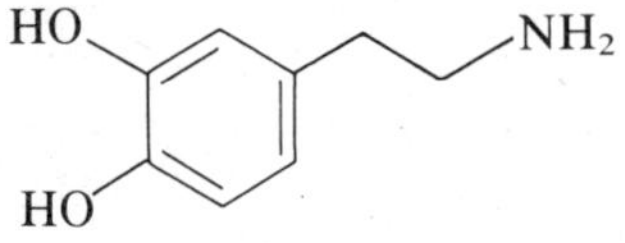

Dopamine

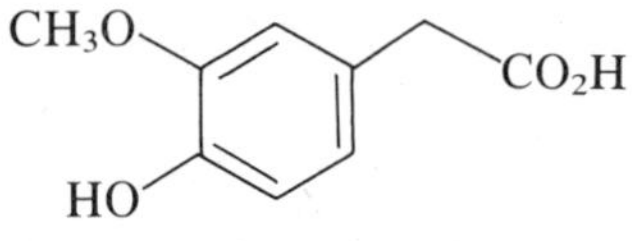

Homovanillic acid

Many biochemical phenomena implicate phenothiazine neuroleptics as modifiers of the permeability of cell membranes. It is believed that cells couple with each other electrically and that this process is regulated at membrane gap junctions by calmodulin, an ubiquitous calcium-ion binding protein that has been described as a mediator of the action of calcium in eukaryotic cells (638). Chlorpromazine and trifluperazine inhibit the function of calmodulin at the gap junctions and thus interfere with intercellular calcium transport (639).

In cooperation with experimental psychopharmacologists, it has been the endeavor of medicinal chemists to separate the many side effects of phenothiazine neuroleptics by molecular modification. Indeed, it has been possible to decrease, although not eliminate, the occurrence of extrapyramidal side effects. As a result of these researches one of the more favorable compounds, thioridazine (640), has become the preferred antipsychotic phenothiazine drug in many psychiatric clinics. The structure of thioridazine serves as a symbol of many of the molecular modifications in the phenothiazine series. The nuclear 2-position has remained substituted in all but two unimportant analogs (promazine, mepazine). The most widely used substituents are Cl (chlorpromazine, prochlorperazine, perphenazine); CF_3 (triflupromazine, trifluperazine, fluphenazine); SCH_3 (thioridazine). A number of other mostly electron-withdrawing substituents (SO_2NH_2, NO_2, $COCH_3$, SO_2CF_3, SCF_3, etc.) have also been used successfully in experimental test compounds.

The second site for molecular modification has been the side chain at-

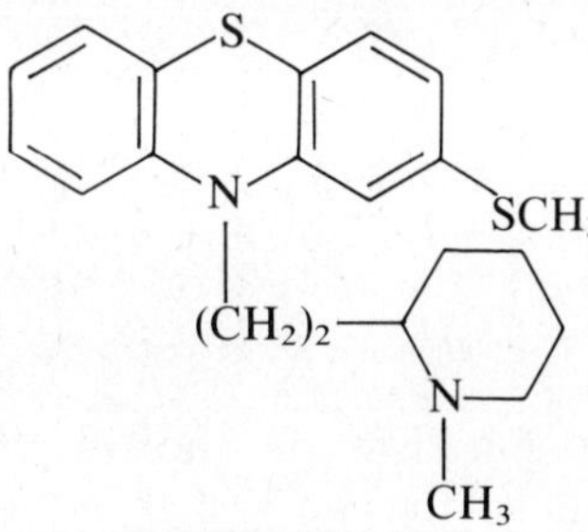

NEUROLEPTIC AGENTS

Thioridazine

tached to the phenothiazine nitrogen (position 10). On the whole, the three-carbon distance from this nitrogen to that of the basic amino nitrogen has proved to be optimal for neuroleptic activity but the arrangement of the side chain has been varied widely. It can be unbranched [$(CH_2)_3$], branched [$CH_2CH(CH_3)CH_2$], or part of a heterocyclic ring.

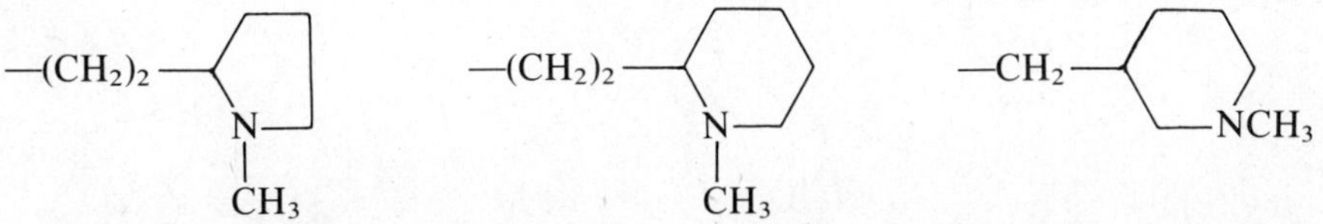

The tertiary amino group has also been varied in many cases. Apart from the dimethylamino group found in the earlier phenothiazine neuroleptics, N-methylpyrrolidinyl and -piperidinyl groups have been of interest. They incorporate one or two carbon atoms of the side chain in their cyclic portion. Of special interest are derivatives in which a piperazine ring represents the basic amine moiety. A balance between added lipophilic bulk and required hydrophilicity is achieved by replacing the methyl group on the second piperazine nitrogen by an ethanol chain. Potency and selectivity are often increased in such derivatives.

3.8.3. Other Tricyclic Systems

Of the more daring molecular modifications, the ring structure of phenothiazine has been altered radically in many cases. One or both benzene rings have been replaced isosterically by thiophene or pyridine; the central thiazine ring has been changed to oxazine, pyridine (acridines), or by exchange of nitrogen and carbon (thioxanthenes). In the latter case, the 9-carbon atom of the thioxanthene system can be unsaturated, introducing a double bond

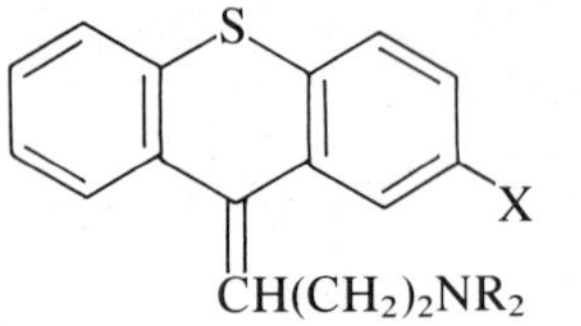

Thioxanthene derivatives

and thereby geometrical isomerism into the side chain, if the 2-position is substituted. Frequently, one of the two isomers is more potent than the other, or may even differ qualitatively in its psychopharmacological activity spectrum. Other types of rigidity (cyclopropane, cyclobutane rings, etc.) placed in the three-carbon side chain also lead to exploitable geometric isomers.

In addition to these modifications, the central ring has been expanded to seven-membered heterocyclic structures, motivated by the possibility that antipsychotic and antidepressant activities might be combined in one molecule. Thousands of molecular modifications have been described and catalogued in monographs (641–644).

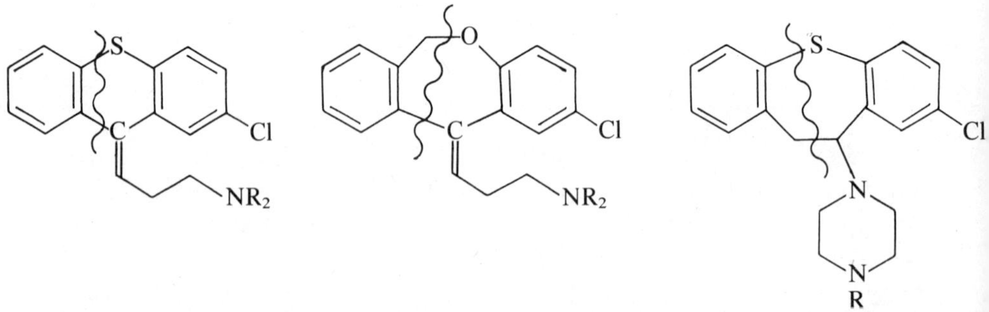

In these tricyclic series, the three schematic formulas illustrate the essential structure–activity relationships in rigid or near-rigid structures. The common features of these three formulas lie to the right of the wavy lines. In the aminopropylidene derivatives, a Z orientation is needed for potent neuroleptic activity. The unsubstituted ring apparently is less specific for binding to the receptors while the chlorine-substituted aromatic ring always lies on the same side as the side-chain amino group. Other suggestions about SAR have been made by Kaiser and Setler (641).

In at least one structural type only vaguely related to the tricyclic antipsychotics and yet derived from them, there was an absence of extrapyramidal side effects characteristic of the phenothiazines and their congeners. They are dibenzo[b, e][1, 4]diazepines, dibenzo[b, f][1, 4]thiazepines, and related compounds. The most widely studied member of this series is clozapine (645). It has an unusual pharmacological profile, and only the occurrence of agranulocytosis during its use interferes with its clinical utility. Its development serves notice that even after thousands of not very successful attempts to increase critical specificity in a given series, the hope of finding

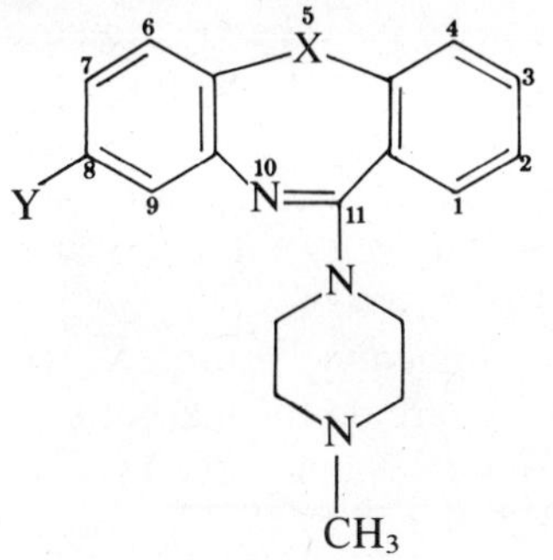

Dibenzodiazepine derivatives: X = NH, Y = H
Clozapine: X = NH, Y = Cl
Dibenzothiazepine derivatives: X = S, Y = H

a suitable analog should not be dismissed. The decision to keep on searching will be one of devotion to the subject, exploitation of acquired experience, and the availability of research funds.

3.8.4. Butyrophenones and Related Compounds

The Belgian pharmacologist Paul A. J. Janssen is an experimental biologist who understands the details of the mainstream of medicinal chemistry. Working on new types of potent analgetic analogs of meperidine, Janssen concentrated on derivatives in which the standard methyl substituent on the piperidine nitrogen was replaced by alkyl aryl keto groups. The "lead" compound, which exhibited morphinelike activity, was screened in 1957 for the

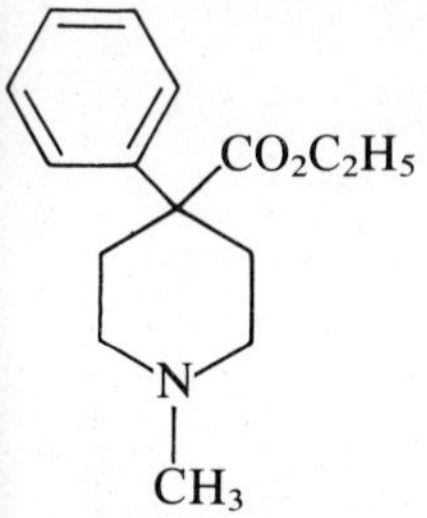

Meperidine

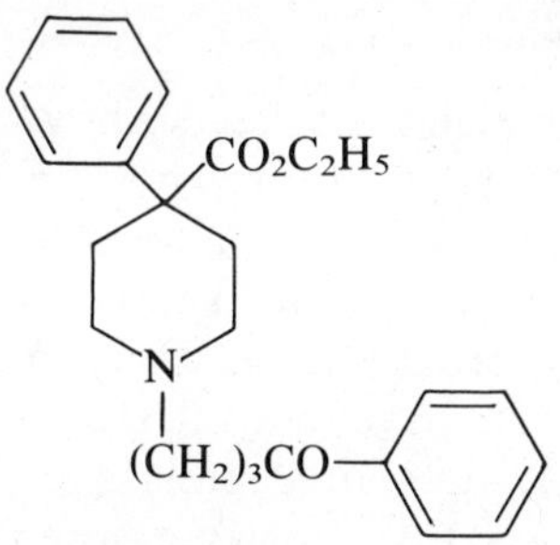

"Lead" compound for butyrophenone neuroleptics

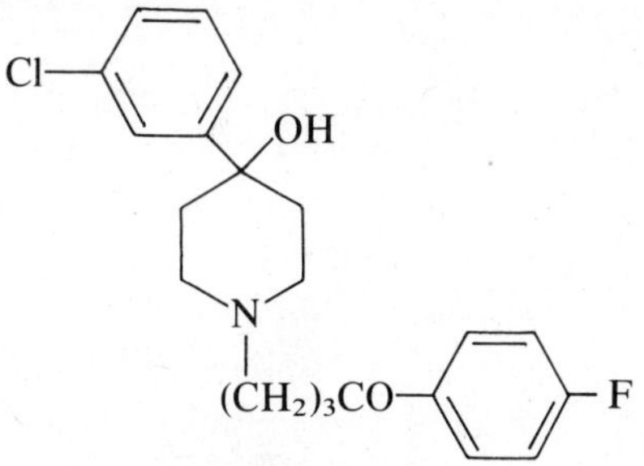

Haloperidol

then fashionable neuroleptic action and was found to produce such a syndrome in animals. The butyrophenone substituent was given preference over alkyl aryl ketones, which were less suitable pharmacologically (646). Substitution of the aroyl nucleus by halogens and pseudohalogens (F_3C) demonstrated that fluorine *para* to the keto group was optimal for neuroleptic potency. Replacement of the carbethoxy group by other functions narrowed down the choice to a tertiary alcohol group, and after synthesizing and testing hundreds of analogs, haloperidol was chosen for clinical tests in 1958 (646). It responded considerably more strongly than chlorpromazine in some assays of the psychopharmacological test battery and became the prototype of over 5000 modifications, of which more than 20 have been studied clinically for antipsychotic properties. In many assay methods the butyrophenones behave just like the tricyclic antipsychotics, whereas in some others there are differences. For example, the butyrophenone neuroleptics are less potent than chlorpromazine in inhibiting dopamine-sensitive adenylate cyclase (647) *in vitro*, but they are more potent *in vivo* (648).

Haloperidol is the only member of the butyrophenone neuroleptics used clinically in the United States, but several others are on the market abroad. Structurally, the molecular modifications comprise many changes in the "lead" structure. The piperidine ring has been replaced by tetrahydropyridyl and piperazinyl in efforts to enhance neuroleptic activity at the expense of analgetic properties. The large amount of SAR data has been reviewed repeatedly (641, 649).

The aminobutyrophenones are δ-aminoketones; other homologs, especially β-aminoketones, also have some neuroleptic activity. The pharmacological "lead" to screening these compounds (Mannich bases) was their α-adrenoceptor blocking activity (650), which had already been implicated

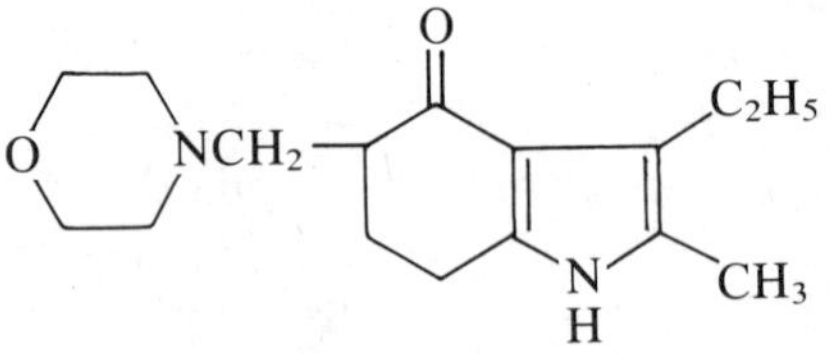

Molindone

in connection with extrapyramidal side effects above. Screening then revealed other CNS activities, including neuroleptic properties. Only one of several types of such compounds will be mentioned here. Molindone (651) is an example of a Mannich base of a pyrrole ketone with a pharmacological profile resembling that of chlorpromazine in certain respects. It has some clinical advantages over other antipsychotics (652).

Another example of using α-adrenoreceptor blockade as predictive of neuroleptic properties goes back to the early α-adrenergic blocking agent piperoxan (385), which had initiated the researches on antihypertensive drugs. Compounds such as pentamoxane, in which the piperidino group has been replaced by a secondary amino group of appropriate chain length, have high neuroleptic activity in animal models (653). Isosteric exchange of O for CH_2 in the alkylamine chain or of CH_2 for O in the dioxane ring also furnish compounds with experimental antipsychotic properties.

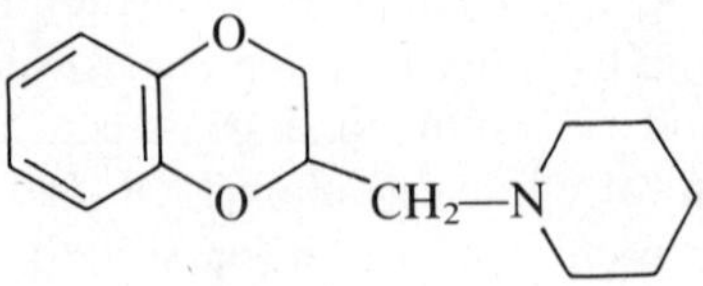

Piperoxan

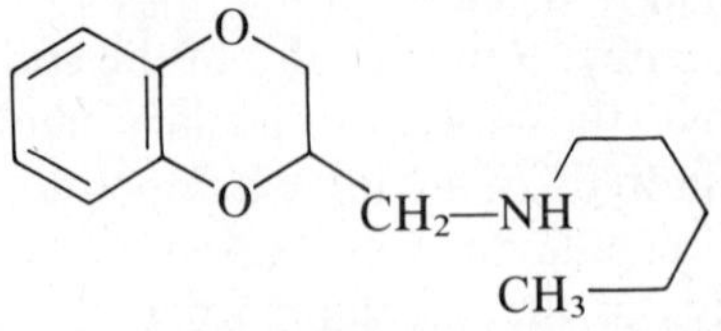

Pentamoxane

3.9. ANTIDEPRESSANTS

The role of medicinal chemistry in the field of antidepressants has been the effort to improve existing "leads" by molecular modification and to study the biochemical mode of action of antidepressants. Medicinal biochemists have contributed to hypotheses trying to explain clinical depression and its pharmacological counterparts in experimental animals. However, pharmacology and other experimental branches of biology have made the fundamental contribution to that research area. The "leads" have been provided essentially by pharmacological and, even more, by clinical observations. These discoveries are presented first in order to put the role of chemical modification into proper perspective.

Many compounds of diverse structure exhibit antidepressant activities in a pharmacological sense, that is, by responding to tests in which artificially depressed animals are restored to a more normal level of behavior. Some of these test methods will be described later. Successful clinical antidepressant activity is subject to human pharmacokinetics superimposed on the intrinsic biochemical activity of the compound. Yet these difficult conditions, made doubly difficult by the vagaries of psychological judgments of human behavior, have opened the door to the psychopharmacological treatment of true pathological depression. Because these fundamental observations occurred in two classes of drugs, the monoamine oxidase inhibitors (MAOI) and the "tricyclic" antidepressants, their discoveries are described first.

3.9.1. Discovery of MAO Inhibitors

Patients suffering from tuberculosis had been treated with isoniazid since 1951. The pharmaceutical firms manufacturing isoniazid were engaged for several years rounding out their patent positions and at the same time studying SAR. During these studies, Fox (187) prepared a large number of l-alkyl derivatives of isoniazid, 4-Py-CONHNHR; the isopropyl derivative, 4-pyridyl-CONHNHCH(CH$_3$)$_2$ (iproniazid), was tried clinically. Many of the patients treated with iproniazid experienced an unanticipated mood-elevating side effect that could escalate to psychotic levels. This phenomenon was too striking to be attributed to the well-feeling following relief from cured tuberculosis lesions.

Biochemists in Zeller's laboratory (261), interested in the inhibition of various enzyme systems by drugs, then discovered that iproniazid was a potent inhibitor of monoamine oxidase (MAO), a flavin-containing enzyme whose mission it is to deaminate biogenic amines and related compounds oxidatively. Biogenic amines have long been associated with various auto-

$$RCH_2NHR' \xrightarrow{MAO} RCH{=}NR' \; (\xrightarrow{H_2O} RCHO + R'NH_2)$$

nomic nervous phenomena, and synthetic analogs of, for example, phenethylamine, such as amphetamine, had been known to stimulate cerebral cortical functions and to elevate mood. In corroboration of its MAO-inhibitory effect, iproniazid was shown to potentiate certain pharmacological actions of some adrenergic amines that must have been spared from the deamination by MAO (654, 655). Iproniazid was then found to cause excitement in animals sedated by reserpine (656); this led to a successful clinical study of iproniazid in depressed patients (657). Hepatotoxicity, often encountered with hydrazine derivatives, soon forced withdrawal of the drug from general use and stimulated research on nonhydrazine MAOI antidepressants.

3.9.2. Discovery of Tricyclic Antidepressants

The history of the prototype drug for the so-called tricyclic antidepressants is equally circuitous but contains more elements of chemical planning. After the recognition of the neuroleptic properties of chlorpromazine, promazine, and other phenothiazines (Section 3.8.2), industry-wide projects were initiated to expand these "leads" in a search for more specific antipsychotic drugs with fewer side effects. Phenothiazine is near-flat, that is, the central thiazine ring is part of its tricyclic aromatic system. Sulfur in aromatic linkage has been known to be isosteric to —CH=CH— (or =CH—CH=), the best known ring equivalents (3) being benzene and thiophene, and in pyridine and thiazole. The phenothiazine neuroleptics arose in the Rhone–Poulenc laboratories (254) from synthetic series designed to yield antihistaminics, analgetics, antiparkinsonism agents, and sedatives. Häfliger and Schindler

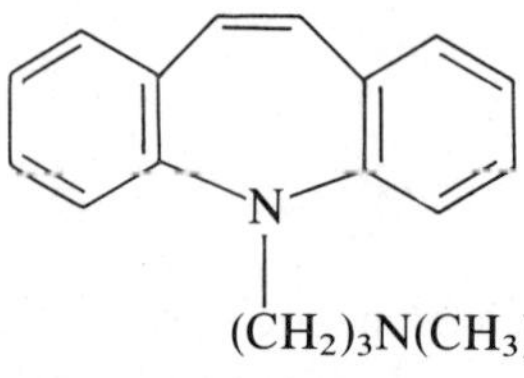

$(CH_2)_3N(CH_3)_2$ $(CH_2)_3N(CH_3)_2$

11-(3-Dimethylaminopropyl)-dibenzazepine 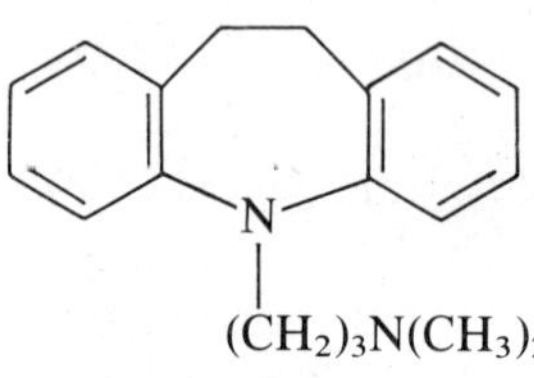   Imipramine

(265), interested in similar drugs, therefore replaced the sulfur atom of promazine with an ethene bridge. As long as this unsaturated dibenzazepine was at hand, it was also hydrogenated to the dihydro compound imipramine. Both compounds were disappointing as neuroleptics in pharmacological tests and were set aside. Imipramine was reinvestigated clinically by Kuhn six years later (264) and unexpectedly elicited good responses, although delayed, in endogenously depressed patients. This observation led to a huge effort in molecular modification of related structures, with 7-membered azepine compounds dominating synthetic thinking. This line of approach was reinforced further when the benzodiazepine anxiolytics were discovered soon after (258–260).

The discovery of the most prominent neuroleptics, antidepressants, as well as the discovery of the antianxiety agents, leaves medicinal chemists with more questions than answers. In the case of the MAO inhibitors and the antianxiety agents of the meprobamate and benzodiazepine types, keen biochemical and pharmacological observations must be credited fully with recognizing the therapeutic potential and type of pharmacological action of these drugs. For the tricyclic antidepressants, clinical pharmacology in the hands of an alert psychiatrist provided the entering wedge into this field (264, 658). The same can be said for the recognition of lithium as a therapeutic agent in manic states (659). In all these cases, pertinent biochemical causative support for the original pharmacological observations has been obtained from the developing biogenic amine theories of mental diseases. But even if these biochemical models had been known before the pharmacological discoveries in animals or in the clinic, they would not have led to the drugs now used in psychopharmacological medicine. None of the phenothiazines, dibenzazepines, benzodiazepines, and carbamates, which constitute the major structural prototypes of psychopharmacological agents, resemble the biogenic amines and other neurohormones and neurotransmitters that have come to be associated with behavioral mental disorders. Attempts to construct connections between these neurohormonal factors and the known drugs by overlaying their two- or three-dimensional formulas over each other have remained inconclusive at best. In other words, the design of psychopharmacological agents has remained empirical, a matter of incredible chance, and of chemical research.

3.9.3. Biochemical Hypotheses

The accumulated wealth of biochemical data relating behavioral disorders to disturbed concentrations of brain chemicals has not yet yielded a clue to how effective drugs for such disorders might be derived rationally. It may well also be a matter of fashion in biochemical researches. During the 1960's, norepinephrine was, as Snyder put it, "the darling of neuropharmacology." A decade later, interest centered on dopamine, probably because more sensitive spectroscopic and immunoassay procedures had made possible a distinction between biogenic catecholamines. We now know about 25 brain and hypothalamic peptides whose combinations can produce subtle variations in behavior. Until SAR of these peptides in behavioral tests will be understood, the design of pertinent drugs affecting enzyme systems in which these peptides play a role may have to remain empirical.

One of the few structural relationships to the biogenic amines shows up in the ability of hydroxylated derivatives of such amines to deplete lastingly the neurotransmitter amines in certain tissues. Thus, 6-hydroxydopamine degenerates neurons in tissues that accumulate dopamine, and 5,7-dihydroxytryptamine has the same effect on tissues that contain 5-HT (660, 661). The 6-hydroxydopamine is converted metabolically to 4,6,7-trihydroxyindoline, and this reacts with nucleophilic groups of tissue components to form covalent materials that can no longer function in the central or peripheral neurons. Additional mechanisms have also been suggested (662). None of these inhibitory and regenerative agents has given rise to medicinal drugs.

The biogenic amine hypothesis of mental disease appears simpler than it really is. Some antidepressants (e.g., fluotracen) differ from imipramine-like drugs in that they are also dopamine receptor antagonists (663). Others (mianserin, danitracen, iprindole) are such weak inhibitors of monoamine uptake that their antidepressant activity must be based on other mechanisms, perhaps on antagonism to serotonin or histamine or an observed decrease in sensitivity of norepinephrine-dependent adenyl cyclase in the forebrain. All this points to the need for comprehensive pharmacological and biochemical screening if SAR are to be interpreted properly.

3.9.4. Test Methods

To be valid, comparisons of antidepressant drugs must be made on the basis of the same or equivalent test procedures. MAO inhibitors are added *in vitro* to MAO's [from brain or tissue homogenates or highly purified enzyme preparations (664)] reacting with a substrate (norepinephrine, dopamine, serotonin, benzylamine, etc.), and the amount of carbonyl compound formed is subtracted from the amount in an uninhibited standard experiment. Pharmacologically, reserpine depression is reversed by MAO inhibitors; reserpine can be replaced by the synthetic tetrabenazine, which acts comparably. Imipramine-like tricyclics are likewise measured by their ability to prevent or reverse tetrabenazine depletion of biogenic monoamines; *in vitro* and behavioral tests are also used (665). These tests are based biochemically on the ability of "tricyclics" to block accumulation of various biogenic monoamines in synaptosomes or brain slices (*in vitro*) or in whole brain (666).

It has become customary to subject each promising candidate compound to a battery of tests that can unveil neuroleptic, anxiolytic, and antidepressant activities as well as peripheral autonomic activities. During the feverish researches on psychopharmacological drugs from 1960 to 1975, many compounds were found whose claim to pharmacological utility depended largely on their action profile rather than on one special activity. The call for such action profiles also had an effect on medicinal chemical planning. It became difficult, if not impossible, to predict what component of this profile would be affected by a given molecular modification. Thus Hansch-type calculations and even less quantitative intuitive modifications could not be applied to a testing program in which multiple tests based on different mechanisms of action decided the direction further structural changes should take.

3.9.5. Monoamine Oxidase Inhibitors

It was relatively easy to select "leads" for MAO inhibitors because here one specific enzyme was involved in the tests. But instead of one specific

ANTIDEPRESSANTS

enzyme, two should really be considered, named monoamine oxidases A and B for lack of better nomenclature. It had been hoped that the discovery of these isoenzymes from different organ sources would lead to more specific inhibitors, and that hope was realized experimentally—but clinically useful MAO inhibitors are, on the whole, unspecific for these isoenzymes.

3.9.5a. Nonhydrazine MAO Inhibitors. The hepatotoxic side effects of several hydrazides and hydrazines prompted a search for structurally different compounds, and two of them, pargyline and tranylcypromine, were selected as clinical candidates. Both have effects on the blood pressure, often paradoxical, and pargyline (667) was introduced as an antihypertensive agent. It inhibits MAO's and potentiates dopa in mice. Some of its derivatives, in which the benzyl group is replaced by 2,4-dichlorophenoxypropyl (clorgyline) or the propynyl group is replaced by cyclopropyl, inhibit MAO-A or B selectively, but this has not been of clinical advantage.

One of the most interesting aspects of the biochemistry of propargylamines such as pargyline is the mechanism of their reaction with MAO (668). The flavin coenzyme reacts with the drug by addition across the 1,4-additive system. This is an example of enzyme inactivation by formation of a covalent adduct with an inhibitor; the enzyme thereby "commits suicide," and the inhibitors have been aptly named suicide enzyme inhibitors.

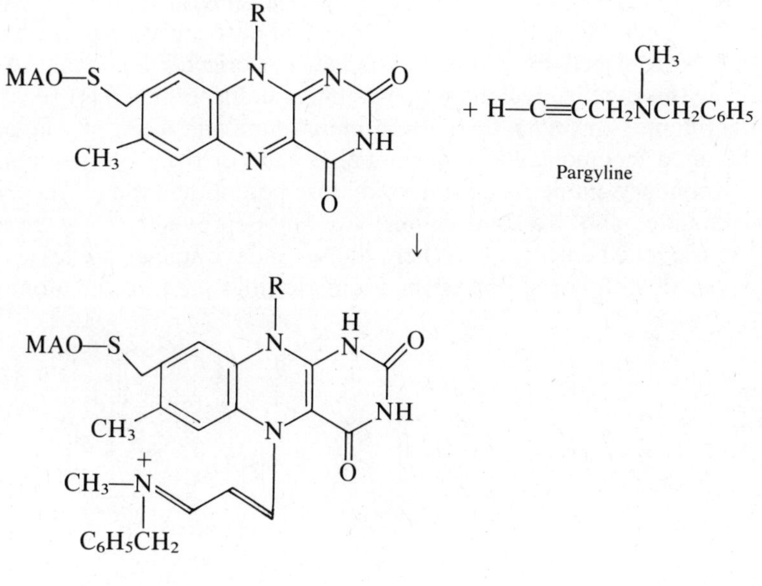

A different fate befalls *trans*-2-phenylcyclopropylamine (tranylcypromine) that had been synthesized as a structural analog of amphetamine (669). It is a potent inhibitor of MAO *in vitro* and *in vivo* and is used as an antidepressant. Patients using tranylcypromine (as well as other MAO inhibitors) may not ingest foods rich in tyramine, such as cheddar cheese, red wine, or bananas, because the deamination of this biogenic amine will be inhibited and tyramine can cause damaging hypertension.

Tranylcypromine is a suicide enzyme inhibitor. It is dehydrogenated by MAO to an imine which then adds to an essential thiol group of the apoenzyme and thereby deprives MAO of further activity (670).

This mechanism cannot be operative in the case of analogs and derivatives of 2-phenylcyclopropylamine, which cannot be dehydrogenated to an imine. Thus, N,N-dimethyl-2-phenylcyclopropylamine and 1-methyl-2-phenylcyclopropylamine are potent inhibitors of MAO (671) but must react by different mechanisms.

Molecular modification of these MAO inhibitors followed traditional lines: nuclear substitution, widening of the cyclopropane ring, changing the acetylenic triple bond to double bonds or small rings, and probing the significance of the various functions for biological activity.

3.9.6. Tricyclic Antidepressants

A wide selection of similar molecular modifications has also been undertaken in the series of tricyclic antidepressants. The most important essential feature of such compounds is their molecular geometry, which sets them apart from tricyclic neuroleptics. The two terminal rings are tilted to one another at considerable angles; this is achieved by seven- or eight-membered central rings. In a few cases, antidepressant-like symptoms are registered in animals treated with compounds whose central ring is six-membered, but 1-chloro substitution, hydrogenation of one terminal ring, or some other device is needed in such cases to interfere with the planarity of the ring system.

The side chain usually has a spacing of three atoms before the amine nitrogen is reached, but this is not a rigid requirement. If the side chain is attached to a central homocyclic ring, the usual nitrogen atom of the azepine ring will be replaced by a double-bonded isosteric carbon atom. A drug containing this dibenzocycloheptane system is amitriptyline; it is used widely as an antidepressant (266, 267). Both imipramine and amitriptyline are N-demethylated metabolically to desipramine and nortriptyline, respectively. These secondary amines are somewhat more potent than the corresponding tertiary amines and are used clinically as antidepressants. This contrasts with the tricyclic neuroleptics, where the secondary amines are less potent. Likewise, substitution of one of the aromatic rings *meta* to the atom (N or

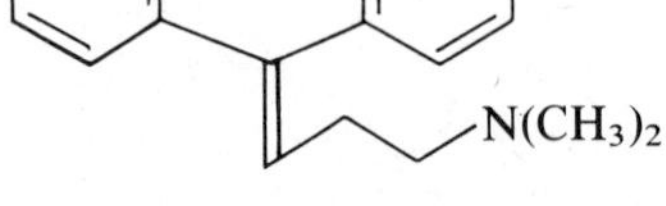

Amitriptyline

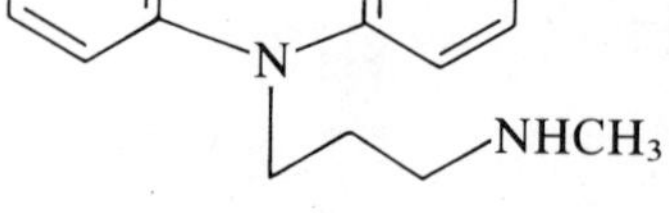

Desipramine

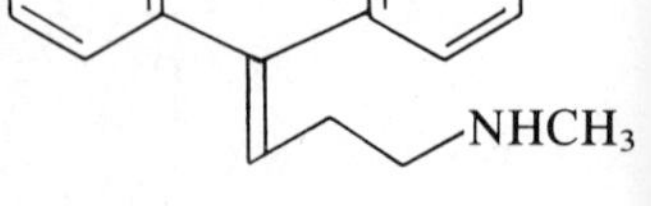

Nortriptyline

C) carrying the side chain has a dystherapeutic effect in the series of tricyclic antidepressants, whereas it is virtually prerequisite for neuroleptic potency.

Devising new compounds with puckered central rings, and devising side chains in which the participating atoms are incorporated in heterocyclic rings and even fused ring systems, has taxed the synthetic ingenuity of medicinal chemists. Much of this work has been done for commercial competitive reasons but it also reflects the inclination of the chemists to study new ring systems. In the course of their extensive investigations, they encountered compounds with profiles combining all types of psychopharmacological emphasis and minor activities, useful or unwanted. For details, the reader is referred to an article by Kaiser and Setler (672). That review also lists miscellaneous compounds of unorthodox and unrelated structure for which systematic screening tests in laboratory animals indicated a potential for antidepressant activity.

3.10. ANTIANXIETY AGENTS

The two principal classes of antianxiety drugs, the urethans of propanediols and the benzodiazepine derivatives, were discovered and developed empirically. Medicinal chemical thinking has had little influence on these methods except to guide molecular modification to a small extent. Even here empiricism has prevailed. The numerous biochemical data on the causation of intense pathological anxiety have not had an impact on the design and understanding of the mechanism of action of these drugs. Pharmacological tests have relied on adjunct activities such as anticonvulsant and muscle relaxant properties. It is a compliment to pharmacological astuteness that effective antianxiety drugs have been recognized, developed, and modified on the basis of hard-to-interpret animal and clinical test methods. Some of these have been reviewed (673).

Severe anxiety can cause psychosomatic illness, and such illnesses aggravate anxiety states in turn. Anxiety has been obtused with alcohol for millenia, with sedative–hypnotics in this century. Low doses of neuroleptics and sedative antihistaminics have also been used, and overt somatic symptoms such as palpitations have been treated with β-adrenergic blocking agents. None of these classes of drugs has been prescribed as much as the modern antianxiety agents. Attempts by medicinal chemists to modify the drugs oriented toward somatic symptoms in order to elevate antianxiety properties *per se* have not had a noticeable success.

3.10.1. 1,3-Propanediols

The starting point for researches in this series was mephenesin, $o\text{-CH}_3\text{C}_6\text{H}_4\text{OCH}_2\text{CHOHCH}_2\text{OH}$, the α-(o-tolyl) ether of glycerol, originally investigated as an antibacterial drug. In the course of pharmacological workup, the antispasm and muscle relaxant properties of the compound were noted (234). Mephenesin suffers from a short duration of action, being metabolized rapidly by oxidation of the primary alcohol group to carboxyl and by phenolic hydroxylation. Therefore it was hoped that esters including carbamates would prolong activity, but this was only partly successful (674). An extensive search for modified analogs revealed that muscle relaxant activity was independent of the aryl glyceryl ether portion and that 2,2-branched 1,3-propanediols were most potent (675). A similar series of diols and their carbamates was studied by Berger for CNS depressant, muscle relaxant, and anticonvulsant activities (676). The dicarbamate of 2-methyl-2-n-propylpropane-1,3-diol [meprobamate, $CH_3C(n\text{-}C_3H_7)(CH_2OCONH_2)_2$], was chosen as a clinical candidate drug on the basis of its activity against electroshock (677). It is a short-acting muscle relaxant and is useful in alleviating tension, anxiety, muscle spasms, and petit mal.

3.10.2. Benzodiazepines

The important drugs in this series were first prepared and modified by Sternbach and Reeder (258); Sternbach has reviewed the events surrounding their discovery (260). The discovery of their activity was made by Randall in the course of pharmacological studies (257, 678–680). Medicinal chemists cannot learn anything from these reports that they could use in drug design in other cases.

Sternbach had been interested as a student in what turned out to be benzodiazepine derivatives and returned to study them some 20 years later. The correct ring structure of these compounds was not understood, but in the Roche laboratories some of them were submitted to pharmacological screening. Randall happened to test new compounds for meprobamate-like profiles and found that the compounds sent over from the chemistry section indeed had the desired properties (257). It took another half year to clear up their structure and begin making derivatives and analogs. Recognition of the antianxiety properties of the new drugs sparked competition in other pharmaceutical companies.

Among the early analogs of the original "lead" were diazepam and oxa-

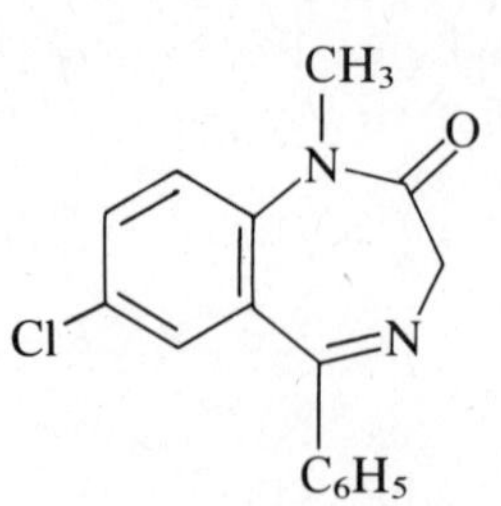

Diazepam

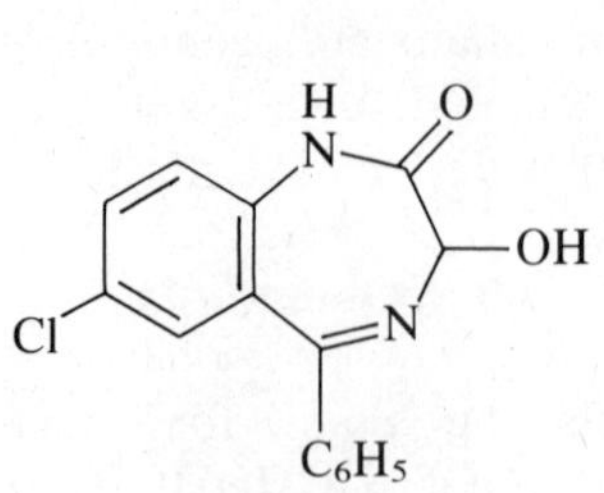

Oxazepam

zepam. Molecular modification followed an opportunistic pattern, making those substitutions, reductions, and analogs that lent themselves to most obvious chemical syntheses. A more thoughtful and medicinal-design-oriented program emerged in later phases. These efforts have been summarized by Childress (673) and by Sternbach (679, 681). Suffice it to say that, as expected, some pharmacological properties of the original muscle-relaxant sedative and anticonvulsant profile became more elevated or less pronounced in some of the analogs. For example, flurazepam (682, 683) has become a widely prescribed hypnotic. The *o*-nitro analog (nitrazepam) is

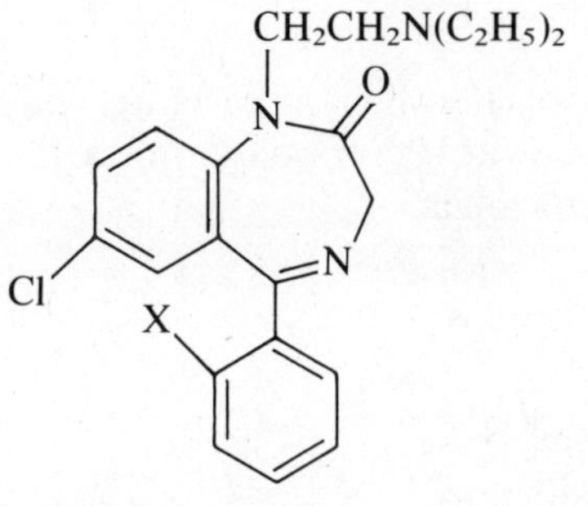

Flurazepam: X = F
Nitrazepam: X = NO_2

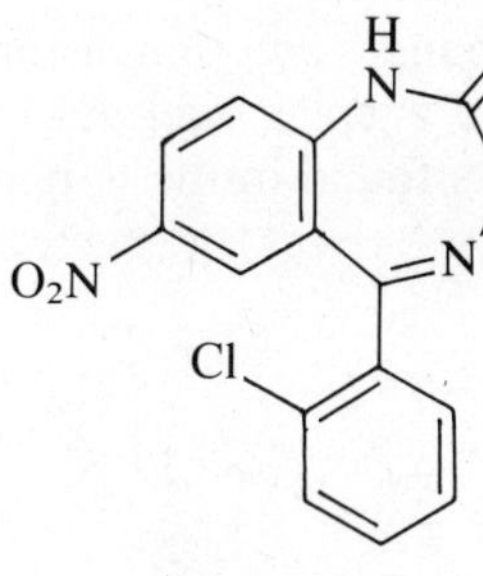

Clonazepam

also a hypnotic. Another nitro derivative, clonazepam, is an effective anticonvulsant.

The molecular modification of large numbers of analogs and derivatives has permitted a limited number of conclusions to be drawn concerning SAR. Among them is that potency decreases if the substituent of the original 2-methylamino substituent (of chlordiazepoxide) becomes too large or if the 5-phenyl group is replaced by other substituents. In the series of 1,4-benzodiazepinones, electronegative substituents are required for highest potency in position 7, NO_2 and CF_3 producing the most potent derivatives. The 5-phenyl group can be substituted in the *ortho* positions or replaced by other aromatic moieties. A number of other substitution regularities have also been noted (673). Compounds in this series have the advantage of being synthesized more easily than the 2-amino derivatives.

When simpler transformations had reached a point of unrewarding returns, chemists turned to synthetic analogs in which both the benzene and the hetero rings of the benzodiazepine system were varied. As could be expected, such innovations forced medicinal chemists to study SAR and the metabolism of derivatives from scratch for each new structural type. These data have been reviewed (673) and little would be learned by discussing them again here. The changes involved have the marks of the standard procedures used in molecular modification. Examples are derivatives of 1,5-benzodiazepinediones such as clobazam (683) and triflubazam, which have antianxiety profiles between those of diazepoxide and diazepam. More far-reaching

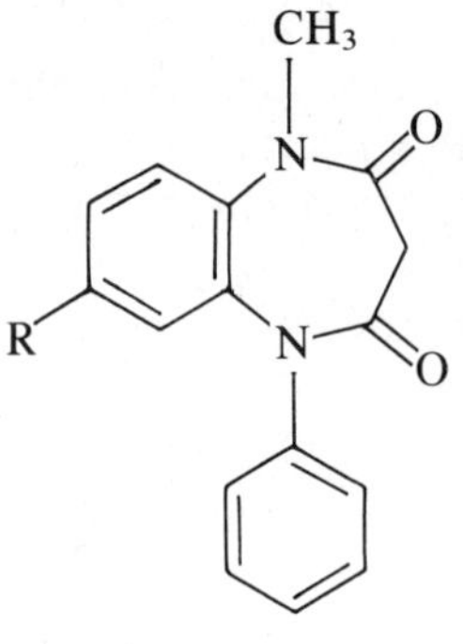

Clobazam: R = Cl
Triflubazam: R = CF$_3$

structural changes are seen in compounds in which the carbonyl and nitrogen atoms of the azepine ring are bridged with the formation of a third fused ring (684), as found in the hypnotic triazolam.

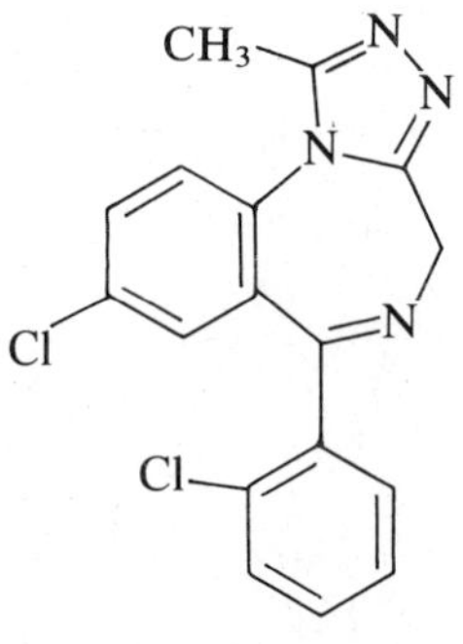

Triazolam

3.11. SEDATIVE HYPNOTICS AND ANTICONVULSANTS

Sedative hypnotics are drugs that induce the state of sleep in varying degrees; some of them sedate excited states, especially at lower doses, while at higher doses they induce and maintain various sleep patterns. Because sleep is a physiological state that occupies one-fourth to one-third of the total lifetime of all individuals, every facet of the sleep process has been studied carefully, especially since the introduction of electroencephalography and related methods. Two states of sleep are generally described. They are rapid eye movement (REM) sleep and nonrapid eye movement (NREM) sleep, characterized by specific brain waves (685). During NREM sleep, a number of biosyntheses take place and the products, prolactin, luteinizing hormone, growth hormone, and biogenic amines, are secreted into the circulation. In animals, especially the cat (686, 687), depletion of serotonin (5-HT) suppresses NREM sleep (688), but the supporting data could not be confirmed in humans (689). Likewise, if administration of reserpine, which depletes serotonin levels, is followed by injection of dopa (a precursor of dopamine, NE, and epinephrine), REM sleep is induced quickly (690). Large doses of 5-HT precursor amino acids (5-HTP, L-Try) increase REM sleep. α-Methyldopa, a precursor of α-methylnorepinephrine that competes with NE, also increases REM sleep (691). These biochemical observations are anything but conclusive, which is too bad, since the controversial involvement of catecholamines, 5-HT, and other biogenic amines in the sleep states is one of the few postulated mechanisms a chemist could utilize in drug design. A few data on the cholinergic effects on REM sleep are also not yet solidly supported. Thus, the discovery and design of sedative–hypnotics has remained an empirical art. The many structural types illustrate the trial-and-error method that has beset the choice of "leads" in this field. They include barbiturates, benzodiazepines, various nitrogen heterocyclics, amides, urethans, alcohols, aldehydes, antihistaminics, sulfones, and inorganic bromide.

3.11.1. Hypnotics

For millennia and until the middle of the 19th century, ethyl alcohol and opium were used to produce euphoric stupor in patients in preparation for surgery and for therapeutic sleep in disease states. Inorganic bromides (mostly KBr) were introduced as sedatives and anticonvulsants in 1857 but in larger doses they produce intoxication (bromism). Nevertheless, they served a useful purpose in medicinal chemistry by directing attention to the possible role of halogens in organic compounds prepared as sedatives.

The first of these was chloral hydrate, $Cl_3CCH(OH)_2$, a compound made by Liebig in 1832. When Liebreich found in 1869 (692) that alkali decomposed chloral to chloroform, he assumed that this reaction also takes place *in vivo* and that the known CNS-depressant action of chloroform was responsible for the hypnotic action of chloral. This assumption turned out to be erroneous. Chloral is actually bioreduced rapidly to trichloroethanol, Cl_3CCH_2OH, and this compound is as effective as chloral hydrate itself. The bioreduction is carried out by alcohol dehydrogenase under the influence of NADH. The pharmacology of chloral hydrate has been reviewed, including its toxicity and its use in "knockout drops" (693).

It is not only the water adduct of trichloroacetaldehyde that has sedative–hypnotic properties; various alcohols can add to chloral to form hypnotic hemiacetals. Trichloroethanol, although an excellent hypnotic in laboratory animals, is too irritating and inconvenient to be used clinically. Its phosphate ester monosodium salt [triclofos sodium, $Cl_3CCH_2OP(O)(OH)(ONa)$], circumvents these difficulties. Chloral and its adducts have found renewed clinical use recently after half a century of neglect. Chloral is also used in tonnage quantities in the manufacture of the insecticide chlorophenotane known as DDT.

In the molecule of chloral, the principal feature is the quaternary carbon atom. This observation may have contributed to the discovery of the hypnotic disulfones (694), which all contain such a quaternary carbon, $R_2C(SO_2R_2')$. Systematic increase of R and R' from methyl to ethyl increased hypnotic activity, a regularity that can now be attributed to the need for increased lipophilicity in sedative hypnotics. This need has been emphasized for these and many other types of compounds by Overton (128), Meyer (127), and Hansch (129).

A rigid molecule, preferably without an ionizable proton at a heavily substituted carbon atom, seems to be a requirement not only for hypnotic activity but for other CNS depressant activities as well. It is seen in many antispasmodic–analgetics, anticonvulsants, and other agents that block nervous transmission. Historically, a quaternary carbon was chosen by Fischer and Mering (695) in their design of the first useful barbiturate. They had other (noncyclic) amides and ureas as "leads" to go by, such as isovaleryl diethylamide, $(CH_3)_2CHCH_2CON(C_2H_5)_2$, and some halogenated or branched analogs. Their choice of 5,5-disubstituted barbituric acids was patterned on the structure of the hypnotic disulfones mentioned above; the alkyl groups in each derivative were identical (two methyls, two ethyls, etc.) for synthetic reasons. Eight years later the first barbiturate with two unlike 5-substituents was synthesized by a new route (696, 697). This compound, phenobarbital, was both a hypnotic (698, 699) and the first effective anticonvulsant (700). Further synthetic refinement then opened the way to 5,5-disubstituted barbituric acids with unlike aliphatic or alicyclic substituents (701). It was in these series that the most potent, selective, and nontoxic barbiturates have been developed.

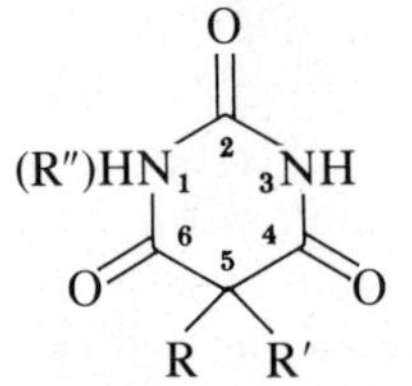

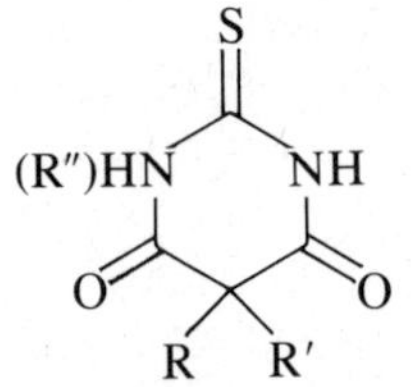

Mostly for competitive reasons, several thousand substituted barbituric acids have been prepared and tested. 5,5-Disubstitution is a *conditio sine qua non* for sedative activity. Tri-substitution at positions 1,5,5 does not abolish activity, and the oxygen atom at C-2 can be replaced by sulfur in the corresponding thiobarbituric acids. All these compounds are capable of proton shifts and existing in tautomeric forms. Depending on the prevalent tautomer, the compound can undergo one or two ionizations, which can be followed potentiometrically and through UV spectra. It takes approximately 40–60% dissociation to enable a 5,5-disubstituted barbituric acid to cross the blood–brain barrier and exert an effect on the CNS. A determination of the pK_a can thus be predictive of CNS activity (702).

The size, character, and shape of the substituents at C-5 are of great importance to biological activity. Hypnotic activity increases until the total number of carbon atoms of the two substituents reaches 6 to 10, and then it declines. This means that lipid solubility must remain within limits. Branching of the alkyl chains leads to greater solubility in water-immiscible solvents, and since secondary or tertiary carbon atoms are more reactive toward degradative metabolizing reagents, the duration of action of branched isomers is shorter. Unsaturation of the substituent groups also improves hypnotic activity; thiobarbiturates have higher potency than the corresponding barbiturates. In other words, greater chemical reactivity leads to higher activity, quicker onset, and shorter duration of action. All this can be predicted for nonpolar substituents. Introduction of any kind of polar groups is detrimental to the activity. This is also expressed in the inactivity of drug metabolites, which carry OH and other hydrophilizing groups, NH in lieu of NR, or oxygen in place of the sulfur in the thiobarbiturates. Stereoisomers of chiral substituents do not differ in CNS activity. All this points to the probability that barbiturate hypnotics must present an optimal balance of lipophilicity and acidity but need not be tailored to specific receptors.

Both the amide and the urea moieties seen in barbiturates have been reproduced successfully in numerous acyclic and cyclic sedative CNS depressants. Among the best-known acyclic derivatives are carbamates, amides, ureides, and similar structures containing —CONH— with quaternary but also tertiary carbon atoms in positions α to the CONH group. Such compounds may contain saturated, unsaturated, halogenated, epoxide, or alicyclic carbon skeletons. A few SAR have been devised, giving unsaturated or electronegatively substituted compounds an edge over saturated ones in

Piperidinediones   2-Methyl-3-aryl-4(3H)-quinazolinones

potency. Among cyclic analogs, 2,4- and 2,6-dioxopiperidines doubly substituted at their 3-position have been used widely as hypnotics. Another system extensively studied is that of 2-methyl-3-(o-substituted-phenyl)-4(3H)-quinazolinones, which partially preserves the oxopyrimidine structure of the barbiturates. For a discussion of hydantoin anticonvulsants, see Section 3.11.2.

The early introduction of chloral (692) and paraldehyde (703) as hypnotics spawned a long series of alcohols and aldehydes to be used for the same purpose. Again, the most potent and pharmacologically suitable compounds in this series of non-nitrogenous substances are tertiary alcohols with one or two unsaturated substituents. Groups such as —C≡CBr and —CH=CHCl are seen in this series, in other words, electron–rich substituents that make the alcohols allylic or propargylic.

Therapeutic applications are often subject to fashions. New series of drugs may be promoted because they are new and not necessarily because they perform better. This is the case, to some extent, with some of the benzodiazepine hypnotics that had originally been studied as antianxiety agents. A number of advantages have been claimed for the benzodiazepine hypnotics but they are in question. Perhaps the most valid advantage is that they are nonfatal when taken in overdoses for suicidal purposes. Barbiturates are frequently fatal under such conditions. The effect of benzodiazepines on different sleep stages has been emphasized as a new development in sleep therapy, and the possibly different mechanism on brain substructures has been noted as an advantage. No doubt the pharmacological profile of the benzodiazepines makes them more acceptable than barbiturates for certain sleep disturbances, but the medicinal chemist can only note the overall hypnotic properties of these compounds and strive for derivatives with more suitable pharmacological acceptability, especially lack of hangover aftereffects.

Almost all the benzodiazepines used as hypnotic–sedatives have a formula in which R_1 is H, F, or Cl; R_2 = H or OH; R_3 = Cl or NO_2; and R_4 = H or a variety of cyclic, basic, halogenated, hydroxylated, or other substituents (704). It seems to have been futile so far to invoke connections between these substituents and improved and selective sedation.

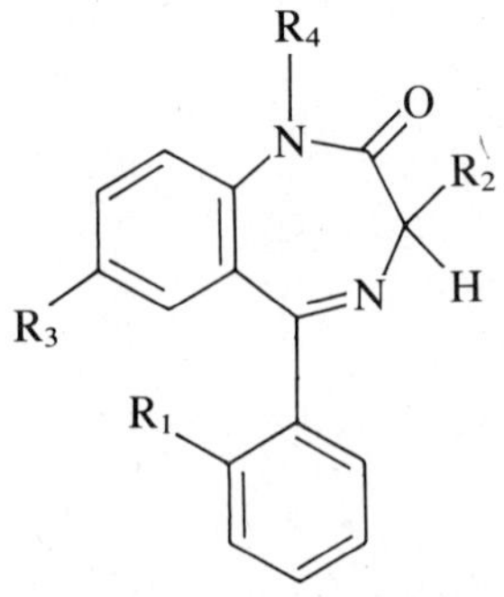

Benzodiazepine hypnotics

3.11.2. Anticonvulsants

When Hoerlein synthesized phenobarbital (696, 697), the compound was tried out as a hypnotic (698, 699) as a competitor to and possible improvement upon barbital. The discovery of the anticonvulsant activity of phenobarbital (700) was a pharmacological bonus. Its N-methyl homolog mephobarbital (5-ethyl-1-methyl-5-phenylbarbiturate) (705) is also an anticonvulsant, but it is largely N-demethylated metabolically, and its activity may therefore be due in part to its conversion to phenobarbital (706). Considering that the most effective agents in the hydantoin series acting on the motor cortex are also 5-phenyl substituted, it may be suggested that aryl substitution α to the ureide group is conducive to the anticonvulsant properties of these cyclic analogs. However, one has to discriminate between the various types of experimentally induced seizures, their response to drugs, and their clinical equivalents. For example, a series of 1,3-bis(alkoxymethyl)-5,5-disubstituted barbiturates was active against electroshock and pentylenetetrazole-induced seizures if there were two phenyls at position 5, but only against pentylenetetrazole if the two 5-substituents were ethyl (707). Another example of the separation of two properties is seen in 1,3-bis(acetoxymethyl)-5,5-disubstituted (R = C_2H_5 or C_6H_5) barbiturates, which are active against electroshock but are not hypnotic (708). Other examples may be found in reference 709.

5-Ethyl-5-phenylhydantoin (Nirvanol) (710, 711) was prepared as an analog of phenobarbital and carried out the expected structure–activity relationship; it exhibits broad anticonvulsant activity but is too toxic for general use in epilepsy. Nevertheless, its activity emphasized the probability mentioned earlier that nitrogen heterocyclics containing ureide or amide grouping would be suitable candidates for CNS-depressant properties. In 1937, Putnam and Merritt (712, 713) studied a series of disubstituted hydantoins that had been prepared in the industry (714), among them the 5,5-diphenyl derivative phenytoin, which had been synthesized 30 years earlier (715). This compound protected cats against electroconvulsive seizures and was introduced clinically as an effective antiepileptic agent with minimum sedation.

As in the series of hypnotics, both ring analogs and open-chain analogs, such as aralkylureas and ureides, have been prepared and tried out in various

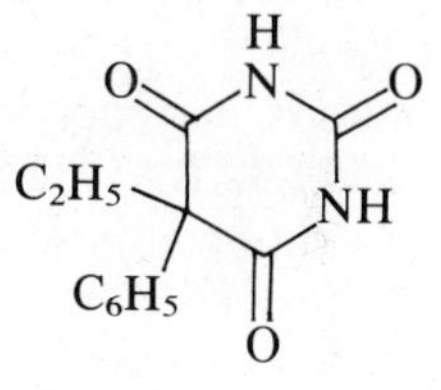

Phenobarbital 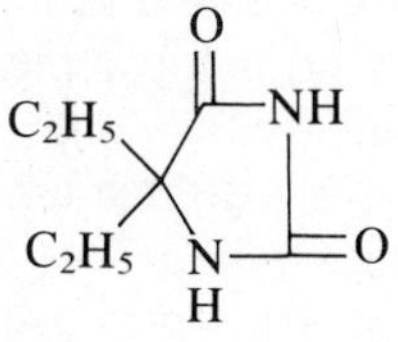   Nirvanol 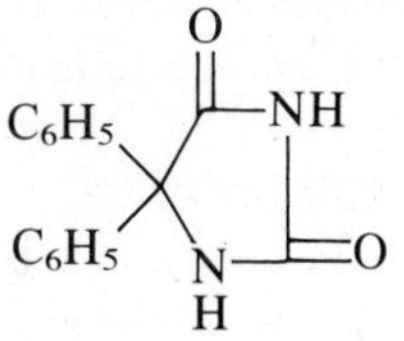   (Diphenylhydantoin) Phenytoin

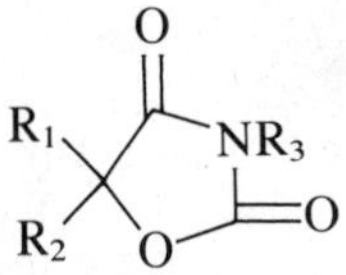

Trimethadione: $R_1, R_2, R_3 = CH_3$
Paramethadione: $R_1, R_3 = CH_3, R_2 = C_2H_5$

anticonvulsant test systems. An example is phenylacetylurea (phenacemide, $C_6H_5CH_2CONHCONH_2$), the optimal although still toxic representative of a large series of analogs (716). Among cyclic analogs, oxazolidine-2,4-diones have been studied most closely (717). Representative drugs in this series are trimethadione and paramethadione, which are effective against petit-mal epilepsy. Interestingly, trimethadione had been synthesized in the course of an analgetic drug research project (718). If the substituents at C_5 are aryl groups, anti-grand mal activity comes to the fore.

The *N*-methyl group of trimethadione is removed metabolically, and the activity of the drug may well be due to dimethadione (719).

Cyclic amides are represented by primidone, the 2-methylene congener of phenobarbital (720); it is a broad-spectrum anticonvulsant similar to phenobarbital in animals. N-Methyl-α-phenylsuccinimide (phensuximide) is used in petit-mal epilepsy. Other succinimides and glutarimides show anticonvulsant activity in experimental seizures.

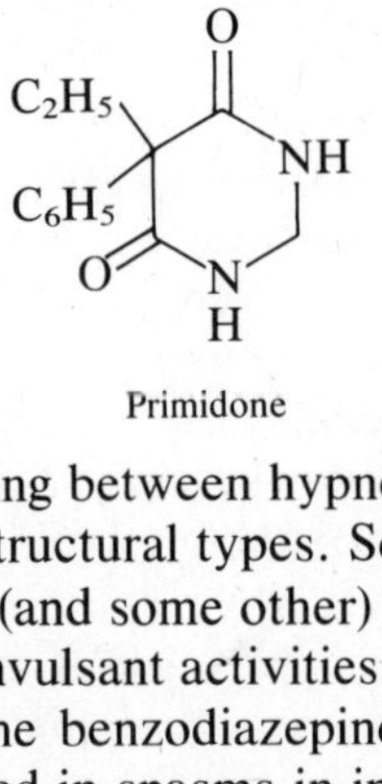

Primidone

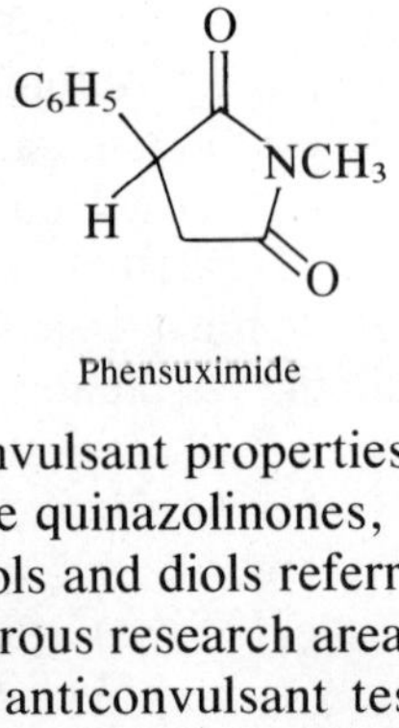

Phensuximide

Overlapping between hypnotic and anticonvulsant properties occurs in a number of structural types. Some of the same quinazolinones, carbamates, and tertiary (and some other) aliphatic alcohols and diols referred to earlier have anticonvulsant activities. The most vigorous research area has been in modifying the benzodiazepine structure for anticonvulsant testing. Nitrazepam is used in spasms in infants, and diazepam as a broad-spectrum anticonvulsant. Trends in SAR may be found in reference 7c, pp. 847–849.

Several branched aliphatic acids show pronounced activity in anticonvulsant test systems. The most active and most acceptable derivative was dipropylacetic acid (valproic acid), which emerged from a random screening program as a fortuitous "lead" (721, 722). These compounds may act as inhibitors of enzymes that destroy gamma-aminobutyric acid (GABA) (723) or they may act by mimicking GABA (724). This is an invitation to study other analogs of GABA for the same purpose.

A number of aromatic and heterocyclic sulfonamides inhibit carbonic anhydrase, and those that can distribute themselves in select brain areas are useful as anticonvulsants. Four such drugs are shown below. Their utility

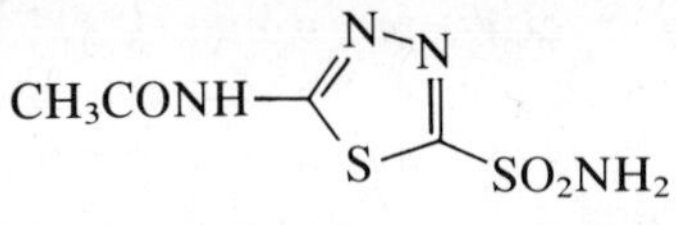

Acetazolamide

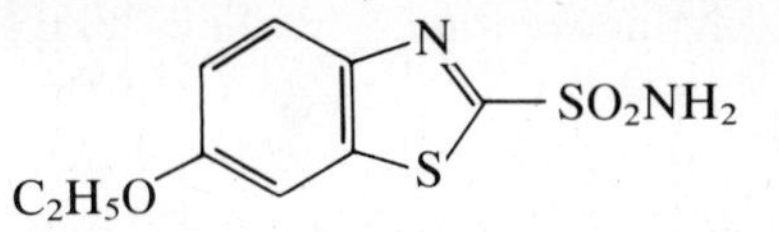

Ethoxzolamide

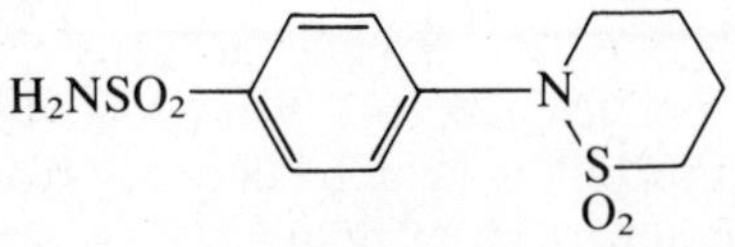

Sulthiame

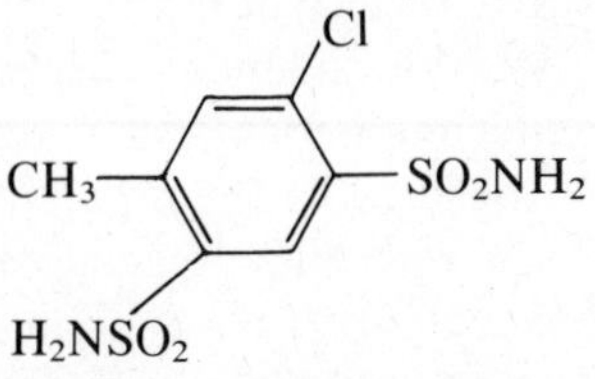

Disamide

in different types of seizures may be due to absorption in special brain tissues. This seems to hold as well for most of the other anticonvulsants described in this section. It may be that the molecular mechanisms of these drugs is the same but is a function of the compound's ability to reach different brain centers and react there with the same biomacromolecules (725). The assumption that GABA deficiency is connected to epileptic seizures is a step in the right direction in identifying a metabolite in this field that may provide a "lead" for analogs in further researches.

Neuronal excitation leading to convulsive seizures in laboratory animals is also caused by dicarboxylic amino acids such as N-methyl-D-aspartic acid. Anticonvulsant action is exhibited by analogs of such convulsants, for example, by 2-amino-7-phosphonoheptanoic acid (726).

3.12. ANALGETICS

3.12.1. Mechanisms of Action

Interest in analgetics centers on their sources and structure–activity relationships, their known and putative mechanisms of actions, and their uses in the therapy of innumerable painful conditions. Analgetic specificity is not high; most analgetics have additional and overlapping activities, desirable or undesirable. The most undesirable side effects and the most difficult to separate are those that result in psychotomimetic symptoms, dependence liability, and the danger of withdrawal symptoms. Some analgetics also exhibit antagonistic properties, ranging from mixed agonist–antagonist to pure analgetic antagonist.

The chemistry of traditional opiate analgetics and their semisynthetic and totally synthetic analogs has been challenging, and this has fueled interest in the study of these compounds. Answering stereochemical questions and formulating synthetic methods have been especially rewarding. Until ten years ago, mystery surrounded the mode of action of the opioid type of analgetics. Adrenergic, cholinergic, serotoninergic, and other mechanisms were invoked based on observations of limited scope. Receptor "shapes" were claimed repeatedly, featuring acidic groups to confront basic functions of analgetics, interatomic distances to account for similar ones in the drug molecules, and "cavities" with geometric dimensions that could accommodate a variety of molecular shapes and sizes. These postulates were supported by biochemical–pharmacological measurements of the ligating rates of many analgetics and analgetic antagonists to several nervous tissues. From these rates the concept of the "opiate receptor" was created (727–729). The pharmacological meaning of this binding can be deduced from the stereospecificity, reversibility, and amount of drug bound. These are the customary data and methods used in pharmacology to measure localization and potency of a drug. Chemically, almost nothing is known about the opiate receptor. Like other postulated receptors, it is supposed to be a lipoprotein. When cerebroside sulfate and phosphatidyl serine were found to bind opiates (730), cerebroside sulfate was used as a model for the putative lipid component of the opiate receptor. The participation of prostaglandin E-stimulated cAMP and adenyl cyclase in bringing about biological activity at the opiate receptors has been demonstrated (731, 732), but the meaning of this finding is not yet clear. There is no doubt that radiolabeled opium alkaloids and their synthetic analogs accumulate at certain sites rich in opiate receptors.

3.12.2. Endogenous Analgetic Peptides

Similarly, endogenous peptides with morphine-like properties concentrate at these receptor regions and react there. These reactions are reversed by naloxone, a pure opiate antagonist. These peptides were discovered, purified, and identified as pentapeptides (733–736). Structurally, they are H-TyrGlyGlyPheLeu-OH (Leu-enkephalin) and H-TyrGlyGlyPheMet-OH (Met-enkephalin). They are a subset of other pituitary opioid peptides with larger molecular weights called endorphins. Met-enkephalin has a sequence of amino acids identical with that of residues 61–65 of the pituitary hormone β-lipotropin (β-LPH) (737). This fragment itself has potent opioid activity. The anatomical distribution of the enkephalins in the CNS and gastrointestinal tract (738) and of β-endorphin mainly in the posterior pituitary (739, 740) excludes a precursor mechanism. The enkephalins and also the β-endorphin found in the CNS are regarded as specific neurotransmitters, whereas β-endorphin in the pituitary has the role of a hormonal modulator in pain perception. It is also apparent that these different peptides are the products of natural molecular modification, and this has lent encouragement to the massive efforts to synthesize peptides with one or several different amino acid groups, with D-amino acids that change the geometry of the flexible molecules, and even with non-amino carboxylic acids, amines and other functions. Over 1000 analogs of the enkephalins have been synthesized. Greater stability towards di- and tricarboxypeptidases, which are mainly responsible for the proteolysis of the enkephalins, can be attained by conversion of the terminal carboxyl to —$CONH_2$, or by inserting a D-amino acid at position 2. The tyrosyl group—which can also be spotted in the morphine alkaloids and provides a link between the enkephalins and the thebaine-derived opiates—is an essential structural feature. A 10.0 ± 1.1 Å distance between the aromatic rings of Tyr^1 and Phe^4 may be important in the design of enkephalins with other aromatic moieties (741).

Systematic SAR have revealed these and a few other predictable regularities, but much empiricism is left in the approach to these peptides. Depending on the tests used to distinguish between receptor activities, enhanced activity has been found for different series of enkephalin as well as endorphin analogs. The potent activity and stability of endorphin with its 31 amino acids has not been matched by shortening the chain. For additional details, see reference 742.

The very large number of peptides mentioned has not produced a clinically useful analgetic. The relatively small yield of active peptides has led to compounds, usually of short duration of action, that have activity in animal tests. Those peptides that are active in humans are limited by various pharmacological drawbacks.

3.12.3. Opioid Analgetics

Similar considerations have overshadowed the clinical utility of opioid analgetics. Even though virtual separation of analgesia and dependence liability has been claimed for a few of these compounds, low levels of dependence liability have barely been overcome. Except for the true narcotic antagonists, respiratory depressant side effects have not been completely eliminated. These are pharmacological limitations, and the medicinal chemist confronts them by trying out other empirical modifications. In few fields of medicinal chemistry has stepwise modification and screening without much rationale been so prevalent as in opioid research. For a systematic survey making use of every intelligent idea, the reader is referred to the contemporary review by Johnson and Milne (742) and the many references to other reviews cited therein.

Beginning with the morphine alkaloids (morphine, codeine, thebaine) and the synthetic morphinans, which lack the oxygen bridge of morphine (743, 744), the same pattern was followed that had led medicinal chemists from other earlier natural alkaloids to synthetic simplified—or at least modified—improved analogs. Codeine, the methyl ether of morphine, is only a tenth as analgetic; this indicates the need of a free phenolic hydroxyl for greater potency. Other functions that can be altered are the alcoholic OH in position 6, the 7,8- double bond, and the N-methyl group, which can be removed (normorphine) or replaced (*N*-alkylnormorphines). The alcoholic hydroxyl can be exchanged for halogen and other nucleophiles, oxidized to carbonyl, or esterified or etherified. Thebaine contains two conjugated double bonds and can undergo Diels–Alder addition reactions. Codeinone can add Grignard reagents (745) (metopone). The 6-OH group can be isomerized to 8-OH (decreased activity), and the C-8 β-halo derivatives are more potent experimental analgetics than morphine (746). A 14-hydroxy group is seen in the potent 14-hydroxydihydrocodeinone (747), in one of the most interesting very early molecular modifications, and in the corresponding morphine derivative (oxymorphone) (748). Substitution of the aromatic A-ring generally impairs analgetic potency. For example, halogen atoms or an acetyl group in the 1-position cause diminished activity with the exception of 1-fluoro substitution. 1-Fluorocodeine is as potent as codeine (749); apparently steric and not electronic factors are decisive in this case.

The list of these cases could be expanded further but would not overcome a very real problem. With the few exceptions of bioisosteric input in later years, molecular modification in the morphine series of alkaloids and the morphinans has been entirely empirical. It has leaned on the organic–chemical interest of enthusiastic alkaloid chemists and ingenious stereochemists. From their synthetic experiments have arisen a dozen or so derivatives and analogs whose pharmacological advantage over morphine and codeine have remained in question. Some of them are widely used but careful inquiry will detect arguments of tradition and not-too-clearly defined and defendable preference. Certainly, dependence liability has not been eliminated in this series. Some pharmacological advantages have been realized, such as reduction of gastrointestinal peristaltic inhibition and metabolic instability. These real but minor advantages have not cancelled out cost factors, and in the series of morphine alkaloids—not their structurally dissected analogs—morphine has been retained by the medical profession over all newcomers.

In the series of thebaine adducts, the idea was explored of a more rigid and more lipophilic molecule that might react more specifically with a receptor molecule. These interesting compounds, which contain a 6,14-*endo*-etheno bridge and a secondary or tertiary alcoholic side chain in position 7,

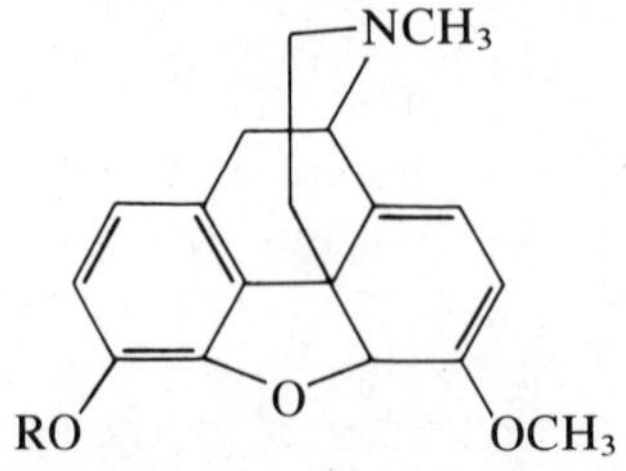

Oripavine: R = H
Thebaine: R = CH₃

can be regarded as derivatives of thebaine or its phenolic parent compound oripavine. Examples of the hundreds of compounds synthesized have been described in Section 2.6. The effect of the lipophilic side chain is so great that the phenolic ring can be omitted without loss of analgetic potency (750).

3.12.4. Simplified Structures

Deletion of sections of a molecule of a naturally occurring prototype drug has always appealed to medicinal chemists for a number of reasons. First, there is the conviction that the metabolism of the natural source organism must have added chains, rings, and functional groups designed to detoxify the metabolite rather than prepare it for therapeutic purposes. Second, experience with many other natural products has shown that molecular moieties responsible for the biological activity can almost always be peeled out of a more complex naturally occurring structure. This has held for alkaloids, peptide hormones, β-lactam antibiotics, and others; it has been and remains the main preoccupation of drug design based on natural compounds. A third reason is that such degradative molecular modification often separates pharmacological properties with which nature may have endowed natural products without regard to medicinal utility.

It is more prudent to remove one small piece of a molecule after another, because this permits one to evaluate the significance of each segment and function. However, more audacious and radical molecular surgery also has its place in paring down unnecessary portions of naturally occurring drugs. In the morphine series, deletion of the furanoid ether oxygen may be classified as an example of stepwise simplification. The resulting morphinans show not only that this ether oxygen is not needed for biological activity but that the steric strain imposed by it on the whole condensed ring system is not essential for analgetic properties (751). The N-phenethyl morphinan derivative is a potent analgetic (752), the N-allyl derivative (levallorphan) a narcotic antagonist without dependence liability (753), and other derivatives occupy intermediate positions similar to those in the morphine series.

A sharp reduction in molecular surface was executed by May (434). Ring C was cut down to two methyl stumps in 2'-hydroxy-2,5-dimethylbenzomorphan without loss of potent activity. Extensive SAR studies in this series of benzomorphans have revealed (754) that retention of the same absolute configuration as in morphine is necessary for highest analgetic potency and

2'-Hydroxy-2,5-dimethylbenzomorphan

lowest dependence liability while the opposite configuration leads to opposite results in both regards. Again, the N-phenethyl derivative has high and pharmacologically valuable potency, the N-allyl derivative resembles levallorphan as an antagonist, and the dimethallyl compound [>NCH$_2$CH=C(CH$_3$)$_2$] (pentazocine) is a widely used narcotic antagonist and analgetic (755).

3.12.4a. Meperidine and Methadone.

The serendipitous discovery of the analgetic properties (428) of meperidine (427) has been described in Section 2.6. A large number of structural modifications has revealed a considerable latitude permissible in retaining and increasing analgetic potency. One of the possibilities is to replace the carboxylate ester group, —COOC$_2$H$_5$, with the bioisosteric group, —OCOC$_2$H$_5$, which is present in prodine. Two diastereomers are known, α- and β-prodine (756), of which

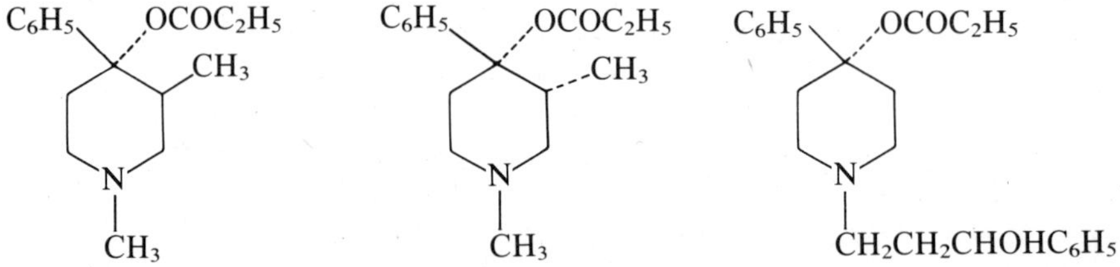

β-Prodine

α-Prodine

1-(3-Hydroxy-3-phenylpropyl)-4-phenyl-4-propionyloxypiperidine

the α-isomer is used clinically. The 3-hydroxy-3-phenylpropyl analog has an activity 3200 times that of meperidine (757). The ester or reversed-ester group can also be replaced by a keto group (—COC$_2$H$_5$). Other more far-ranging modifications include fentanyl in which an anilide group is attached to the phenethyl-substituted piperidine ring (758, 759).

Methadone also arose from a program of studying anticholinergic blocking

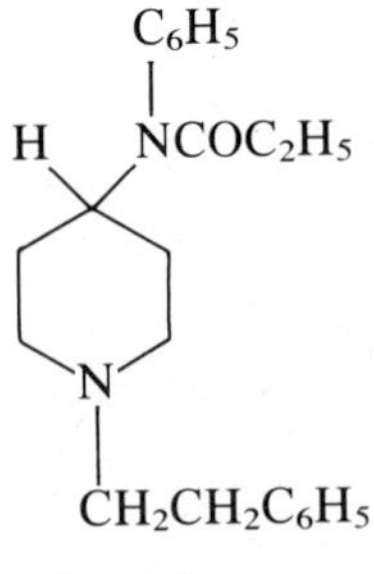

Fentanyl

agents, in this case a keto group instead of an ester group being connected to the amine portion. The drug became known outside the original laboratory through interrogation of German scientists by a team of American investigators (429). The potent analgetic action of methadone was discovered, as in the case of meperidine, by pharmacological screening.

In order to explain the analgetic activity of methadone on structural grounds, a number of nonbonded quasi-cyclic arrangements have been proposed that would bring the methadone structure in line with the piperidine

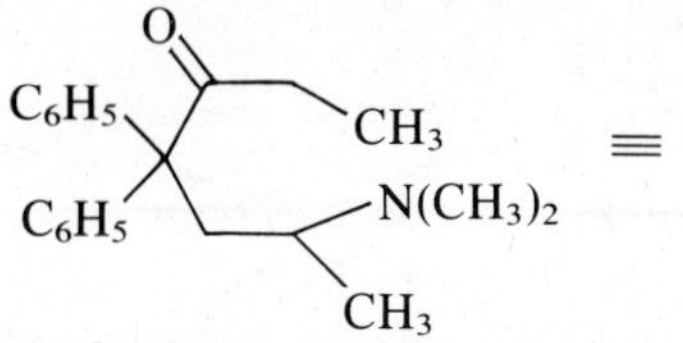

Methadone 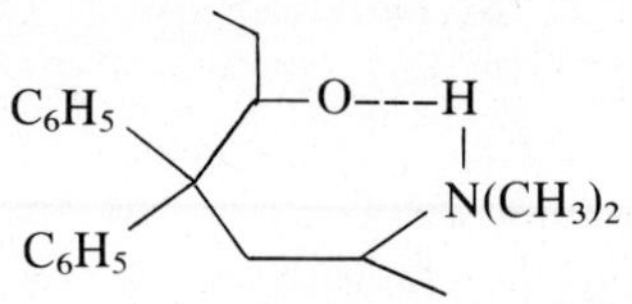   Nonbonded interpretation of methadone (···)

structures of morphinan and meperidine (760–762). In any event, such quasi-cyclic formulations do not appear to be of major influence (762).

Typical molecular modifications of methadone have included homologation and cyclization of the dimethylamino group, reduction of carbonyl to —CHOH—, removal and relocation of the methyl branching, isosteric replacement of one or both phenyls by thienyl, and other standard variations. Replacement of the keto group by an amide group led to dextromoramide; insertion of an ester oxygen between the blocking groups and the carbonyl (as well as use of a benzyl instead of one phenyl) gave dextropropoxyphene (763), a popular analgetic. In thiambutene, the blocking thiophene rings converge on an amine chain, and instead of an electron-rich carbonyl group, a double bond has been introduced.

Among the methadols α-(+)-acetylmethadol is the most potent orally effective analgetic with few side effects (764).

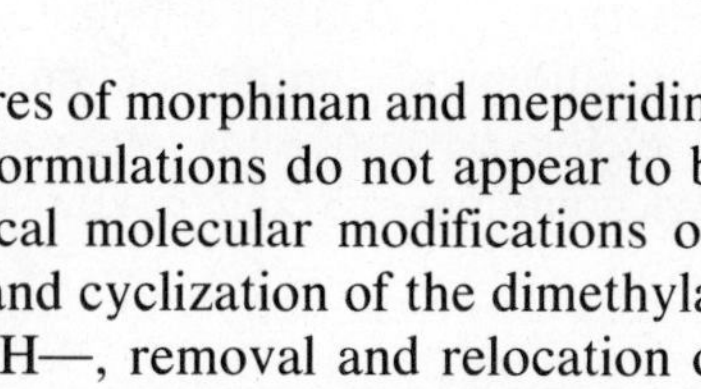

Thiambutene 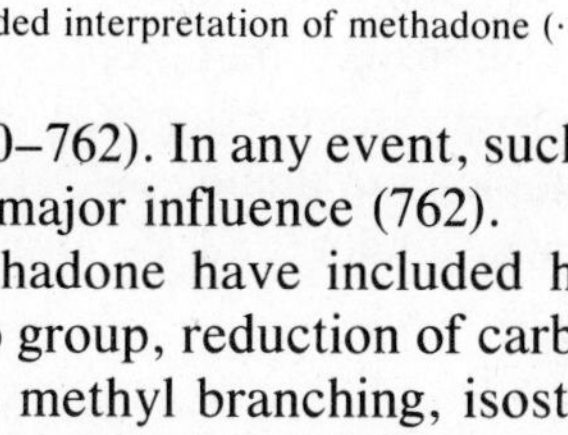   Acetylmethadol

3.12.4b. Other Analgetics. A number of other structural types exhibit morphine-like analgetic activity, albeit usually with more or less dependence liability. They were found to have these properties in the course of routine overall pharmacological screening of compounds made and tested for other purposes, for example, the potent but high-abuse-potential benzimidazoles (e.g., etonitazene) (765, 766). More consistent was the investigation of laudanosine analogs, since this alkaloid occurs in opium (767). For a few additional examples related to other drug types, see reference 742.

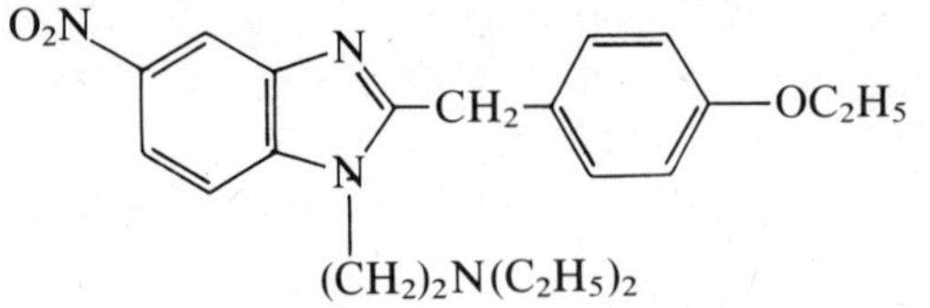

Etonitazene (structure with O_2N, benzimidazole, $-CH_2-$, phenyl, $-OC_2H_5$, and $(CH_2)_2N(C_2H_5)_2$)

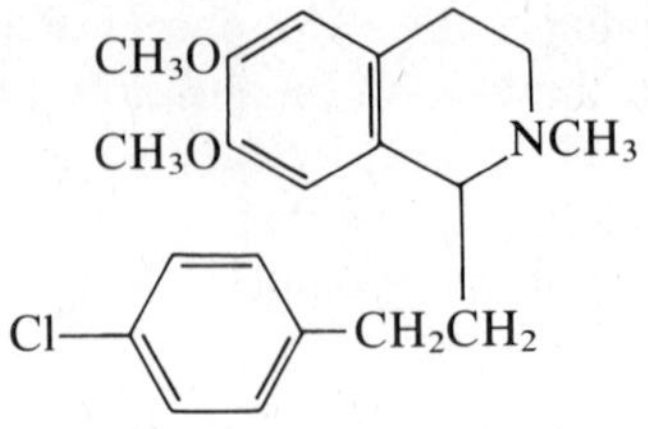

4-Chloro homolog of laudanosine (structure with CH_3O, CH_3O, NCH_3, Cl, CH_2CH_2)

Among ancient drugs with a therapeutic folklore of pain relief are various products from the plant *Cannabis sativa* (768, 769), including crude extracts as well as tetrahydrocannabinols. Their analgetic activity appears to be due to the 11-hydroxy metabolites of Δ^8- and Δ^9-tetrahydrocannabinols (770). These observations triggered extensive studies of synthetic cannabinoids, for example, aza analogs (771). Some of these compounds, such as the *N*-methyl aza analog and its *N*-propargyl homologs, exhibit potent analgetic activity in animals.

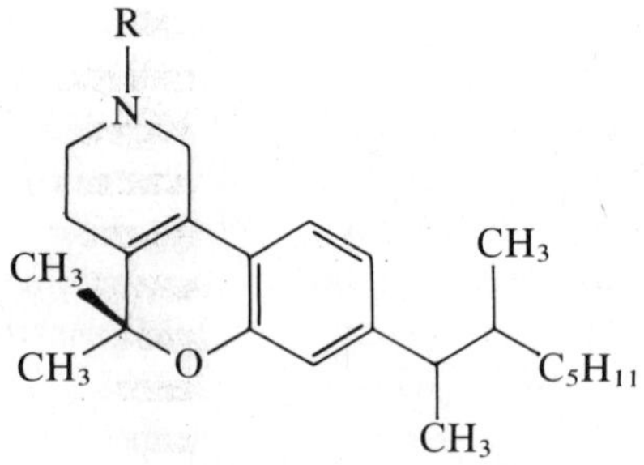

Aza-cannabinoids: R = CH_3, $CH_2C \equiv CH$, etc.

3.13. ANTIINFLAMMATORY AGENTS

Although aspirin, several aminopyrazolones, and a few other synthetic drugs now classified as antiinflammatory agents have been known since the turn of the 20th century, until recently they were thought of primarily as antipyretics and minor analgetics. However, they had been prescribed for rheumatic pain, and aspirin and other salicylates had become the drugs of choice in rheumatic heart disease. The real impetus to antiinflammatory drug research arose from clinical observations of Philip S. Hench that arthritis often undergoes remissions during pregnancy. This drew attention to steroidal hormones and led to the discovery of the antiarthritic properties of cortisone.

3.13.1. Steroidal Drugs

The synthesis of cortisone and cortisol and the clinical application of megadoses of these hormones for a variety of arthritic conditions represented major advances in the chemistry of natural products, in the art of organic chemical synthesis, and in the therapy of inflammatory diseases of unexplained etiology. But the initial euphoria about these innovations was short-lived; the side effects produced by corticosteroids soon dampened clinical enthusiasm (347). In 1953 Fried and Sabo (348) found that synthetic 9α-fluorocortisol [9α-fluoro-11β,17,21-trihydroxy-pregn-4-ene-3,2-dione] was 10 times more potent than cortisol in several test methods. Since then, hundreds of other steroids have been synthesized as potential antiinflammatory agents. Some of them succeeded in separating a few of the multiple pharmacological effects of the natural hormonal prototypes. This holds especially for decreases in retention of sodium ions and loss of potassium ions. Stimulation of CNS processes and of appetite are harder to separate. Other

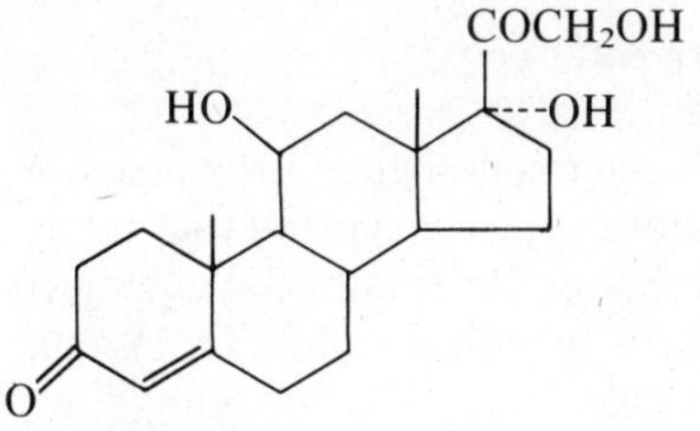

Cortisol

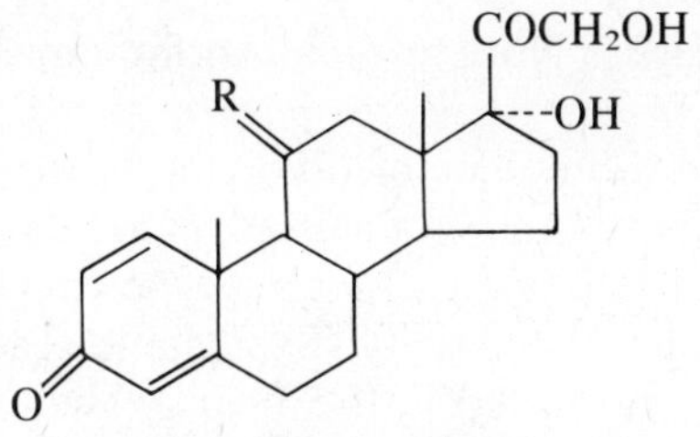

Prednisone: R = O
Prednisolone: R = —OH,···H

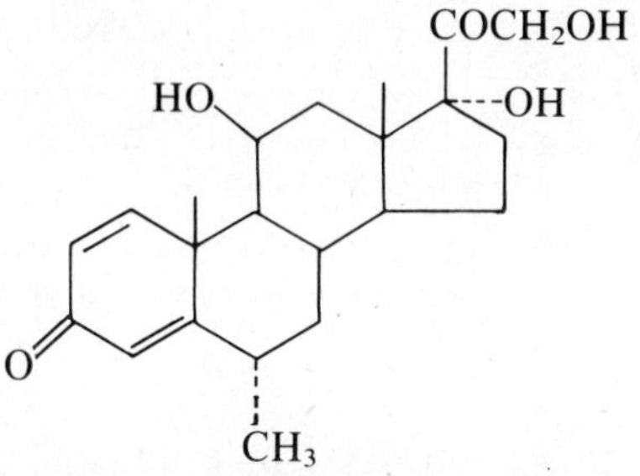

6α-Methylprednisolone

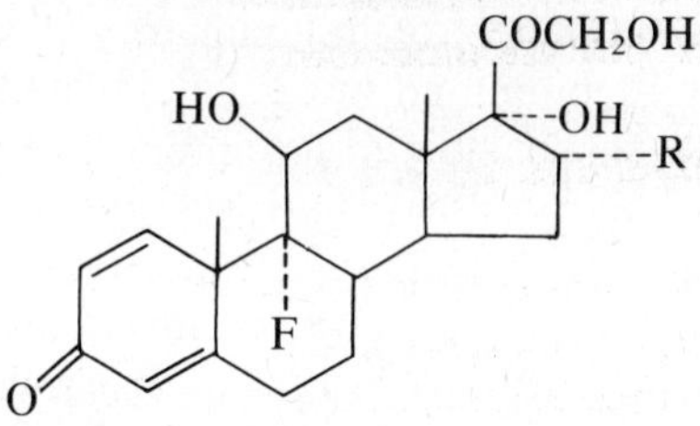

Triamcinolone: R = OH
Dexamethasone: R = CH_3

side effects have not been enhanced or reduced adequately, and this has limited the number of clinically useful drugs in this series. Cortisol is now sold without a prescription. Other clinically widely recommended antiinflammatory steroids are prednisone, prednisolone (772, 773), 6α-methylprednisolone, triamcinolone, and dexamethasone (774), with several others closely following suit.

None of these steroids is perfect: they possess multiple biological activities and the structural requirements for these activities overlap. If the actions of antirheumatic corticosteroids could be localized, many of the complications would be eliminated. This is said to be the case in esters of prednisolonic acid (775), which are nearly as active as prednisolone in the

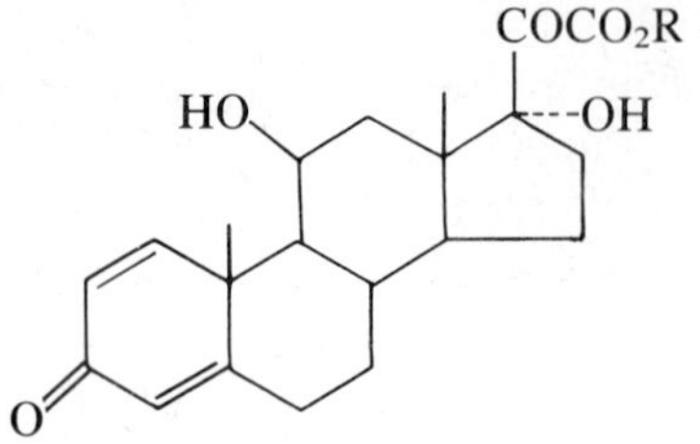

Prednisolonic acid esters

cotton pellet granuloma assay but are much more weakly active systemically. Furthermore, these esters do not suppress pituitary–adrenal function or cause liver glycogen depletion in rats.

3.13.2. Nonsteroidal Antiinflammatory Agents

The difficulties of divorcing the various pharmacological properties in the activity profile of steroidal antiinflammatory drugs suggested that other, totally unrelated structural types might be better candidates for enhanced biological specificity. Systematic searches for nonsteroidal antiinflammatory agents started in the 1950's. They depended on slowly emerging concepts of the etiology of inflammation as they arose from ongoing biochemical, immunochemical, and pharmacological studies. In turn these led to assays of test drugs in more meaningful, although by no means conclusive, laboratory models of inflammatory disorders. Thus it was no longer required that a compound lower an elevated body temperature, as aspirin does, for it to be considered an antiinflammatory agent (776). The huge market for antiinflammatory drugs, estimated at $600 million in 1981, and the continued quest for the elimination of unwanted, stubbornly adherent, side effects (777), have made nonsteroidal antiinflammatory drugs one of the most active reseach areas in medicinal research.

3.13.2a. Salicylates. Naturally occurring carboxylate esters of salicylic acid, discussed in Section 2.2.1, have been used in the therapy of febrile, rheumatic, and inflammatory diseases for centuries. It is ironic that it took more than 70 years for one facet of the molecular action of the salicylates to be elucidated. In 1971, it was demonstrated that therapeutic doses of aspirin–as well as of the potent antiinflammatory drug indomethacin–inhibit the enzymatic biosynthesis of two prostaglandins, PGE_2 and PGF_2 (778–780), and that this accounts for the analgetic and antiinflammatory activity of the two drugs. Indeed, many other antiinflammatory compounds (see, e.g., refs. 781–795) may act by similar mechanisms. Aspirin transacetylates the lysyl group of the enzyme, cyclooxygenase (796, 797) which plays a role in the arachidonic acid cascade (798).

Of the thousands of analogs and derivatives of salicylic acid synthesized and tested, a few of the more imaginative molecular innovations deserve

ANTIINFLAMMATORY AGENTS

attention. Many *o*-hydroxy aromatic carboxylates have been tried in which the benzene ring of salicylic acid is replaced by other aromatic or quasiaromatic nuclei. The group of oxicams (799, 800) represents *N*-arylcarboxamides of 4-hydroxy-1,2-benzothiazine-1,1-dioxides of which piroxicam (R = CH_3, Ar = 2-pyridyl) is a typical example. It acts at the cyclooxygenase

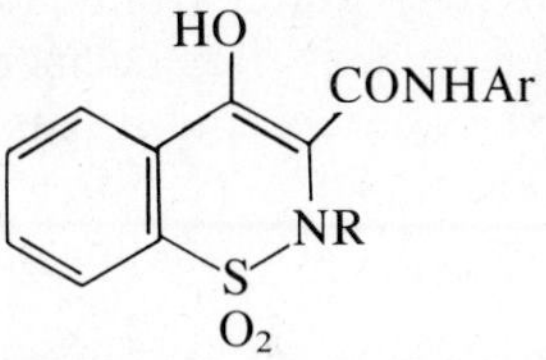

4-Hydroxy-1,2-benzothiazine-1,1-dioxide-N-arylcarboxamides
Piroxicam: R = CH_3, Ar = 2-pyridyl

step as a reversible inhibitor of the conversion of PGH_2 to PGE_2 and PGF_2 and is effective in rheumatoid arthritis at well tolerated doses.

3.13.2b. Fenamates. Another successful modification consists of replacing the phenolic hydroxyl of salicylic acid by an aryl-substituted imino

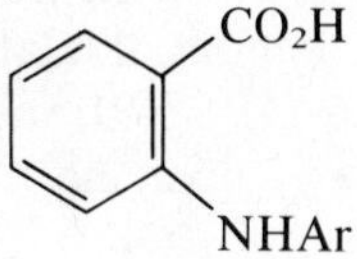

Fenamates

group, which results in isosteres of aryl ethers of salicylic acid. These compounds, *N*-arylanthranilic acids, are *o*-carboxydiphenylamine derivatives with a weakly basic anilino group. They have been given the collective name of fenamates (801). A number of side effects limit their use (802); this is perhaps explainable by their potent inhibition of cyclooxygenase (803).

3.13.2c. Pyrazolinones. A number of *N*-phenyl-*N*-alkyl-substituted pyrazoline-5-ones with additional functional groups such as 4-dimethylamino (aminopyrine) (804, 805) or without further functional substitution (806) date back 80–90 years, when their antipyretic action was discovered by a chemical error: they were conceived as structural analogs of quinine whose structure was still in flux at that time. In spite of their tendency to cause frequent

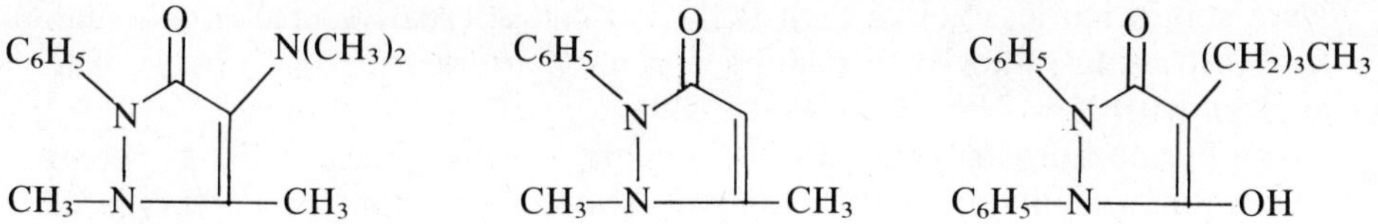

Aminopyrine   Antipyrine   Phenylbutazone

hematologic damage, they have remained in use in Europe until recently. That also meant that research for improved analogs continued at a gingerly pace, and phenylbutazone (783) surfaced during that work as a long-lasting drug. Its N^1-p-hydroxyphenyl analog (783) is a metabolite of phenylbutazone and is also synthetically available; it is pharmacologically slightly superior to phenylbutazone but both drugs suffer from causing gastrointestinal disorders and occasional agranulocytosis. Replacement of the n-butyl group by $(CH_2)_2S(O)C_6H_5$ increases the activity and uricosuric activity to clinically practical levels in sulfinpyrazone (807). The story of these findings emphasizes the potential of following up and extending old and apparently defunct research areas.

Both experimental and clinical pharmacologists have contributed "leads" to medicinal chemistry by observing antirheumatic and antiarthritic properties of drugs tried for entirely unrelated conditions. Thus, the antimalarial chloroquine (808), the anthelmintic and immunostimulant levamisole (809–811), and the sulfhydryl amino acid penicillamine (812–814) have been used in the treatment of arthritis. Even gold compounds, originally tested in tuberculosis, have experienced a renaissance in this field (815, 816).

The suspected relationship of the etiology of arthritis to a breakdown of immune responses has led to experiments not only with immunostimulants but also immunosuppressants. Among the latter are alkylating agents (817, 818), antimetabolites such as 6-MP (819), methotrexate, and azathioprine. Microtubular inhibitors such as colchicine (820) have also been tried, but not with very promising results.

In a field as empirical as antiarthritic drugs it is understandable that natural products with a therapeutic folklore have been given attention. Peptides such as pepstatin (821), muramyl dipeptide (822), and the hydrogen-bonded cyclosporin A (823) may serve as examples. The groping for "leads" is seen in the mostly disappointing testing of flavonoids (824), the triterpenoid escin (825), and the constituent of licorice, glyzyrrhetinic acid (826).

The application of biochemical metabolite models to the design of medicinal protagonists and antagonists should have served well in the case of antiinflammatory agents as progress was slowly made in unraveling the events leading to the inflammatory process. The present ideas concerning inflammation have been reviewed (827). The acute inflammatory process involves blood platelets that ultimately restore normal circulation. The prostacyclins act as inhibitors of platelet aggregation, and this is used in a test of potency of analogs of prostacyclin (PGI_2). A number of these analogs have been tested (828). Several of them with X = S, SO, SO_2, and CH_2 were somewhat less active than PGI_2 and had alternative properties, such as constricting rather than dilating coronary arteries (X = S) (829). The analogs with X = N< mimic PGI_2 (830).

6,9-Pyridazaprostacyclin, an "aromatic" analog, also was a potent smooth-muscle dilator. These examples, chosen at random from a very large

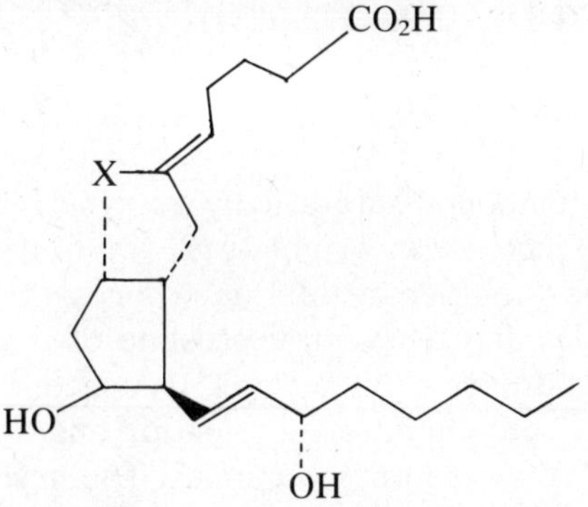

PGI₂: X = O

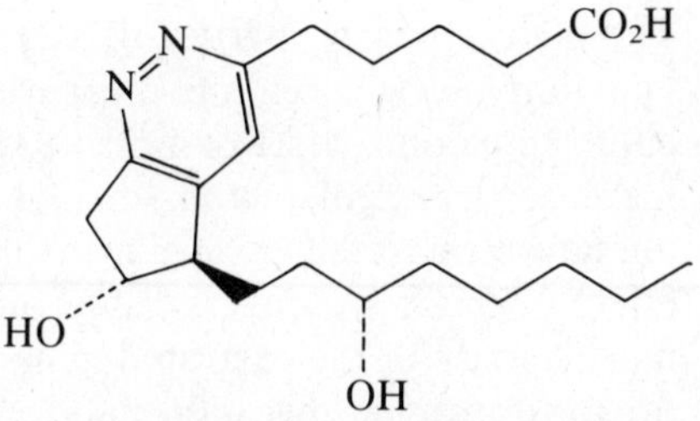

6,9-Pyridazaprostacyclin

variety of such acids and their esters, illustrate elegant bioisosteric replacements of these complex structures. Similarly, a number of analogs of thromboxanes (TXA$_2$ and TXB$_2$) have been synthesized and tested. For example, pinane-TXA$_2$ is a selective inhibitor of thromboxane synthetase but does not inhibit cyclooxygenase or prostaglandin synthetase. It also inhibits TXA$_2$ and other, more stable, endoperoxide analogs from aggregating platelets and contracting smooth muscle (828, ref. 39).

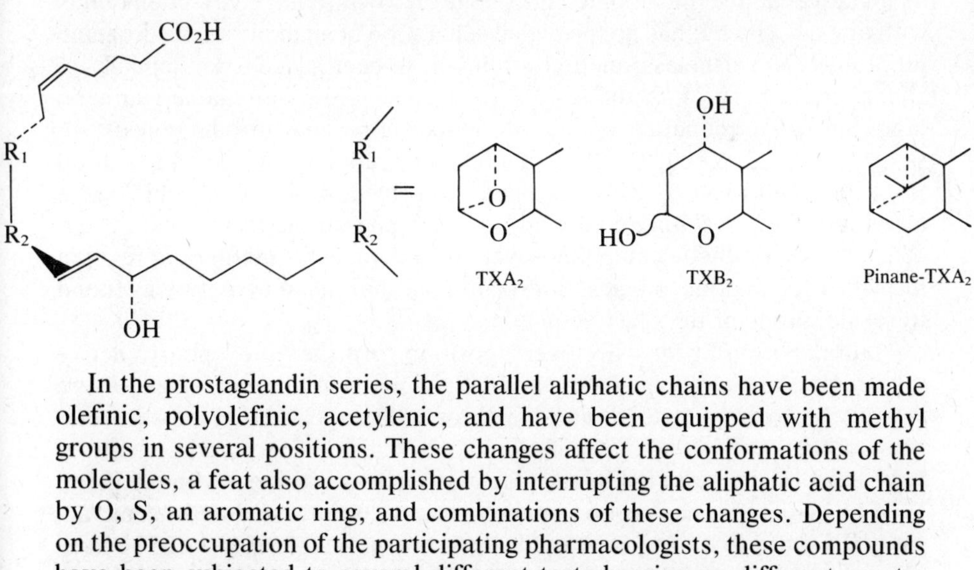

TXA$_2$ TXB$_2$ Pinane-TXA$_2$

In the prostaglandin series, the parallel aliphatic chains have been made olefinic, polyolefinic, acetylenic, and have been equipped with methyl groups in several positions. These changes affect the conformations of the molecules, a feat also accomplished by interrupting the aliphatic acid chain by O, S, an aromatic ring, and combinations of these changes. Depending on the preoccupation of the participating pharmacologists, these compounds have been subjected to several different tests bearing on different prostaglandin activities (827, pp. 1174–1188). In regard to antiinflammatory action, several antagonistic effects could be interpreted as part of the activity profile of such drugs, but it is a long way from enzyme inhibition to practical antiinflammatory action.

3.14. STEROIDAL HORMONES AND THEIR ANALOGS (831)

3.14.1. Androgenic–Anabolic Compounds

The androgenic properties of steroidal compounds are usually assayed by injection into a capon and measuring the increase in weight or growth of its comb (coxcomb test) or both (832, 833). Another method determines the increase in weight of the seminal vesicles and the ventral prostate of the immature castrated rat as a measure of androgenic potency (833). A second effect of those steroids is based on an increase in nitrogen retention and/or muscle mass of the castrated male rat or other laboratory animals. The best organ to provide this information is the levator ani muscle (834). Comparisons of the weight of this muscle and the seminal vesicles and ventral prostrate gives a ratio of anabolic to androgenic activity (834, 835). This ratio is of importance in judging the potential value of the steroid under investigation. For a review of clinical, biochemical, biosynthetic, and pharmacological studies as a background to SAR see (831).

3.14.1a. Structure–Activity Relationships. If one reads summaries of SAR in this series, very few defendable trends can be seen. Introduction of a 17α-methyl group (larger alkyls are disadvantageous) into testosterone or its putative active metabolite, 5α-dihydrotestosterone, gives compounds with oral activity but has not provided separation of anabolic and androgenic properties. Nevertheless, methyl groups have been placed synthetically in almost every position of the testosterone molecule, a formidable synthetic task. Similarly, reduction of the 4,5 double bond and introduction of unsaturation at 1,2, 6,7, 9,10, 11,12, or 14,15, and reduction of the 3-keto group to CHOH and even to CH_2 has been performed, with variable biological activities of the resulting compounds, depending on the test method used. When totally synthetic compounds were tested, racemic mixtures were often furnished for pharmacological study although hormonal activity was found to reside in one of the diastereomers.

Hydroxylation at virtually every position furnished only poorly active myotropic and androgenic compounds. Exceptions to this generalization are 4- and 11β-hydroxytestosterones, with or without 17-methyl substituents; some of these exhibit a fair separation of activities. Halogenated derivatives have been very uneven in biological response, but occasionally one of them, such as halotestin, has an oral anabolic effect 20 times and androgenic activity 9.5 times that of 17α-methyltestosterone in rats (836) and can be used clinically.

It would be gratifying to be able to report that these and many other molecular modifications have had a rationale for their design, but with the few exceptions to be mentioned later this has not been the case. The prodigious effort poured into the synthesis of these many analogs, isomers, and derivatives was motivated by the hope that some working hypothesis could

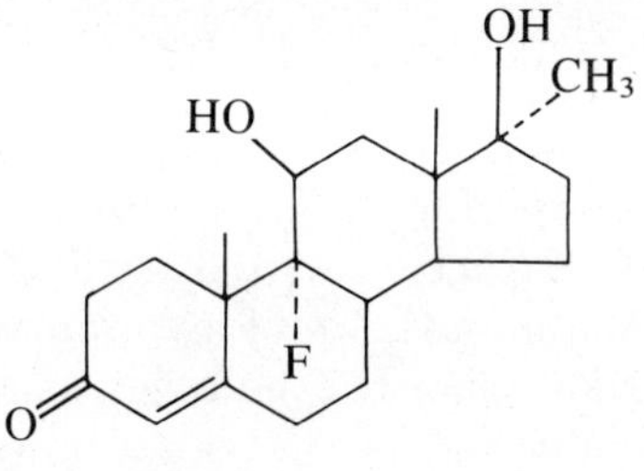

Halotestin

be evolved from sheer numbers of substances. Commercial competitive thought also played a prominent role but there were preciously few intelligent conclusions to be drawn. The reason these trials and errors are mentioned here is that in this compact series of analogs, limited to some extent by the chemical and steric conditions of the steroid system, these efforts demonstrate the groping for rationales in molecular modification that still besets many areas of medicinal chemistry. Deleting many further purely empirical attempts to increase potency and separate androgenic and myotropic properties, a few of the more defendable modes of reasoning should be mentioned.

One of these is the application of bioisosteric principles to the steroid skeleton. A CH=CH group is classically isosteric with sulfur, and indeed 2-thia-A-nor-5α-androstane has high anabolic and androgenic activity, indicating that steric rather than electronic factors are important at C-2 and

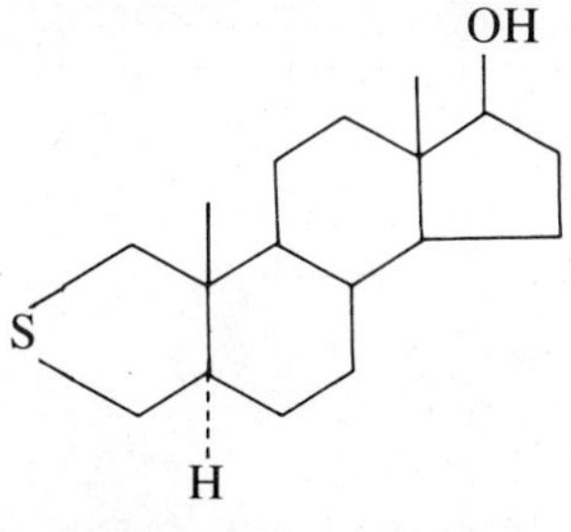

2-Thia-A-nor-5α-androstane

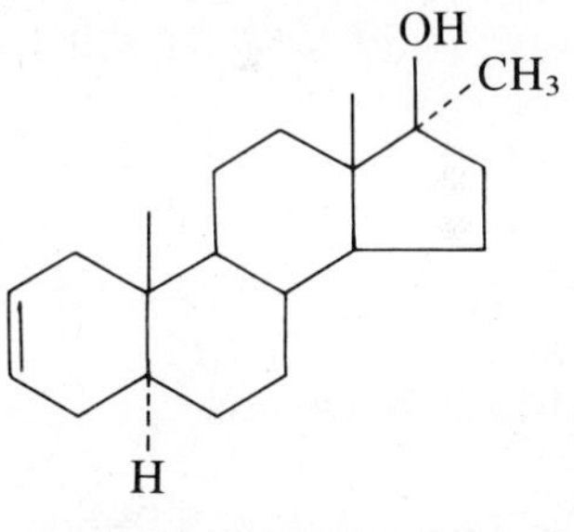

17α-Methyl-5α-androst-2-en-17β-ol

C-3 (837). This is in accord with the activity of the steroid prototype of the compound, 17α-methyl-5α-androst-2-en-17β-ol (838). Quite commonly, 17-methyl derivatives have higher (and oral) activities throughout these series.

The 2,3-unsaturated steroids have offered the opportunity not only to replace the double bond but to add to this reactive linkage. Epoxides, episulfides, and numerous heterocyclic 5-membered rings as well as cyclopropane analogs have been prepared and tested, after a few early compounds of this type had shown interesting activities and often high potency. Contractions and expansions of the A and D rings have also been studied (reference 831, p. 892).

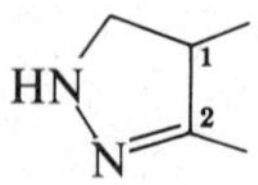

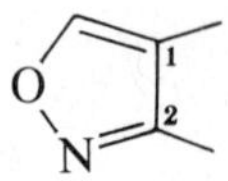

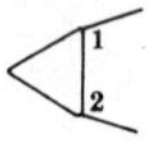

The interesting activity profiles observed for 18- and 19-nor derivatives in the progestational and estrogenic steroids spilled over also into the male sex hormone series. Separation (1:10) of myotrophic and androgenic activities was realized in nandrolone. On the other hand, the isomeric 18-nor-testosterone was essentially inactive in both tests (839). A dozen esters of

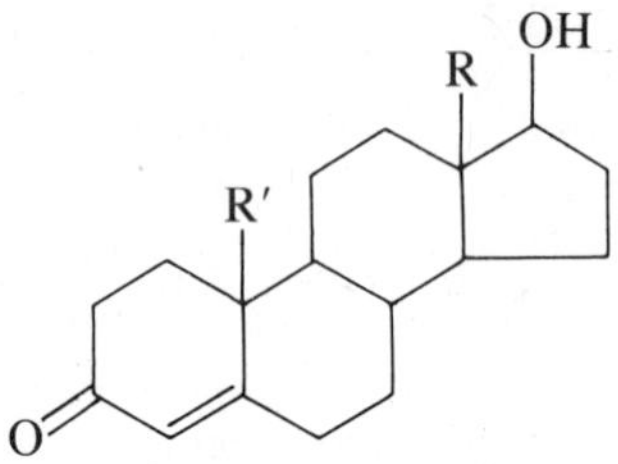

18-Nortestosterone: R = CH$_3$, R' = H
19-Nortestosterone (nandrolone): R = H, R' = CH$_3$

nandrolone with long-chain aliphatic and arylaliphatic acids have been in use as latentiated long-acting anabolic agents based on an early observation (840) that esterification of testosterone prolongs activity. Apparently, androgenicity is proportional to lipophilicity and the rate of hydrolysis by liver esterase (841). The latter relationship could explain the low androgenicity of sterically hindered esters. Similarly, 17β-alkoxy ethers are at least a little active as long as the ether group can be cleaved *in vivo*, emphasizing the need of a free 17β-hydroxyl for androgenicity (842). Some of these results resemble those in other series of biologically active alcohols. Attempts to construe receptor surfaces that would complement these and other SAR have remained inconclusive for lack of detail.

3.14.2. Estrogens

The estrogenic steroid hormones (843) all contain an aromatic A ring (no substituent at C$_{19}$) with a 3-OH or 3-OCH$_3$ group, but estrogenic properties are found in a variety of other structures as well. After the early discoveries of estrone (844, 845), estriol (846, 847), and 17β-estradiol (848), a number of further urinary estrogens were found, indicating that hormonal activity is not restricted to a very few similar structures. Compounds with 2-OCH$_3$, 18-CH$_2$OH, and 15- and 16-OH groups and 6,7- and 7,8-double bonds were among these derivatives. These variations should have given medicinal chemists ideas about alternative molecular modifications but the time for that had not yet arrived: No total or partial syntheses were available that would have made possible modified structures, and the quantities of estrogens isolated from ovaries or pregnancy urines were inadequate to serve as

STEROIDAL HORMONES AND THEIR ANALOGS

starting materials for synthetic steps. Nevertheless, enough estrone accumulated to serve as a ketone in a reaction that had been of interest to Inhoffen (849). Estrone was treated with potassium acetylide in liquid ammonia to furnish 17α-ethynylestradiol, which turned out to be 20 times more potent orally than estrone in the Allen–Doisy (850) rat test and in humans (851). This drug and its 3-methyl ether, mestranol, are still being used in the field of oral contraception.

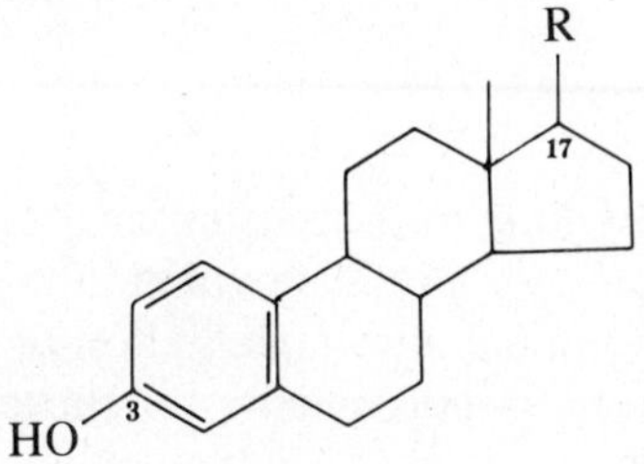

Estrone: R = =O
Estradiol: R = H, OH
17α-Ethynylestradiol: R = HO, C≡CH

Estrogenic activity is found not only among steroids but also in totally synthetic compounds such as di- and triphenylethane, naphthalene, and phenanthrene derivatives (852). Dodds devised the first orally active synthetic estrogen, diethylstilbestrol (853). It had been conceived as an analog of estrone (estradiol made its appearance three years later) in which the B and C rings had been cut open and the D ring had to become phenolic. On this basis the *trans* configuration was assigned to diethylstilbestrol; later, by chemical deduction, it was proved to be correct.

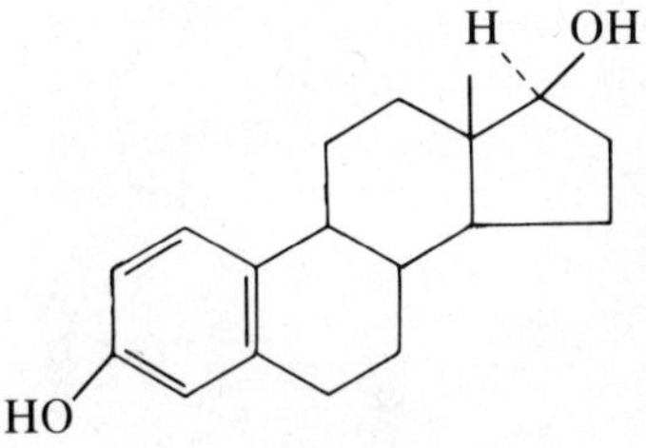

17β-Estradiol

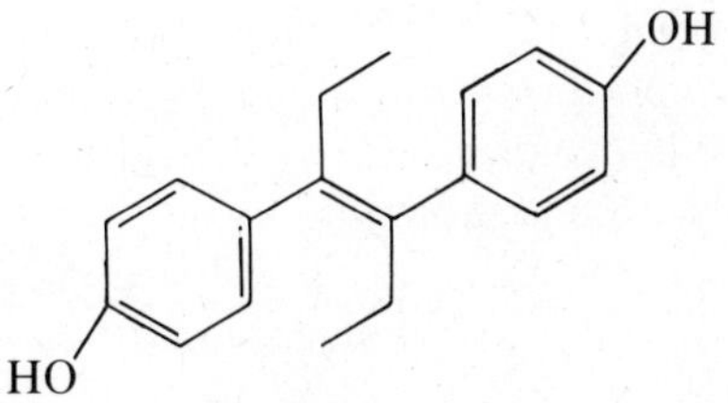

trans-Diethylstilbestrol

As long as activity had thus been located in *trans*-stilbene derivatives, other compounds in this series were also tested because of chemical interest in their structures (854–856). The medicinal-chemical rationale for the design of such drugs as clomiphene is not very convincing (857).

Some mono- and diesters of 17β-estradiol with carboxylic acids are very potent estrogens by the parenteral route (858).

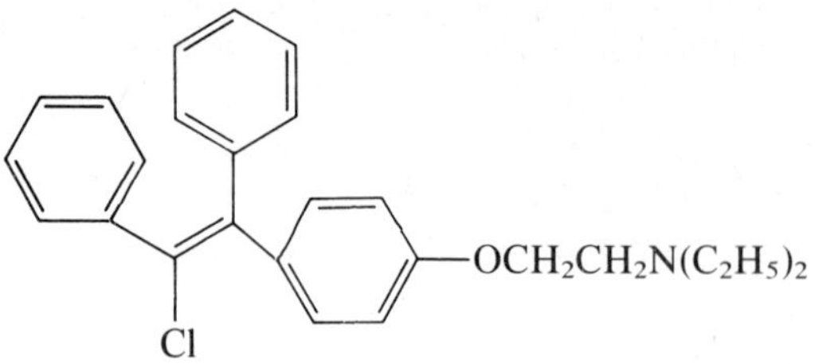

Clomiphene

3.14.3. Progestational Agents

The hormone of the corpus luteum, progesterone, regulates a number of tissue changes; among them it suppresses ovulation in pregnant female animals and aids in the development and implantation of the fertilized ovum. The structure of progesterone is not specific for these activities. Deletion of the angular methyl group (or groups) is advantageous; rings A and B can be further unsaturated, methyl or chlorine can be substituted at C-6, and position 17 can be substituted by an α-acetoxy group. Moreover, androstanes carrying a 17-ethynyl group are orally active progestationals, especially if the angular CH_3 at C-19 is deleted or lengthened (to ethyl) and methyl is introduced at C-6. This does not exhaust all the possible modifications that were attained by ingenious syntheses and by guesswork and good fortune.

The biosynthesis, metabolism and hormonal mechanisms of action are of great interest but have not provided any ideas leading to the design of more potent and specific agents. On the other hand, antagonists may well be constructed based on the inhibition of enzymes involved in the biosynthesis of steroidal hormones, but such studies do not seem to have been published. Antagonists would be of value in hyperandrogenism, in the control of sex hormone-dependent cancers, and in other sex hormone disorders. The inhibition of the conversion of testosterone to its active dihydro derivative has been studied for such purposes. A complex derivative of progesterone, called cyproterone acetate (859) has been found to compete with testosterone and dihydrotestosterone for androgen receptors (860) and can be used clinically as an antiandrogen.

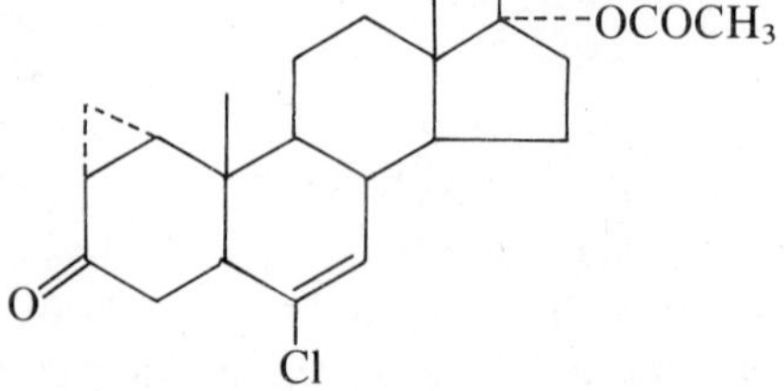

Cyprosterone acetate

The most important use of estrogens and progestogens is as antifertility agents. Considering an estimated number of 200 to 400 million women who are using such agents worldwide, this has been and remains a highly active research area.

Almost all reviews of antifertility agents stress the clinical aspects of these hormonal compounds, whether they are used simultaneously, sequentially, on different days of the menstrual cycle, orally, or parenterally. These important techniques have not been able to guide medicinal chemists, however, in recommending to their biological colleagues which compounds to use in a particular way. Screening in laboratory animals (usually rats) has remained the guiding technique for subsequent clinical trials. These agents are used voluntarily by women and not for therapeutic reasons unless therapeutically necessary abortions and similar contingencies arise. The voluntary aspect of these drugs has led to stringent demands for both acute and long-term safety and freedom from toxicity and carcinogenicity. These agents have been reviewed (861–863). Here, only the structures of widely used oral contraceptives are listed (Fig. 2).

These and a few further analogs (esterified at alcoholic OH groups, etherified at C-3, etc.) represent the arsenal of steroid contraceptives. Little is to be learned from this array of compounds concerning SAR, especially since practical results heavily depend on the correct combination of such agents as determined by trial and error and on the method and timing of administration.

In addition to these steroids and diethylstilbestrol, numerous other compounds have been tested as contraceptives in animal experiments and in a few clinical trials. Some of them, to be found in reference 864, may ultimately become clinically useful based on their different mode of action, relative lack of long-term side effects, and lower cost—an important factor in a drug taken voluntarily by low-income populations. At this time, the variety of structures ranging from basic triphenylethanes and reserpine to prostaglandins, and the different test methods used in determining the activity of these compounds, do not permit medicinal–chemical deductions or predictions.

Some of the female steroid contraceptives also have an effect on male sperm count, potency, libido and other qualities affecting male fertility. Again, a variety of unrelated structures also produce some of these effects, and speculations about the choice of any of these "leads" appear premature (35).

3.15. ANTIINFECTIOUS AGENTS

3.15.1. Topical Antibacterials and General Antimicrobials

In their search for nutrients, microorganisms attach themselves to plants, animals, and even inanimate objects and metabolize the tissues or fibers of

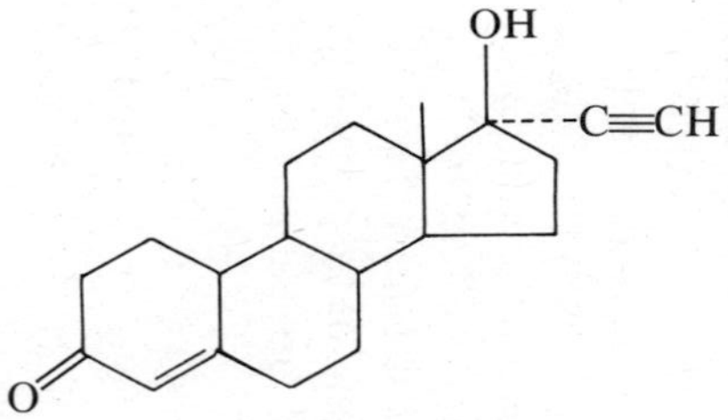

Ethynodiol diacetate

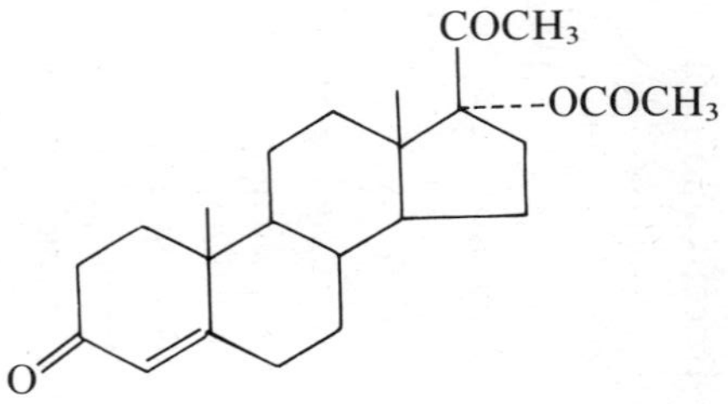

Medroxyprogesterone acetate

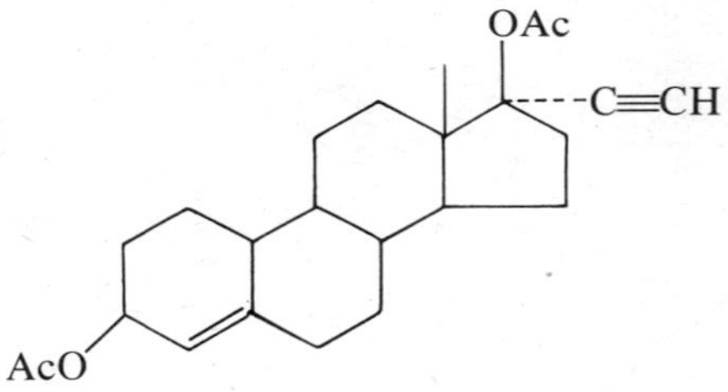

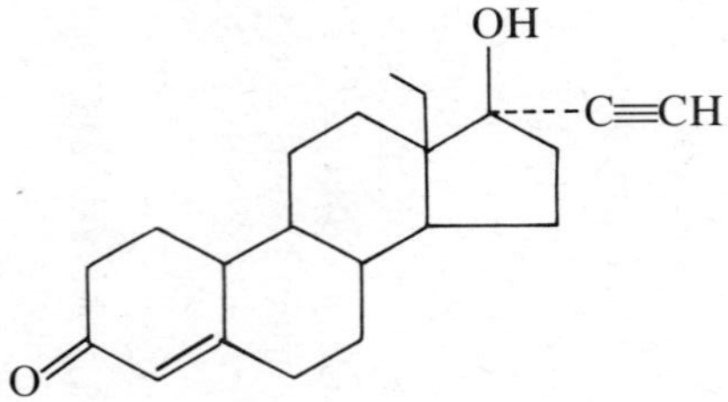

Superlutin

Norethinodrol

Mestranol

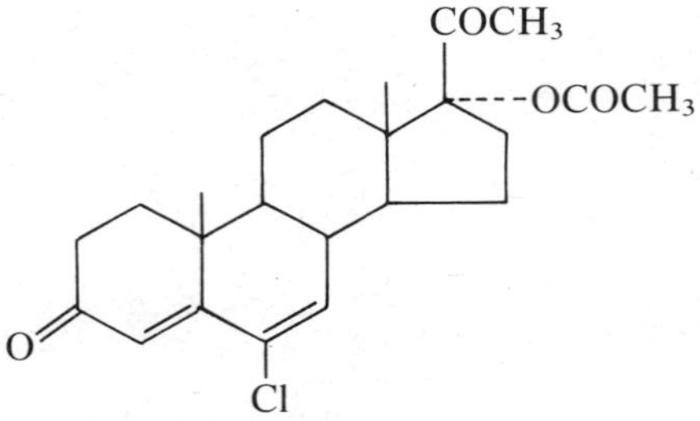

Chlormadinone acetate

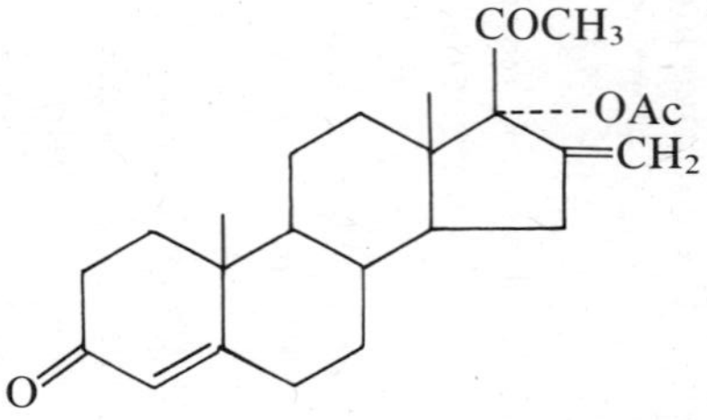

Norethindrone
(or its 17-acetate)

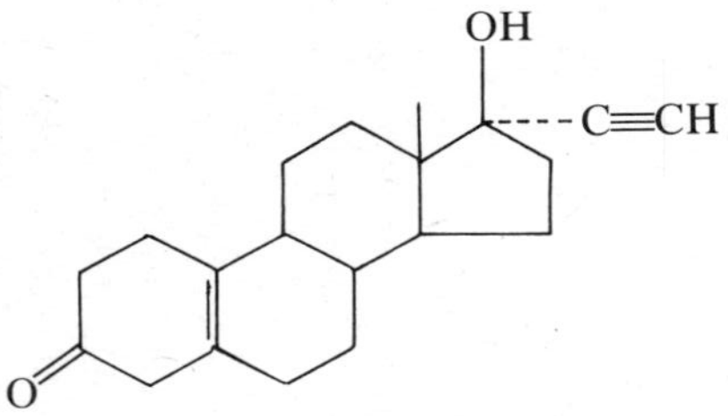

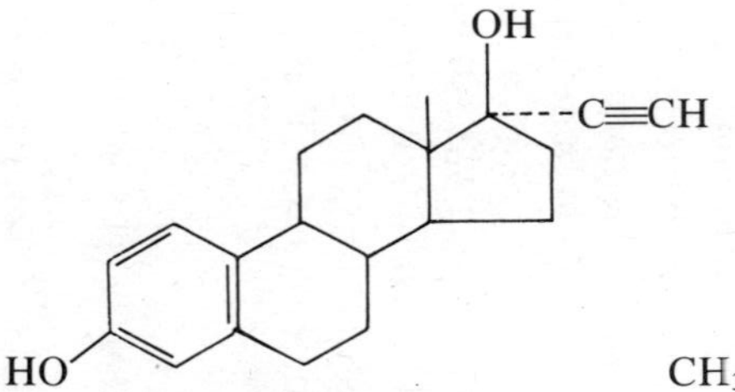

Ethynylestradiol

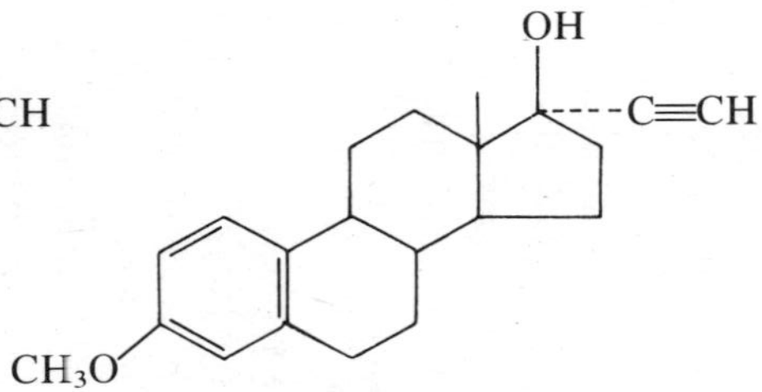

D,L - Norgestrel

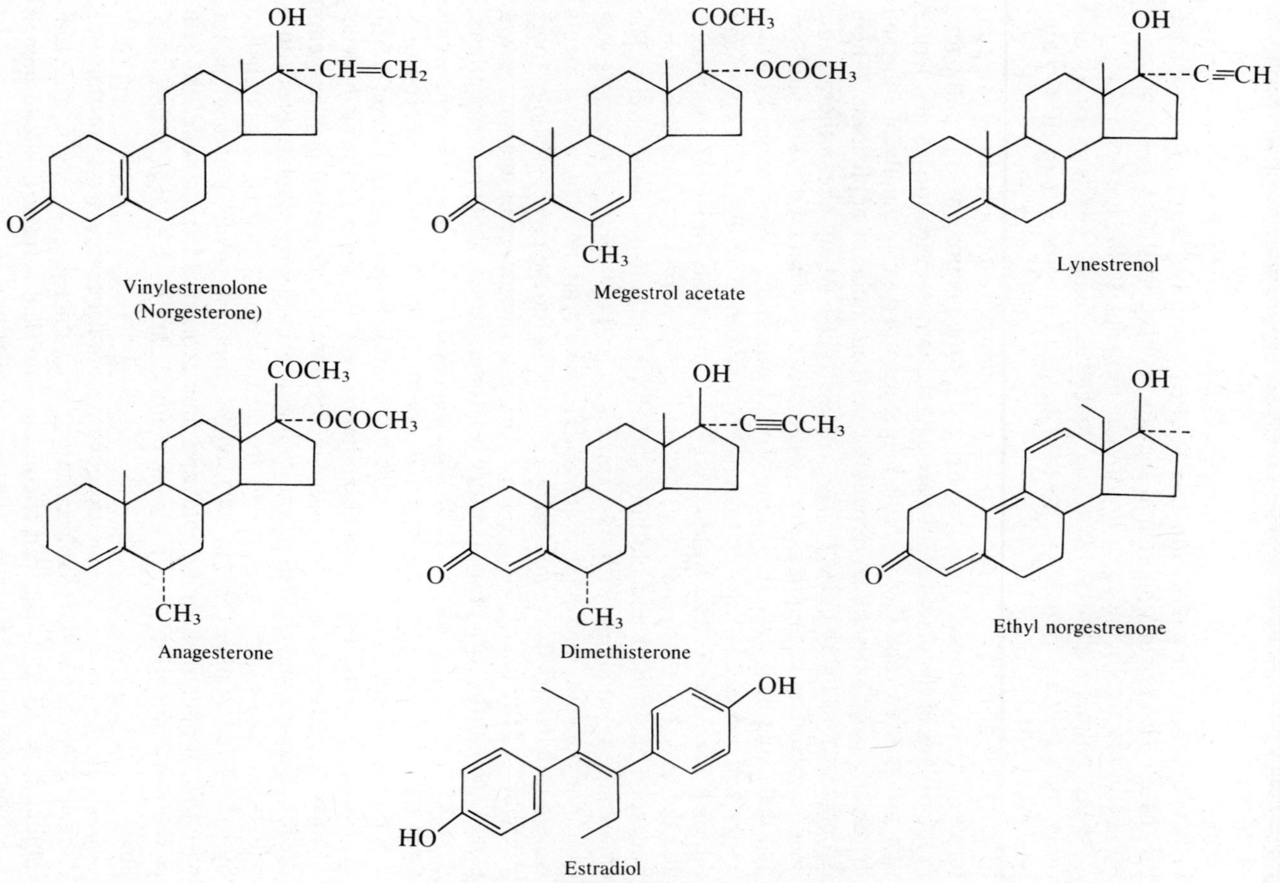

FIG. 2. Clinically used contraceptives.

these hosts. In some cases, a symbiosis is set up whereby metabolic products of the microbes are utilized by the host. The digestion of cellulose by special bacteria in the intestinal tract of ruminant animals or the bacterial synthesis of K vitamins in the gut of humans and other mammals are examples of such processes. Each plant and animal species also harbors microbes that are of no symbiotic significance nor are they pathogenic to the host.

Many animal parasites, protozoa, bacteria, fungi, yeasts, viruses and other microbes are pathogenic to their hosts. The products of these pathogens are either outright toxic or they are antigenic, and they call for detoxification or mobilization of the host's immune responses. Drugs that interfere with biochemical reactions needed for the life processes of pathogens in the tissues of the diseased host are called chemotherapeutic agents. Drugs that are applied topically without being absorbed systemically are termed disinfectants or antiseptics, depending on their mode of application. There is considerable overlapping between these types. Many chemotherapeutic agents (including antibiotics) are used topically on infected tissue surfaces and in dermatology, whereas some topical antiseptics are absorbed to some extent and can exert systemic toxicity. Nonsystemic antibacterial chemicals are also used to sterilize water, surgical instruments, hospital equipment, and the like. For thousands of years, people have tried to combat microbes that spoil their food and clothing and rot their houses and shelters. The embalming of cadavers with oils, spices, and other vegetable products was essentially an antimicrobial practice. The explorations of new continents were motivated by the search for spices needed for the preservation of foodstuffs before the era of refrigeration. The intrinsic nature of antiseptics was established at the turn of the century by Robert Koch, Delepine, and others, who proposed early tests for such materials (865). Some colored compounds stain parasitic cells and thereby make them visible under the microscope. Koch presented his ideas about staining tubercle bacilli in a lecture in Berlin on March 24, 1882 (866). In the audience was Paul Ehrlich, who had described selective bacterial staining in his doctoral dissertation four years earlier. He returned to his laboratory and set up Koch's staining procedure with tuberculous sputum (867). The first enduring test method was published by Rideal and Walker (868) in 1903, who compared the effect of a substance with that of phenol (carbolic acid) and thus expressed the phenol coefficient of the test compound. Other testing methods are of more recent date (865, 869).

Early in the studies of antimicrobial agents it became apparent that one could not predict with reliable certainty what microbes would be affected, inhibited, or killed by a given chemical. Empirical screening has remained a deplorable necessity in the utilization even of well-designed antimicrobial drugs. In the examples in this section, we will encounter many such compounds, originally expected to perform well against certain pathogens only to be found more effective against entirely different organisms when screened broadly for antimicrobial activity.

In a few instances this activity could be rationalized to some extent. As an example, urea was noted to be antibacterial in 1906 (870), perhaps, as we understand it now, because of its ability to break hydrogen bonds in proteins. Many simple and complex compounds containing urea moieties possess similar properties but their real utility may lie in other microbial areas.

3.15.1a. Halogens. Among the earliest antiseptics were the halogens. Iodine was introduced by the French surgeon Chatin in 1839 for the treatment of battle wounds (12), and chlorine and hypochlorite salts have been employed for 80 years to purify drinking water and as household sanitizers. Less irritating and longer acting sources of hypochlorite ion (chlorophors) were found when protons of aromatic sulfonamides, cyclic imides, and amidines were replaced by "positive" chlorine. Chloramine T [p-$CH_3C_6H_4SO_2N^-Cl\cdot Na^+$] was the first of these chlorophors (871, 872). These compounds probably act as oxidizing agents and enzyme inhibitors (873). 3-Chloro-4,4-dimethyl-2-oxazolidinone is regarded as one of the best synthetic N-chloramines (874).

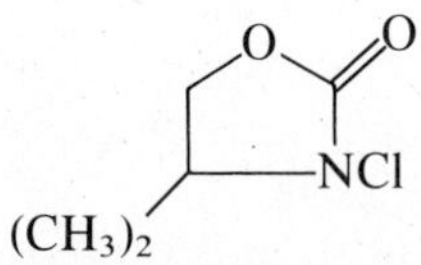

Among iodophors, a complex of iodine and polyvinylpyrrolidone (povidone-iodine, or PVP-I) (875) has been used widely for topical treatment. While its low toxicity has made it popular, none of the iodophors is markedly superior to 1% iodine in 70% ethanol (876).

3.15.1b. Phenols. The earliest practical successes in disinfecting surgical areas, instruments, and the then ungloved hands of the operating team were achieved with phenol by Joseph Lister in 1867 (47), although Kuchenmeister had pointed out the antimicrobial properties of this material before. Lister's studies were based on his acceptance of Louis Pasteur's germ theory of disease and met with fierce opposition by hospital administrators who denounced the suspicion of uncleanliness in their wards. One of these physicians was James Simpson, who had introduced chloroform into obstetrics a few years earlier. He in turn had been harassed by the clergy, who claimed that anesthetics were Satanic decoys (877).

The necrotizing effects of phenol called for less aggressive and if possible more effective analogs with a greater margin of safety. The cresylic acids, a crude mixture of isomeric cresols from coal tar, were among the first disinfectants of this class (878). Halogenation of phenols increased antibacterial potency (879, 880) and decreased mammalian toxicity, as attested by the wide use of chlorophene (4-chloro-2-hydroxydiphenylmethane). Diphenolic compounds in which the two phenolic nuclei were connected directly (879) or by various bridging atoms were also found to comprise effective antibacterials (881); these studies culminated in hexachlorophene [2,2'-dihydroxy-3,4,6,3',4',6'-hexachlorodiphenylmethane], synthesized by Gump (882). It remained the most widely used multipurpose antiseptic but it had to be limited to prescriptions when toxic effects on the CNS of infants were discovered (883). Bisphenols inhibit D-glutamyl ligase (884), which catalyzes the incorporation of D-glutamic acid into uridine 5'-diphosphomuramyl-L-alanine, a component of bacterial cell walls. Other mechanisms of action of phenols have also been suggested, among them chelation of iron (885). This possibility has been studied especially for 8-quinolinol (oxine, chinosol), which readily chelates trace metals (886, 887). Halogenated 8-quinolinols possess broad-spectrum antibacterial and antifungal activity (888). 5-Chloro-7-iodo-8-quinolinol (chinoform, clioquinol) (889, 890) has been employed for the treatment of amebic dysentery and bacterial diarrheas, but these effects have not been confirmed (891, 892) and the drug is no longer available for these purposes.

Salicylanilide is an antifungal agent (893) in the textile industry. Its halogenated derivatives ("halosalans") had been in use as antiseptics (894, 895) but they cause photosensitization. Benzyl p-hydroxybenzoate (benzyl paraben) is a constituent of several gargles for infected throats (896). One of the most popular polyphenolic antiseptics is n-hexylresorcinol, similarly used in oral preparations (897, 898).

Structurally simple hydroxy antibacterials include ethanol and 2-propanol, which act upon many bacteria by protein denaturation but are not sporicidal (899). Among aldehydes, formaldehyde solutions are used for the sterilization of surgical equipment, and the undiluted compound has been employed as a fumigant. Hexamethylenetetramine (methenamine) depends on the release of formaldehyde for its activity as a urinary antiseptic. Formaldehyde itself has been supplanted by ethylene oxide as a gaseous chemosterilant (900, 901). Another nonspecific gaseous alkylating agent is β-propiolactone (902).

3.15.1c. Quaternary Ammonium Salts.
Quaternary ammonium ions with at least one long-chain hydrophobic (lipophilic) substituent ("invert soaps") have antiseptic properties; some of them are used for general disinfection and as oral antiseptics. These properties were mentioned first in 1908 (903) but the introduction of antiseptic "quats" had to wait until 1935 when Domagk (904) found that all compounds of this series killed bacteria provided one radical was aliphatic, with 8 to 18 carbon atoms, 14 being optimal. In low but effective concentrations they are quite nontoxic. They affect cell membranes but their molecular mode of action still needs clarification.

Of the thousands of quats studied, only a few are commercial products. They are all used as topical antiseptics for the skin, for burns, and as an udder wash in the dairy industry. They are prepared by quaternization of

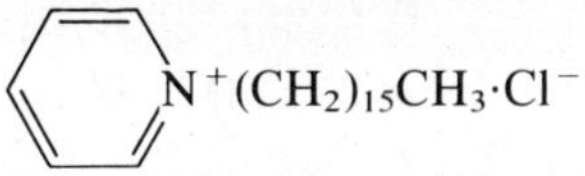

ANTIINFECTIOUS AGENTS

Cetylpyridinium chloride: pyridinium-$N^+(CH_2)_{15}CH_3 \cdot Cl^-$

Cetrimonium bromide (905): $(CH_3)_3N^+(CH_2)_{15}CH_3 \cdot Br^-$

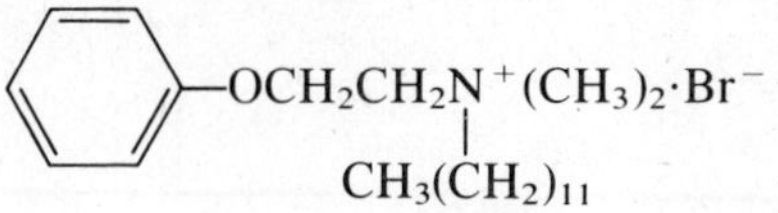

Domiphen bromide (906): Ph–$OCH_2CH_2N^+(CH_3)_2 \cdot Br^-$ with $CH_3(CH_2)_{11}$ substituent on N

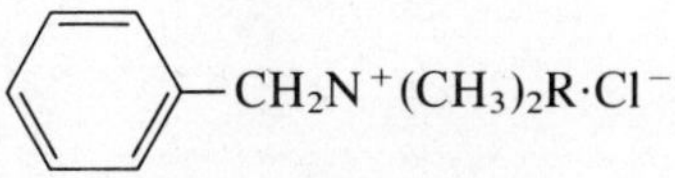

Benzalkonium chloride: Ph–$CH_2N^+(CH_3)_2R \cdot Cl^-$
R = C_8H_{17} to $C_{18}H_{37}$

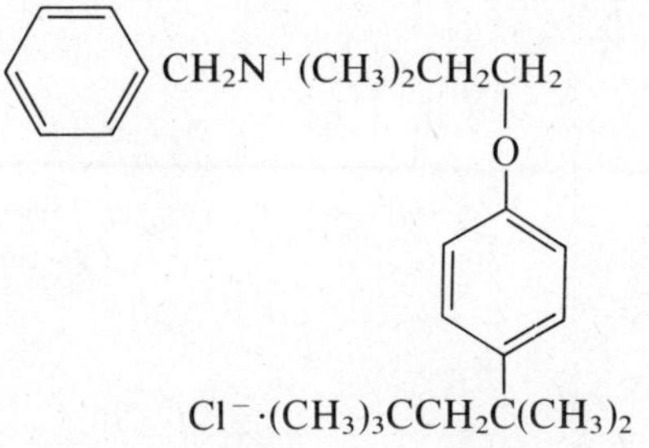

Octaphonium chloride: Ph–$CH_2N^+(CH_3)_2CH_2CH_2$–O–C$_6$H$_4$–$C(CH_3)_2CH_2C(CH_3)_3 \cdot Cl^-$

the respective tertiary amines with long-chain alkyl halides; this operation is very popular because it does not require much skill or imagination.

3.15.1d. Ureas, Amidines, Biguanides. Several compounds related to early trypanocidal azo dyes were designed to contain urea groups in an effort to minimize the dyestuff character of the prototype azo dyes in the proposed chemotherapeutic agents (60). Afridol Violet and Chlorazol Fast Pink BK were urea–azo hybrids, and the success in protecting animals from trypanosomal infections encouraged the final step, the exchange of the remaining

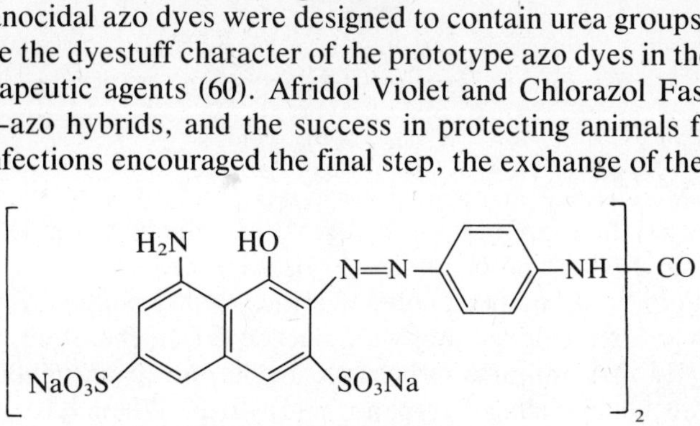

Afridol violet

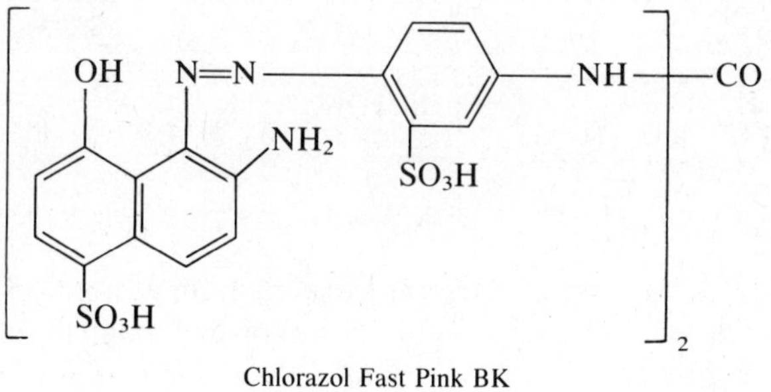

Chlorazol Fast Pink BK

azo chromophore with a bioisosteric amide linkage. This led, after many variations of substituent positions and insertion of further anilide moieties, to suramin, first prepared in 1917 by O. Dressel and R. Kothe (907). The structure, kept secret by the Bayer Farbenfabrik for which they worked, was elucidated later by Fourneau at the Paris Pasteur Institute (64, 65).

In other structural series, diarylureas (carbanilides) were found to have

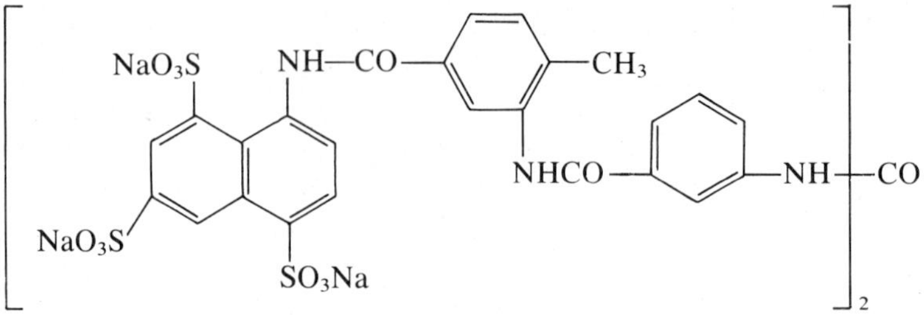

Suramin

potent topical antibacterial activities. Triclocarban (908) and cloflucarban (909) have been studied in depth. Triclocarban breaks down to chloroanilines, which have been implicated as the cause of the methemoglobinemia neonatorum noted with this antiseptic.

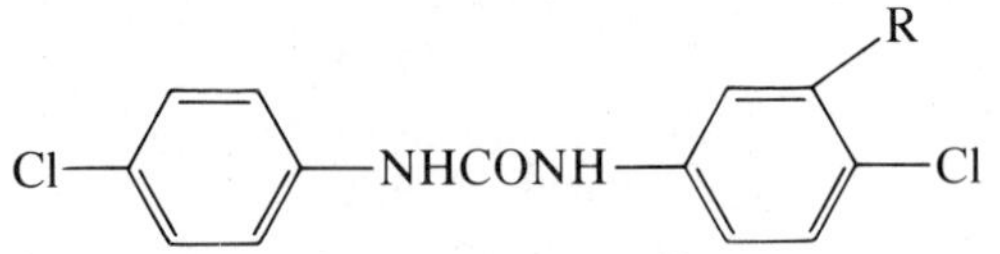

Triclocarban: R = Cl
Cloflucarban: R = CF$_3$

Aromatic amidines are more basic than ureas; the first compounds in this series, exemplified by propamidine (910), were designed on a premise of shaky serendipity. It had been noted that the motility of most trypanosomes depends on an extracellular supply of glucose (911). Therefore, the trypanocidal activity of suramin was believed to be due to an inhibition of the carbohydrate metabolism of trypanosomes (912). When 1,10-diguanidinodecane (synthalin) was singled out as the most suitable diguanidine that, in metabolic studies, lowered the blood sugar level in animals (913), it was tested for trypanocidal activity and turned out to be active (912, 914, 915).

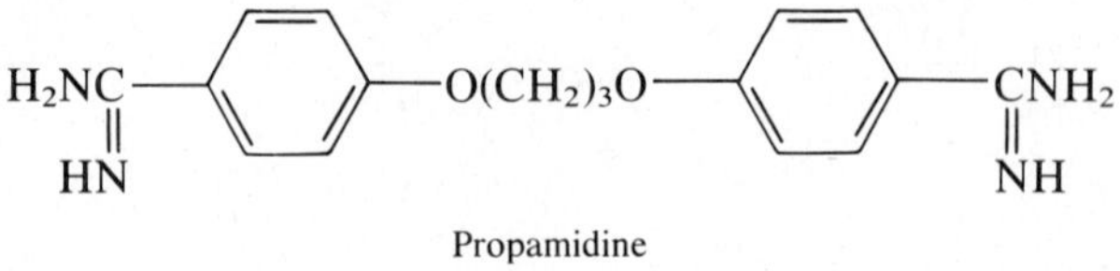

Propamidine

Synthalin kills trypanosomes in a glucose medium at nonhypoglycemic concentrations and could therefore not act by this mechanism; nevertheless, its trypanocidal activity could not be denied. Now molecular modification took over and led to the study of diamidines, diisothioureas, and diguanidines whose two functional groups were located at the ends of alkylene chains or at aromatic nuclei connected by saturated and unsaturated alkylenes, diethers, and so on (916, 917). Several of these diamidines are useful trypanocidals. Propamidine is also bactericidal for gram-positive bacteria and some fungi (918).

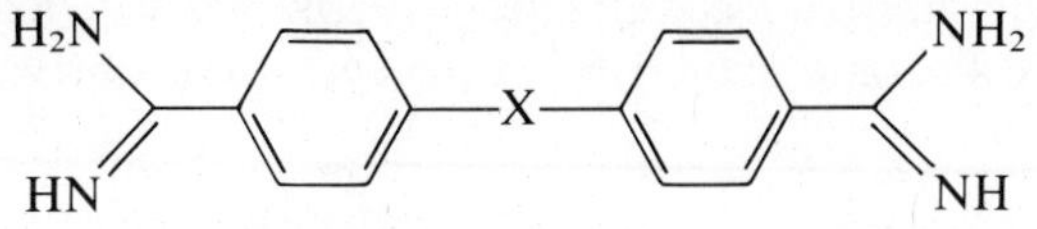

Stilbamidine: X = CH=CH
Propamidine: X = O(CH$_2$)$_3$O
Pentamidine: X = O(CH$_2$)$_5$O
Diminazene (917): X = N=N—NH

An extension of amidine chemistry is reflected in guanidines. Biguanides had been derived from aminopyrimidines by ring opening during studies to design antimalarials on a logical basis (919). The original antimalarials of the chlorguanide type contained one biguanide group, but at the time of that work the doubling of functionalities ("bis-ing") had become popular, particularly in ganglionic blocking agents (920, 921). When this concept was applied to the then novel biguanides, in analogy to the propamidine congeners, alkylenebis(biguanides) were developed (922) of which chlorhexidine

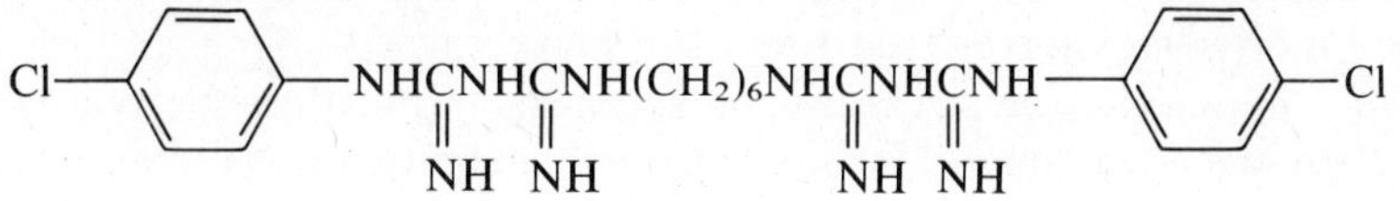

Chlorhexidine

was the choice (923) pharmacologically. On antimicrobial screening, it was selected as a topical antibacterial agent. It inhibits many gram-positive and gram-negative bacteria but not mycobacteria, spores, or viruses (924). Apart from the usual antiseptic applications in the surgical theater, in veterinary medicine, on burns, and in mouthwashes, chlorhexidine seems of value in preventing dental plaque (925).

3.15.2. Antimicrobials Used Topically and Systemically

3.15.2a. Nitroheterocycles as Antimicrobials. As mentioned before, there is no sharp dividing line between topical and systemic antibacterials. Compounds used as systemic antibacterials can usually be applied for therapy on body surfaces and even to unrelated objects: thus, oxytetracycline arrests the lethal yellowing of coconut palms.

Nitro compounds have been eyed suspiciously by pharmacologists because nitrophenols, such as picric acid, that had been employed as antiseptics on burned skin, turned out to be dangerous systemic poisons. When the presence of a nitro group in chloramphenicol was established by IR spectroscopy, research on this antibiotic was almost abandoned. Since then, other nitroheterocyclics have been discovered among antibiotics, for example, the antitrichomonal azomycin (926, 927). Many of them had to be shelved when mutagenicity and carcinogenicity were found in their activity spectrum, apparently caused by reduction of NO_2 to a reactive intermediate that interferes with normal DNA functions (928). Nevertheless, several such compounds have become established drugs for their unrivaled special activities.

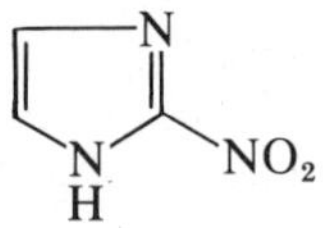

Azomycin

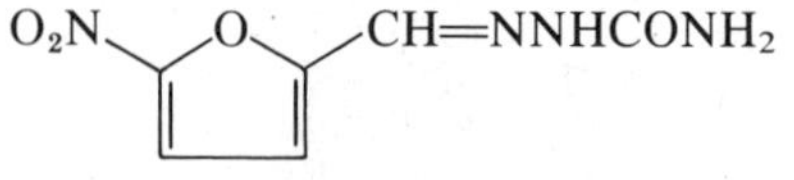

Nitrofurazone

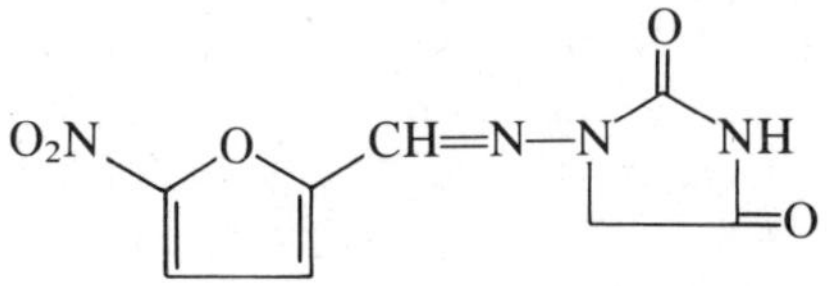

Nitrofurantoin

Broad-spectrum topical antibacterial activity was first seen in nitrofurazone (5-nitro-2-fural semicarbazone) (221, 929) and nitrofurantoin (930). Other nitrofural derivatives followed later; in several cases their side effects caused their removal from the American market.

The simplicity of the structure and synthesis of nitrofural derivatives led to extensive molecular modification and testing of nitro compounds in related ring systems. The first really successful analog was metronidazole (931, 932), which has become the standard drug in the treatment of trichomoniasis and giardiasis (933).

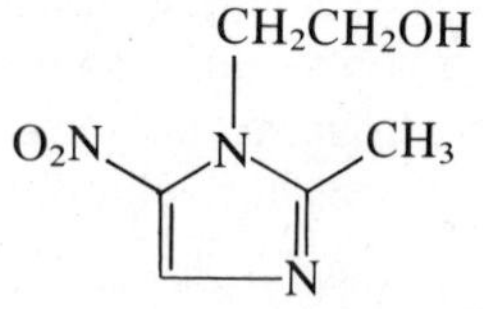

Metronidazole

3.15.2b. Condensed Pyridonecarboxylic Acids. In the sequence of testing compounds for antimicrobial activity, one hopes that the pharmacodynamics and metabolism of the drug will permit application as a systemic as well as topical agent. Failing this, topical use is usually the least damaging if systemic absorption produces unacceptable toxicity. In between is the possibility of using the drug as a urinary-tract or intestinal-tract antiseptic. This is the case for nalidixic acid (934) which inactivates an essential com-

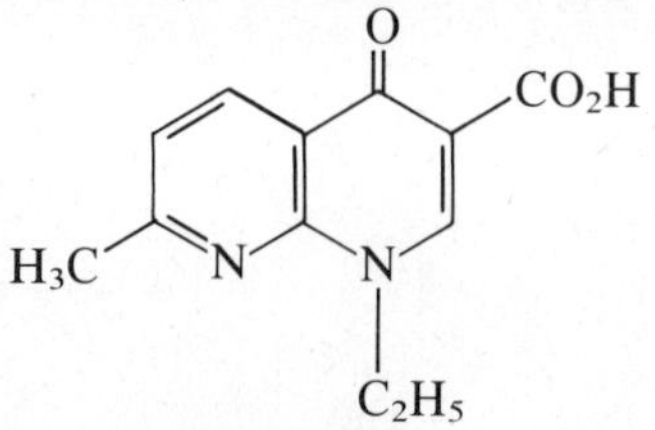

ANTIINFECTIOUS AGENTS

Nalidixic acid

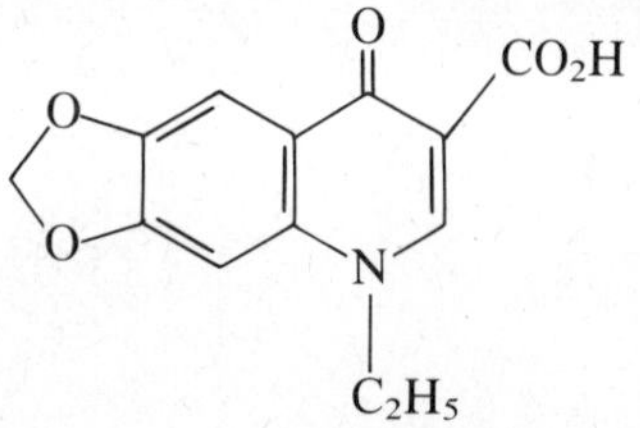

Oxolinic acid

ponent in DNA replication (935, 936). Similarly, oxolinic acid (937), a 4-quinolonecarboxylic acid, was chosen as a suitable urinary-tract antiseptic.

3.15.2c. Dyestuffs. Captivating the imagination of microbiologists at an early date was the working hypothesis that dyes might be toxic to cell processes and selective enough to stain mainly cells to be affected and, if possible, not the surrounding tissue areas. Microscopic studies of selectively toxic stains were started with protozoa, and bacterial staining followed after a few years. Churchman (938) used the triphenylmethane dye, crystal violet (gentian violet), and in the course of his observations coined the term "bacteriostasis." Gentian violet (939) has remained a useful rosaniline dye; it has topical antibacterial, anthelmintic, and antifungal value. Malachite green (brilliant green) and rosaniline (fuchsine) (940) are other members of this group but their use has declined; they have been replaced by nonstaining, broad-spectrum antibacterials.

These dyes probably intercalate with DNA and promote ribosomal breakdown (941, 942), a view compatible with an early assumption that they formed complexes with amphoteric cell substituents (943).

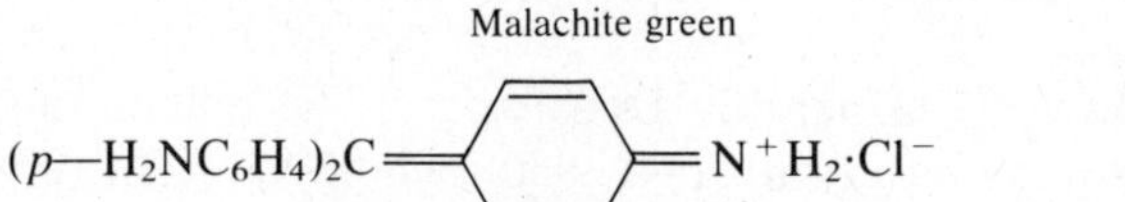

Gentian violet

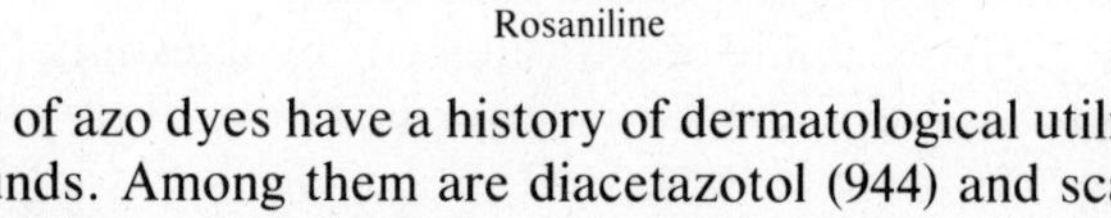

Malachite green

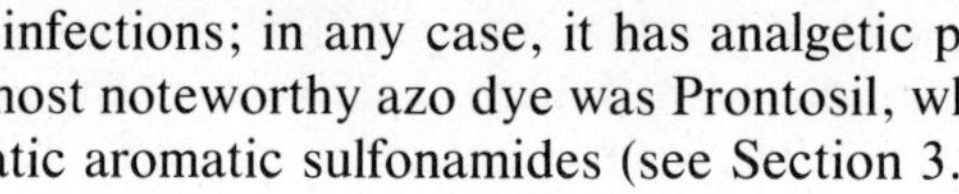

Rosaniline

A number of azo dyes have a history of dermatological utility in the treatment of wounds. Among them are diacetazotol (944) and scarlet red, used mostly in veterinary medicine. Phenazopyridine (945) was thought to act on urinary-tract infections; in any case, it has analgetic properties in such diseases. The most noteworthy azo dye was Prontosil, which heralded the era of bacteriostatic aromatic sulfonamides (see Section 3.16).

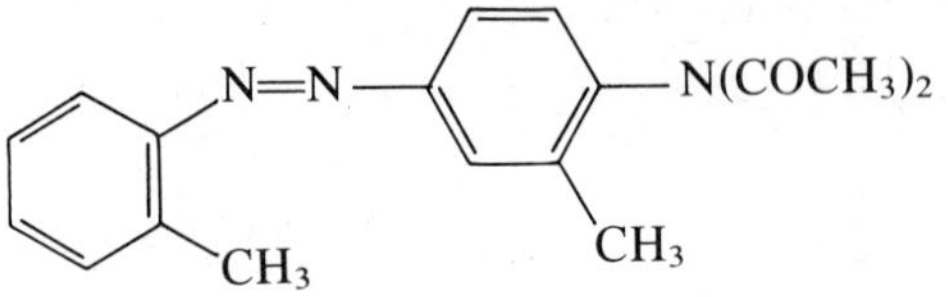

Diacetazotol

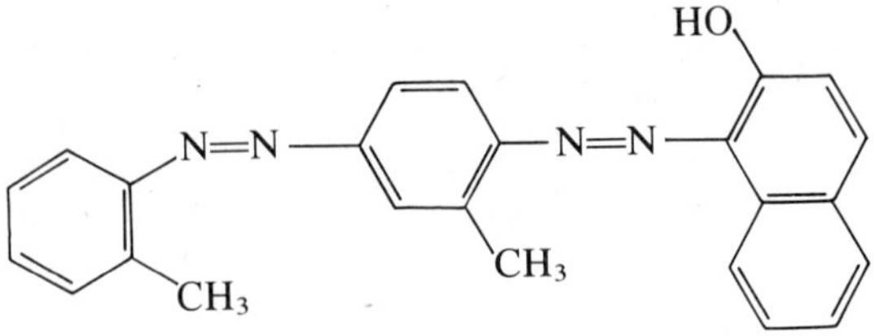

Scarlet red

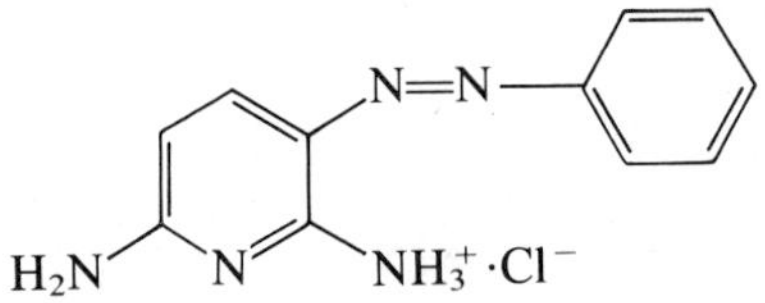

Phenazopyridine

Aminoacridines are colored by virtue of resonance; acriflavin was tested by Browning in the treatment of trypanosomiasis (946). It is a mixture of proflavin (3,6-diaminoacridine) and its quaternary methochloride, and is used topically for minor fungal and bacterial infections. Similarly, ethacridine (2-ethoxy-6-aminoacridine), prepared commercially in 1922, and aminacrine (9-aminoacridine) have antibacterial properties (947). These dyes inhibit DNA synthesis (948).

3.16 CHEMOTHERAPEUTIC ANTIBACTERIALS

3.16.1. Sulfanilamides and Diaminosulfones

3.16.1a. Early Sulfonamide Dyestuffs. The edifice of chemotherapy was erected mainly by Ehrlich (949, 950), who lived from 1854 to 1915. One of its working hypotheses, proposed in his doctoral dissertation in 1878, was based on the selective staining of parasite cells by dyestuffs (951). Various pathogenic microbes were stained more or less selectively on microscope slides, and dyestuffs toxic to these microbes could be expected to interfere with the cellular functions of the pathogens or even kill them. The success of azo dyes *in vitro* and *in vivo* encouraged Ehrlich's students and their contemporaries to continue with similar experiments. The side chain of a derivative of methylene blue, a phenothiazine dyestuff with unconfirmed antimalarial activity, was extended to create the first 8-aminoquinoline antimalarials. Similarly, when he embarked on a program of antibacterial research, Domagk turned to azo dyes again as preferred candidates in a vast screening pool of 30,000 chemicals. It was one of the achievements of the bacteriostatic aromatic sulfonamides that they exploded the myth of chemotherapeutic dyestuffs and thereby opened up antimicrobial drug design to other structural types.

The conception of azo dyes, which contain a sulfonamide substituent *para* to the azo linkage, appears in retrospect like a final sequel to the emergence of some of the European pharmaceutical industries. Two types of firms opened their R & D laboratories during the second half of 19th century. One type was founded by individual physicians and pharmacists who experimented on new—almost always naturally occurring—drugs in laboratories attached to their offices or apothecary stores. As they expanded, these laboratories became the backbone of the manufacturing units in which the production of drugs was scaled up to meet the growing demand.

The second type of pharmaceutical firm was the result of diversification of existing chemical plants, especially in Britain, France, and Germany from 1870 on. The synthetic dyestuff industry established itself after W. H. Perkins' introduction of "Mauve" and soon looked for opportunities outside purely chemical manufacturing to use and sell their intermediate and synthetic products. One such opportunity seemed to lie in the manufacture of drugs. Kolbe's synthesis of salicylic acid from phenol, a dyestuff intermediate, was one of the earliest examples of these efforts to break ground in a new industry (952).

The earliest dyestuffs were used mostly on wool, hides, and silk, the fibers available at that period. Cotton (cellulose) and semisynthetic fibers such as nitrocellulose, cellulose acetate, viscose, and the purely synthetic polymers of our day followed suit and offered the science and art of dyeing increasingly difficult problems because of the relative insensitivity of functional groups repeated on the polymers. Even in wool dyeing some problems were hard to overcome, especially fastness to laundering and resistance to shrinking. It was therefore a step forward when Hörlein, Dressel, and Kothe of the Bayer works of the I. G. Farbenindustrie found that a sulfamyl group imparted fastness to acid wool dyes (953, 954). This was interpreted as a firmer linkage of the sulfonamide-containing azo dye to animal proteins, and could have constituted a "lead" for affinity to other proteins as well, for example, microbial proteins. But the time for this comparison had not yet come in the industrial environment of a dyestuff factory.

In the late 1920's the need for antibacterial agents to combat both gram-positive and gram-negative bacterial infections increased in urgency. There was no other "lead" but the selective staining with dyestuffs, which had furnished several antiprotozoal drugs, and this approach was therefore transposed to *in vitro* antibacterial tests. One of the earliest trials was with hybrids of antiplasmodial-dyestuff structures, obtained by coupling diazohydrates with dihydrocupreine (955), a compound derived from the antimalarial cinchona alkaloids.

Among the assorted coupling components was diazotized sulfanilamide, yielding 5'-p-sulfamylazodihydrocupreine (R = SO_2NH_2). This derivative had a low *in vitro* antibacterial activity and was therefore not investigated

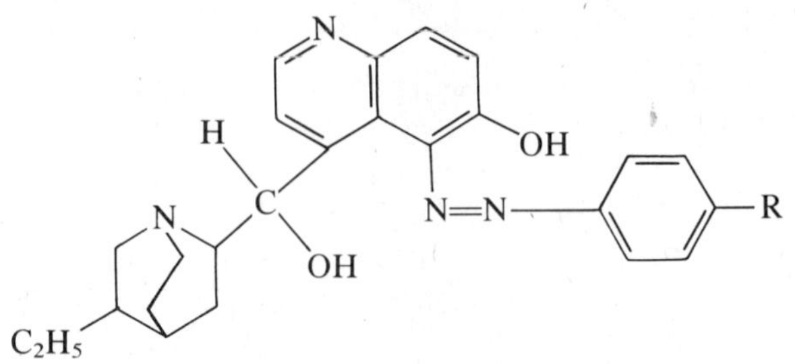

5'-p-Sulfamylazodihydrocupreine: R = SO_2NH_2

further. If the metabolic reductive cleavage of this azo dye had been suspected then, the discovery of the antibacterial properties of sulfanilamide could have occurred 16 years earlier and prevented untold deaths from coccus infections. A suggestion that sulfanilamidochrysoidines might be degradable to sulfanilamide (956) was not followed up.

In a letter written in 1972, Heidelberger said (957):

> Like everyone else at the time, Walter Jacobs and I thought that a substance had to be directly bactericidal in order to be useful in combating bacterial infections. We had been successful with trypanosomiasis by applying Jacob's idea of changing the —OH of —COOH to NH_2, as – $CONH_2$, in order to get an organic arsenical past tissue barriers. Accordingly, when we tried to get something better than optochin against pneumococcal and streptococcal infections, we started first with amides and then, by analogy, went to —SO_2NH_2. The possibility that any substance as simple as sulfanilamide could cure bacterial infections never entered our heads, nor did our microbiologist even ask to test it. We even improved Gelmo's [963] method of preparation and went on to convert sulfanilamide into highly bactericidal substances which killed infected mice faster than the infections alone! As slaves to an idea, we missed the boat in 1915, losing the chance to save many thousands of lives, and the development of the sulfonamides was delayed twenty years.

Systematic studies on azo dyes with potential antibacterial activity were initiated by Mietzsch and Klarer in the late 1920's; the first announcement of their researches did not appear until 1935 when patent rights to these compounds had been secured by the German dye trust. They selected a reputed 10,000 dyestuffs from the archives of their company for this test series but soon concentrated on wool dyes with a sulfamyl group because these promised to have higher affinity to microbial proteins as well. Their hopes were confirmed: several dyes had high bactericidal activity *in vitro*, but this did not carry over to survival times in infected mice. Since a clinically promising agent would have to show such *in vivo* activity, the pharmacologist Domagk, who collaborated on this test program, decided to shortcut the test procedure and screen the compounds directly *in vivo* (958). In 1932, he found that Prontosil rubrum, a red sulfamyl azo dye, protected mice against streptococcus and rabbits against staphylococcus infections. Pneumococcus infections were not affected, and Prontosil was inactive *in vitro*.

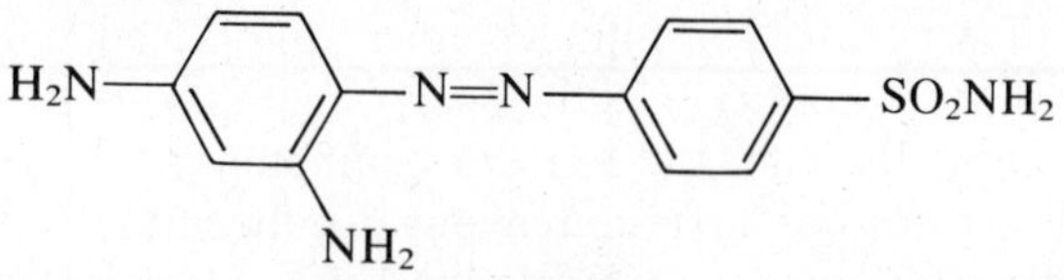

$$H_2N-\bigcirc(NH_2)-N=N-\bigcirc-SO_2NH_2$$

Prontosil

We now know the cause of the *in vitro* inactivity: the dye had to be reduced first by the mammalian host before the released sulfanilamide could act as a dihydrofolate synthetase inhibitor. But this was not understood at that time. Moreover, commercial restrictions stood in the way of more fundamental studies.

The first patient to receive Prontosil was Hildegarde Domagk, the daughter of its discoverer. Among the first patients to receive sulfanilamide was Franklin D. Roosevelt, Jr., the son of the President. His recovery from a streptococcic throat infection helped to overcome early doubts of the medical value of antibacterial chemotherapy.

3.16.1b. Sulfanilamide. Dr. and Mme. J. Tréfouël of the Pasteur Institute in Paris secured a sample of Prontosil and determined its structure by hydrogenolysis to sulfanilamide and 1,2,4-triaminobenzene. Still under the influence of Ehrlich's theory of selective staining, and after reviewing the comprehensive claims of the German patents in this field, the Tréfouëls and Dr. and Mme. Daniel Bovet (959) embarked on an SAR study of sulfamyl-containing azo dyes by coupling diazotized aromatic aminosulfonamides with a variety of phenols. Changes in the structure of the phenolic coupling components did not affect antibacterial activity, whereas any deviation from the *p*-sulfamylbenzeneazo structure abolished the activity. Bovet concluded that the phenolic moiety did not matter, that the host organism reduced the dyes to sulfanilamide, and that this compound accounted for the antibacterial activity of the parent dyestuffs. This belief was supported further by the isolation of colorless 4-acetamidobenzenesulfonamide from the urine of mice dosed with the intensely colored Prontosil. A test with sulfanilamide itself on mice infected with streptococci confirmed this hypothesis. In addition, sulfanilamide was active *in vitro* against a number of gram-positive organisms. Further antibacterial data in mice and in patients appeared soon (960–962). Twenty-seven years after a graduate student, Paul Gelmo in Vienna, had first prepared sulfanilamide (963), this simple compound became the first recognized clinically effective and curative antibacterial agent.

This history of sulfanilamide takes an ominous turn if one considers German Patent 226,239 of May 18, 1909, issued to Heinrich Hörlein of the Bayer Farbenwerke. This patent, issued one year after Gelmo's publication (963), describes a synthesis of sulfanilamide. Therefore, sulfanilamide was no longer patentable in 1930 when Domagk, Klarer, and Mietzsch (956, 958) began their work on sulfonamide-containing dyestuffs as bacteriostatic agents. Prontosil was tested in infected mice in 1932 but Domagk did not publish his results before 1935. With the protagonists of this drama gone, one will never know whether the I.G. Farbenindustrie realized the activity of sulfanilamide but for commercial reasons disguised the compound—then in the public domain—in the form of dyestuffs. Hörlein (954) has stated officially that the sulfonamide dyes were made for their fastness on wool and that Domagk's discovery of their action against streptococci "directed our subsequent work into a new channel. Numerous new azo dyes containing sulfonamides were prepared but the test object was no longer the wool fiber but the mouse infected with streptococci."

Although still used in veterinary medicine, sulfanilamide was abandoned in human clinical practice after a few years when the toxic side effects of the drug became too objectionable. Also, its antimicrobial spectrum was narrow and its metabolism led to crystalluria of its N^4-acetyl derivative in the kidney. In an effort to improve these adverse factors, over 10,000 structural analogs, derivatives, and other molecular modifications were synthesized and tested, mostly in the pharmaceutical industry worldwide. It was soon found the p-aminobenzenesulfonamide ion, $H_2NC_6H_4SO_2NH^-$ (964), could not be altered without loss of potency, but substitution at N^1 was often advantageous and capable of wide variation. In this contiguous series, great improvements in therapeutic properties were achieved by molecular modification.

3.16.1c. *N*-Substituted Sulfanilamides. Substitution at N^1 of sulfanilamide by 2-pyridyl (965) gave sulfapyridine; it was reported in the medical literature in 1938 (966) as the first drug useful in the therapy of pneumonia, but it was too toxic. It is still used occasionally to treat dermatitis herpetiformis. At that time, two events increased the urgency of medicinal research in this field. World War II had begun and antibacterial agents were needed in quantity for the treatment of combat wounds. The other factor was the appearance of the first rationale in drug design in the application of isosteric alterations of existing drug structures. Pyridine and thiazole are classical isosteres (967–969), and the replacement of 2-pyridyl in sulfapyridine by the isosteric 2-thiazolyl group to give sulfathiazole was a logical development in medicinal chemistry (970). Sulfathiazole (971, 972) was much more successful than sulfapyridine both antibacterially and toxicologically, and it has remained a useful drug (964).

H₂N—⟨benzene⟩—SO₂NHR: R = [thiazole] [1,3,4-thiadiazole-CH₃] [3,4-dimethylisoxazole]

Sulfathiazole · Sulfamethizole · Sulfisoxazole

In sulfathiazole, two hetero atoms are present in the N^1-ring substituent, and this encouraged the synthesis of other heteroaromatically N^1-substituted sulfanilamides with five- and six-membered ring systems. In the 1,3,4-thiadiazole series, sulfamethizole (973) had greater antibacterial potency and solubility and exhibited better pharmacodynamic performance (974, 975). Sulfisoxazole is highly soluble and is excreted rapidly, and so is its N^4-acetyl derivative. It has activity against *Proteus vulgaris* and *Escherichia coli,* and it is used in urinary infections (976–978).

In the 6-membered heteroaromatic series, the isomeric diazines—pyridazines, pyrimidines, and pyrazines—exhibit the best antibacterial activity. Sulfadiazine was the first of these compounds to be accepted for clinical use, followed by sulfamerazine and sulfamethazine (979–983). Each of these

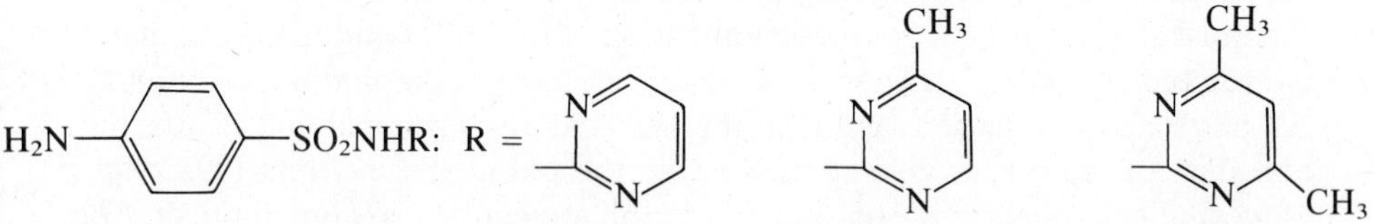

H₂N—⟨benzene⟩—SO₂NHR: R = [pyrimidin-2-yl] [4-methylpyrimidin-2-yl] [4,6-dimethylpyrimidin-2-yl]

Sulfadiazine · Sulfamerazine · Sulfamethazine

pyrimidyl derivatives is endowed with certain pharmacodynamic advantages that make it more suitable for special therapeutic tasks. Some methoxy-substituted pyrimidyl and pyrazinyl derivatives are distinguished by very long biological half-lives in clinical practice, which makes it possible to administer these drugs once a day or even less frequently. Their formulas and half-lives are listed here with their nonproprietary names and selected references.

H₂N—⟨benzene⟩—SO₂NHR: R = [5-methoxypyrimidin-2-yl] [5-methyl-4,6-dimethoxypyrimidin-2-yl]

Sulfamethoxydiazine, 36 h (984) 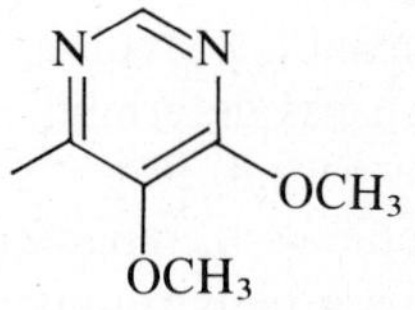 · Sulfamethoxine, 150 h (988, 989)

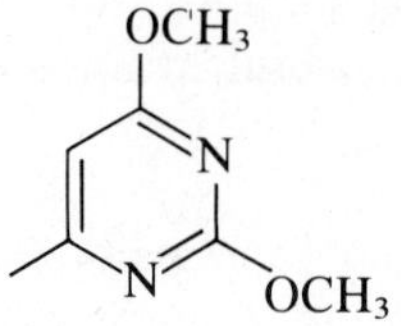

[4-methoxy-6-methylpyrimidin-2-yl-OCH₃] [6-methoxypyridazin-3-yl] [3-methoxypyrazin-2-yl]

Sulfadimethoxine, 40 h (985–987) 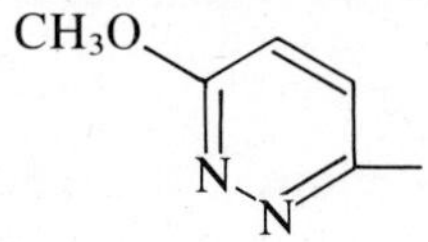 · Sulfamethoxypyridazine, 37 h (990) 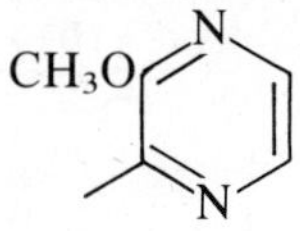 · Sulfamethoxypyrazine, 65 h (991)

With the advent of broad-spectrum antibiotics, efforts were made to modify the structure of anti-gram-positive sulfanilamides so that gram-negative organisms would also be affected. There was no way to design such analogs *de novo* but small "leads" could be strengthened by patient molecular modification. For example, sulfaclomide, a broad-spectrum derivative, was found in this way (992). No rationalization for this finding has been offered. Shepherd has reviewed these structure–activity relationships (993).

$$H_2N-\underset{}{\underset{\text{Sulfaclomide}}{\bigcirc}}-SO_2NH-\underset{}{\bigcirc}\begin{smallmatrix}Cl & CH_3\\ & N\\ N & CH_3\end{smallmatrix}$$

A good deal of thought has been given to the relationship of the physical properties and the chemotherapeutic activity of sulfonamide drugs. One of the most striking properties is the ionization of the sulfonamide group, which is affected by the N^1 substituent as well as by the *p*-aminophenyl moiety. All physical comparisons point to the anion as the principal active species, and this, of course, is also controlled by the pH of the medium (994, 995). Nevertheless, the nonionized portion of the molecule also contributes to the activity (528). More recent studies have used spectroscopy (996, 997) and molecular orbital calculations to correlate polarizability and inhibitory activities, and on the whole have established a linear relationship (998–1002). From a multiparameter linear free energy calculation, optimal values for ionization constants and hydrophobicity, which affect maximum bacteriostatic activity for a given organism, have been derived (1003). Another property that affects the activity of sulfonamide drugs is their binding to proteins that are not concerned with active enzyme sites and receptors (1004–1007).

Finally, the metabolism of the drugs by different organs, mainly the liver, is a limiting factor in different hosts. The main course of metabolism of sulfonamides involves N^4-acetylation, with the solubility of the N^4-acetyl derivatives determining the rate of renal excretion—and thereby some uses of the parent drug (987)—and of damaging crystalluria (1008–1011). Rates of metabolism range all the way from near zero for sulfisomidine to almost complete glucuronidation for sulfadimethoxine. N^4-Phthalyl and N^4-succinylsulfathiazoles remain practically unabsorbed after oral administration and are very slowly deacylated in the intestine. They are useful in intestinal infections (1012, 1013).

3.16.1d. Mechanisms of Action. Medicinal chemistry and general biochemistry overlap in the exploration of mechanisms of drug action at the enzyme level and, if the enzyme structure is known, at the molecular level. Biochemists have been, and will be, busy elucidating normal chemical cellular events but frequently they can separate out a specific step more readily by interference or blockade with a traceable enzyme inhibitor. Furnishing chemicals whose structure might predispose them to act as potential enzyme inhibitors has been one of the greatest services of medicinal chemistry to biochemistry and a force unifying these two sciences. The mechanism of action of sulfanilamide and its analogs is a case in point; the structural simplicity of the *p*-aminobenzenesulfonamide unit has made possible a rather precise explanation of the biochemical reactions involved in bringing about bacteriostasis. Only five years after Domagk's initial publication of Prontosil rubrum (958), various substances were found to antagonize the action of sulfanilamide, for example peptones (1014), bacterial constituents (1015), and others. Woods added yeast extracts to this collection (1016) and indicted *p*-aminobenzoic acid (PAB) as the probable *in vitro* antagonist in these materials. The same was shown *in vivo* (1017, 1018), and PAB was isolated in the form of derivatives (1019, 1020) and as a pure substance (1021–1023). With this evidence at hand, Woods (1016) postulated that the structural similarity of PAB and sulfanilamide made it possible for the drug to interfere with bacterial enzymes dependent on the metabolic PAB. Fildes, in a paper entitled "A Rational Approach to Chemotherapy" (1024), elaborated from this hypothesis a general theory of drugs as metabolite antagonists.

The prophetic insight of Woods and Fildes was fully rewarded by later experimental findings. Tschesche (1025) suggested that PAB or its glutamic acid amide (PABG) condenses with a diphosphorylated pteridine component to give pteroylglutamic (PGA) (folic) acid and that sulfanilamide competes with PAB in this biosynthesis, inhibiting the enzyme, PABG synthetase, or another synthetase. This was actually observed later in bacterial culture and cell-free enzyme preparations. Indeed, the competition was confirmed by isolating products analogous to PGA in which ^{35}S-labeled sulfanilamide had been incorporated (1026, 1027). Details of these events have been summarized by Anand (1028).

Since sulfonamide drugs block the biosynthesis of folate (and thereby of dihydrofolate) in pathogens that depend on their own biosynthetic supply of this metabolite, the therapeutic action of sulfonamides is increased by combination with inhibitors of dihydrofolate reductase (1029–1031). As in other cases of a multipronged attack upon a biosynthetic pathway, the sequential blockade of the availability of one-carbon transfer systems in nucleoside base synthesis by sulfonamides plus dihydrofolate reductase inhibitors has greatly extended the utility of both types of drug. By mutually aiding each other's range of activity, the drugs in combination have made infections such as toxoplasmosis, chloroquine-resistant falciparum malaria, salmonellosis, chronic bronchitis and others more accessible to treatment.

3.16.2. Bis(4-Aminophenyl) Sulfone

This sulfone may be regarded as a vinylogue (arylogue) of sulfanilamide in which the mutual influence of sulfone and amide groups on each other has

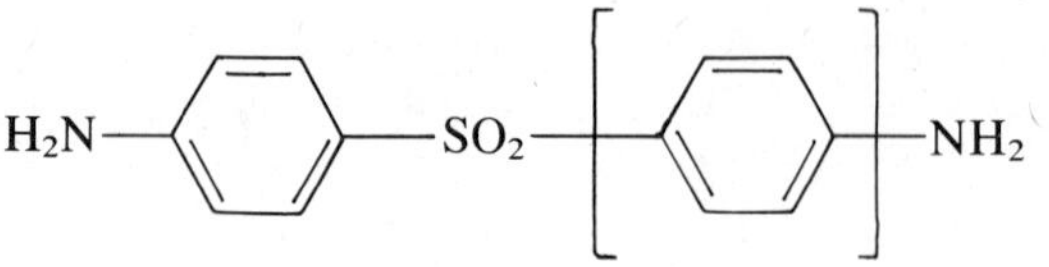

$$H_2N-\langle\bigcirc\rangle-SO_2-\left[-\langle\bigcirc\rangle-\right]-NH_2$$

been modified by an interspersed aromatic ring. Apparently the benzene ring represents an optimum for this vinylogy since analogs with other aromatic or quasi-aromatic systems exhibit decreased biological activity.

Bis(4-aminophenyl) sulfone (4,4'-diaminodiphenyl sulfone, dapsone) was screened for antibacterial activities and found to have a chemotherapeutic effect in experimental tuberculosis (1032). The possibility that other mycobacterial infections might also respond to this drug or its didextrose sulfonate (glucosulfone) (1033, 1034) was substantiated by their effect on rat leprosy (1035) and in clinical cases of human leprosy (1036). The N,N'-diacetyl derivative of dapsone, and the bisbenzal Schiff base of this compound are used as repository forms of the parent drug with a long duration (up to 80 days) of action (1037, 1038). The discovery of dapsone has revolutionized the therapy of leprosy.

In view of the vinylogy of sulfanilamide and dapsone, it is not surprising that both drugs inhibit the multiplication of susceptible bacteria by a similar mechanism. In bacteria (1039) and mycobacteria (1040, 1041) the action of dapsone is antagonized by PAB *in vitro* and *in vivo*. Overproduction of PAB leads to the emergence of drug-resistant bacterial strains (1042, 1043). The occurrence in certain bacteria of an isoenzyme form of dihydrofolate synthetase that is less sensitive to sulfanilamide has also been blamed for resistance (1044).

Like other drugs, the sulfonamides exhibit a variety of side effects, some of them unrelated to their antimicrobial activity. Among them is inhibition of carbonic anhydrase (1045), which led to the development of such clinically useful analogs as acetazolamide, a diuretic and antiglaucoma agent. Further, molecular modification (1046) provided saluretics of the hydrochlorothiazide (1047) and furosemide (1048) types. A hypoglycemic side effect of a thiadiazole-substituted sulfanilamide was observed in 1942 (1049). Drastic molecular modification, especially deletion of the *p*-amino group and replacement by methyl, methoxy, chloro, and so on, in an empirical search led to carbutamide (1050) as a prototype for other oral antidiabetic agents. Another observation was the decrease in the renal clearance of drugs and metabolites by some aromatic sulfonamides (1051), which led to probenecid, useful in extending the half-life of penicillin in the body.

3.17. ANTIMYCOBACTERIAL AGENTS

The chemotherapy of tuberculosis, leprosy, and other mycobacterial infections has often been an applied specialty of medicinal scientists and bacteriologists whose expertise spans antiinfectious drugs and antibiotics and who concentrate their efforts on the slowly infecting mycobacteria. These organisms, with their characteristic membranes, have threatened humans with widespread and dreaded epidemics. Before the advent of chemotherapy, these disabling, deforming, and slow-killing plagues constituted severe public health hazards and therefore invited a fertile, diversified, and highly regarded field for investigation.

As in other areas of antibacterial chemotherapy, the progress of screening programs depended on the elaboration of suitable laboratory methods in animal models. The chronic character of the clinical mycobacterial diseases is not reproduced in acute infections of laboratory animals. The most widely used screening method for experimental tuberculosis is infecting mice with the human virulent strain of *Mycobacterium tuberculosis* H37 Rv and evaluating the results of test drugs in terms of ED_{50}, bacterial count, survival time, and lung pathology. Active compounds are then retested in the rhesus monkey, *Macaca mulata* (1052), whose tuberculosis parallels the human disease.

Experimental leprosy could not be established with the pathogen, *M. leprae,* before 1960, when local infection in the mouse footpad was accomplished (1053, 1054). Another more exotic animal model is the armadillo, which slowly develops disseminated leprosy after inoculation with human leprosy bacilli (1055) and can be used in test procedures.

The earliest antimycobacterial drug to make an appearance was dapsone, bis(4-aminophenyl) sulfone, which has been discussed in Section 3.16.2. Dapsone and its bis(glucose sodium bisulfite) adduct (glucosulfone) were first shown to suppress experimental tuberculous infections (1032, 1033), but this effect did not carry over well to clinical tuberculosis. The discovery of its effect on experimental rat leprosy (1035) opened the way to its use in human leprosy, where it has become the standard chemotherapeutic agent.

3.17.1. Antimycobacterial Antibiotics

In 1944, Waksman and his coworkers announced the isolation of streptomycin and its use in tuberculosis (179). The history of streptomycin and other antituberculous antibiotics, cycloserine (1056), viomycin (1057), and kanamycin (1058) are discussed in Section 3.18.3. Capreomycin, a complex of cyclic polypeptides isolated from *Streptomyces capreolus* (1059), contains an iminodihydropyrimidine ring similar to the one in viomycin; this antibiotic is used in chronic tuberculosis with resistant bacilli. The antituberculous rifamycins, obtained from *Nocardia mediterranea,* are a mixture of several antibiotics (1060) whose *ansa* structure involves an aromatic (naphthalene) moiety spanned by an aliphatic bridge; in rifamycin B this starts with an amide group at position 2 and goes via a 14-carbon triene chain to a lactone in position 5 (1061, 1062). Several hundred derivatives of rifamycin B, the first clinically acceptable antituberculous member of this series (1063), furnished SAR data which after some tortuous empirical modifications (1064), including analogies to isoniazid, led to the N-amino-N'-methylpiperazine-hydrazone of 3-formylrifamycin SV (1065) (rifampicin, rifampin) (1066). Rifampin is a relatively nontoxic highly active antituberculous agent that is also active in many other gram-positive and gram-negative bacteremias (1067) and in human leprosy (1068).

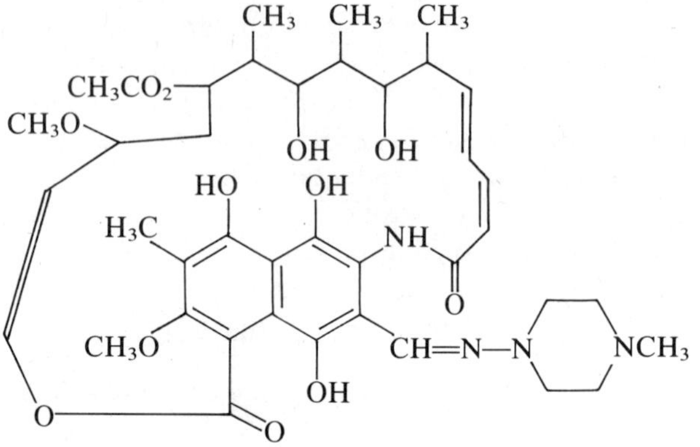

Rifampin

The mechanism of action of rifampin (and the other rifamycins) involves inhibition of bacterial DNA-directed RNA-polymerase; the mammalian enzyme is resistant to even high concentrations of the drug (1069, 1070).

3.17.2. Synthetic Agents

Two years after the announcement of the antituberculous properties of streptomycin, two structurally simple compounds were introduced into the chemotherapy of acid-fast mycobacterial infections, namely, p-aminosalicylic acid and thiacetazone. They were followed in 1951 by isoniazid and in 1952 by pyrazinamide. The history of the discovery of these drugs included both experimental logic and serendipity.

3.17.2a. *p*-Aminosalicylic Acid. The biochemist Bernheim, studying the oxygen consumption of tubercle bacilli in the Warburg apparatus, observed that benzoate and salicylate ions increased this bacillary "respiration" (1071). The "growth" of the bacilli in *vitro* is inhibited effectively by 2,3,5-triiodobenzoic and 3,5-diiodosalicylic acids (1072), and this was interpreted as antagonism to the promoting effect of benzoate and salicylate. Therefore, additional iodinated aromatic acids, sulfonic acids, as well as dialkylaminoalkyl ethers of polyhalogenated phenols were tested for inhibitory activity *in vitro* and *in vivo* (1073, 1074). Although some of the latter ethers were very active they could not be followed up further because of CNS side effects. Looking for other antagonists to salicylic acid, Lehmann

ANTIMYCOBACTERIAL AGENTS

(180) conceived the analogy between salicylic and *p*-aminosalicylic acid and compared it to another case of antagonism, that of PAB to sulfanilamide. Indeed, *p*-aminosalicylic acid (PAS) turned out to be strongly tuberculostatic although it does not inhibit the oxygen consumption of tubercle bacilli. Since PAS was a well-known compound (1075) many isomers, analogs, and derivatives were tested with the hope that patent difficulties could be circumvented. These hopes were not fulfilled.

Interestingly, the antibacterial activity of PAS is antagonized by PAB, probably by competition for an active cell uptake process. However, it is more likely that PAS activity is due to the inhibition of the biosynthesis of the iron ionophore, mycobactin, which occurs in the mycobacterial cell wall (1076, 1077).

3.17.2b. Thiosemicarbazones and Thioureas.

The success of aromatic heterocyclically substituted sulfonamides in antibacterial chemotherapy encouraged many investigators to synthesize and test similar derivatives of additional ring systems. Among these were sulfathiadiazoles (224) which, however, were of no interest chemotherapeutically. They were synthesized by cyclizing the appropriate aryl thiosemicarbazones with ferric chloride to the corresponding aminothiadiazoles and condensing these with a protected aminobenzenesulfonyl chloride (1078). In order to feed additional compounds into this screening program, the intermediates in these syntheses were also tested. One of them, *p*-acetamidobenzaldehyde thiosemicarbazone [Ar = *p*-AcNHC$_6$H$_4$], proved to be highly tuberculostatic. It was the first active drug in this series and therefore was called TbI (tibione). An epidemic of tuberculosis in defeated Germany in the aftermath of World War II was held in check with this rather toxic agent.

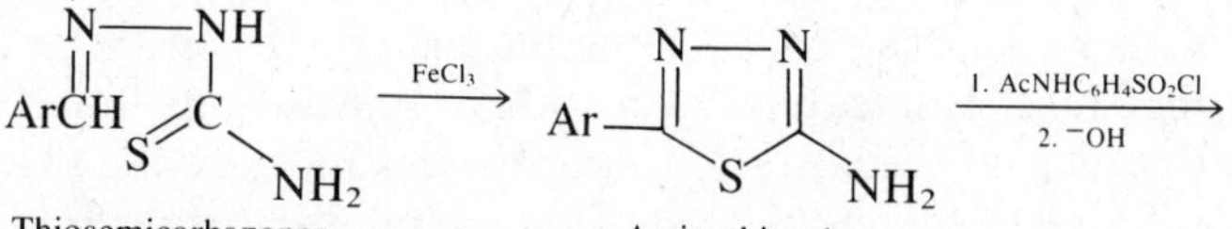

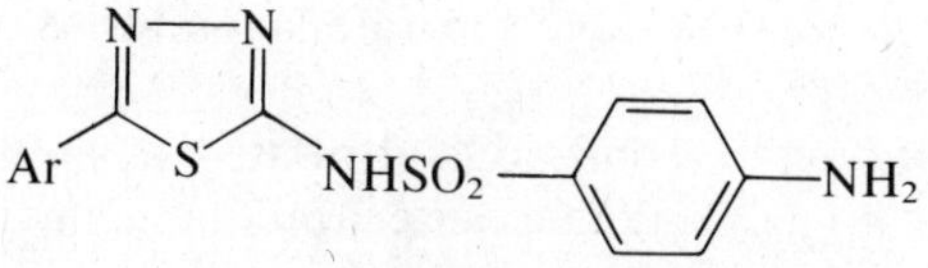

Thiazetazone, as tibione was called later, is used only rarely now because of serious clinical side effects. It has been modified extensively, especially because the art of synthesizing analogs from aldehydes and thiosemicarbazide lent itself to undergraduate programs. Several diarylthioureas have had a slightly better therapeutic profile (1079) but only one, thiambutosine, *p-n*-C$_4$H$_9$OC$_6$H$_4$NHCSNHC$_6$H$_4$N(CH$_3$)$_2$-*p*, has been able to maintain itself in limited cases of antileprotic therapy (1080). These thioureas had first been suggested for antimycobacterial tests because of an alleged morphological relationship between mycobacteria and fungi (1081) and because thioureas had been known to have antifungal activity.

M. leprae uses an enzyme, diphenol oxidase, to oxidize phenolic compounds. This enzyme is inhibited by sodium diethyldithiocarbamate, a chelating agent used to remove copper in Wilson's disease. It also inhibits *M. leprae* on mouse footpads (1068).

3.17.2c. Amides and Thioamides. In 1945, Chorine (182) and Huant (1082) reported that niacinamide was tuberculostatic, and this was confirmed three years later (1083, 1084). After a debate about whether the vitamin activity and the antituberculous activity of niacinamide were interdependent, the question was laid to rest by the discovery that compounds in the isomeric isonicotinic acid series were also tuberculostatic without having vitamin properties (1085). An SAR study followed, and led, *inter alia,* to thioisonicotinamide (184), which has pronounced tuberculostatic action; however, it is too toxic for clinical use. Increased activity was found among 2-alkyl derivatives (1086). The 2-ethyl derivative has been introduced clinically under the name ethionamide.

Amides in other ring systems have also been investigated; of these pyrazinamide [2-carboxamidopyrazine] has aroused some clinical interest (183). Its use is limited by its hepatotoxicity.

3.17.2d. Isoniazid. In the course of these studies, the thiosemicarbazone of pyridine-4-aldehyde was to be prepared (187) from isonicotinyl hydrazide by the McFadyen–Stevens reaction. This was a slow, old-fashioned way of synthesizing pyridine-4-aldehyde, and cooperating pharmacologists filled the time lag of obtaining the material by testing all the synthetic intermediates. Isonicotinyl hydrazide (isoniazid) was found to be one of the most potent tuberculostats, 16 to 55 times more active than streptomycin. This is one of the most famous accidental discoveries in the history of medicinal chemistry. It is noteworthy that the same type of screening program unearthed isoniazid in three separate laboratories during the same year (188, 189, 1087).

The antibacterial action of isoniazid is attributed to an inhibition of mycolic acid synthesis with a concomitant effect on the nonmycolic acid lipids (1088). In this regard, isoniazid acts similar to ethionamide.

3.17.2e. Ethambutol. During one of the many screening searches for antituberculous compounds, the therapeutic activity of N,N'-diisopropylethylenediamine was discovered (1089, 1090). Molecular modification singled out (+)-N,N'-bis(1-hydroxy-2-butyl)ethylenediamine, $HOCH_2CH(C_2H_5)NHCH_2CH_2NHCH(C_2H_5)CH_2OH$ (ethambutol) as the clinically most suitable derivative in this series. The high steric and structural specificity of eth-

ANTIBIOTICS

ambutol denies the working hypothesis that alkylenediamines might act by metal chelation; their antimycobacterial activity should, in this case, not be so sensitive to minor structural and steric changes. The mode of action of ethambutol is still uncertain.

3.17.2f. Clofazimine.

When a solution of 2-aminodiphenylamine is treated with ferric chloride, a red crystalline substance precipitates that inhibits multiplication of human H37 tubercle bacilli *in vitro* and is not inactivated by human serum (1091, 1092). Its toxicity is low, its *in vivo* activity moderate. Its structure was shown to be a riminophenazine. SAR modifications selected the bis(*p*-chlorophenyl) isopropyl derivative (clofazimine) as the most interesting member of this series. Although highly active in murine tuberculosis, clofazimine is of no value in the human pulmonary disease. However, atypical mycobacteria and *M. leprae* respond well to this compound, and in clinical leprosy it is about equivalent to dapsone. The

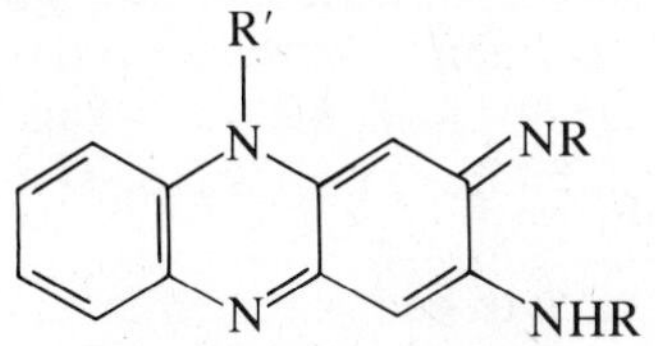

Riminophenazine: R = R' = H
Clofazimine: R = *p*-ClC$_6$H$_4$-; R' = (CH$_3$)$_2$CH-

riminophenazines depend on their *p*-quinoid system for their activity; in anaerobic media, the mycobacteria reduce the quinone system. Clofazimine as well as rifampine would be drugs of choice in leprosy if they were not so expensive (1068).

3.18. ANTIBIOTICS

3.18.1. Introduction

An estimated 4200 or more antibiotics were known around 1980, their structures covering the whole range of organic chemistry from simple compounds to compounds of extreme complexity. It is therefore no wonder that their classification leaves much to be desired. Antibiotics are defined as chemicals produced by microorganisms that inhibit the "growth" (multiplication) of other organisms. Synthetic compounds that act the same way are often not classified as antibiotics. But what about chloramphenicol, originally isolated from several streptomycetes (222) but soon synthetized (223) and now a wholly synthetic broad-spectrum antibiotic all over the world? What about the many penicillins that are synthesized and manufactured by acylation of (biosynthetic) 6-aminopenicillanic acid? Some purely synthetic compounds inhibit the life processes of certain microbes by the same biochemical mechanisms as do materials obtained from microbial fermentation processes.

Quite a few higher plants also produce antimicrobial substances, and these ought to be included in the definition of antibiotics. Finally, several antibiotics have primarily antineoplastic properties; should their antibiotic status be abrogated since neoplasms are not microorganisms?

The current definition of antibiotics was proposed by Selman Waksman in 1942 and, like the nonmetric system of weights and distances in the United States, it has not been changed to a more logical format. Before 1942 antimicrobial substances biosynthesized from microbes were called toxins, lysins, mycoins, and so on. These impure products had been used in folk medicine for over 2500 years. For example, the Chinese applied a moldy curd made from soybean decoction to boils and carbuncles, the mold being an apparent source of antibiotics. In 1877, Pasteur and Joubert (1093) observed that cultures of anthrax bacilli could be killed by mixing with "common bacteria" and that laboratory animals could be protected from the effects of the deadly anthrax bacilli if "common bacteria" were injected simultaneously. Shortly thereafter Garré noted the inhibition of *Staphylococcus* in gelatin cultures by other bacteria. In Germany, an impure product from *Pseudomonas aeruginosa* was used clinically from 1890 to 1905. It contained an enzyme, pyocyanase, and an antibiotic, pyocyanine, whose structure was not elucidated until 1929 (1094).

Medicinal chemists working on antibiotics have to be interested in methods of structure elucidation and synthesis of natural products. The frequently encountered complexity and uniqueness of antibiotics requires greater agility in organic chemistry than perhaps in any other field of biologically active substances. For these reasons, investigators of antibiotics chemistry must be able to meet the challenge of unexpected structural problems and synthetic difficulties.

3.18.2. β-Lactam Antibiotics

3.18.2a. Penicillins. In 1929 the British bacteriologist Alexander Fleming forgot a petri dish containing a culture of *Staphylococcus aureus* on the windowsill of his laboratory, and a green mold grew on it (174). From there on, a series of chance events occurred that has been described by Beveridge and is here repeated verbatim (1095). Although it has no direct bearing on medicinal chemistry, it does show the role unexpected occurrences play in drug discovery.

> It had been assumed that what Fleming saw was a not uncommon phenomenon that others must have seen and ignored. The truth is quite different. Fleming himself tried many times to "rediscover" penicillin, but never succeeded. Ordinarily penicillin does not lyse staphylococcal colonies; it prevents them from developing, but if added after they have developed it has no apparent effect. Hare found that it was very difficult indeed to reproduce lysis of staphylococcal colonies, even when the cultures were deliberately inoculated with a known penicillin–producing mould. Only after a long series of experiments was he

able to discover the very special conditions required to produce the phenomenon. He showed that the rare oddity that Fleming observed could only have arisen as a result of a chain of events that produced just the right, unusual conditions. Fleming must have inoculated a culture dish which at the same time became accidentally contaminated with a mould spore, and instead of putting it in the incubator as is normal practice, it seems he must have got it mixed up with old cultures and inadvertently left it on the bench. The dish lay there during Fleming's vacation and it so happened that there was an unseasonably cold snap in the weather that provided the particular temperature required for the mould and the staphylococci to grow slowly and produce lysis of the staphylococci colonies near the mould colony. Hare remarks that the odds against this combination of unlikely events happening just by accident, which of course it did, must be astronomical.

Yet another extraordinary circumstance was that the particular strain of the mould on Fleming's culture was a good penicillin producer although most strains of that mould, *Penicillium,* produce no penicillin.

The mold was identified as *Penicillium notatum* (174), and the chemical suspected as the lysing principle was named penicillin. Fleming suggested that the brown powdery material from the aqueous extractions of the fungus might be used as a topical antiseptic. It took eleven more years for the systemic value of penicillin to be fully appreciated in animals (1096) and in humans (1097).

One reason for this delay was probably the emergence of the sulfonamide antibacterials, which, from 1935 on, intensely centered the attention of chemotherapists on these new drugs. The clinical systemic importance of penicillin was not realized until two years after the outbreak of World War II, and the pharmacology, production, and clinical application of penicillin was not revealed until after the termination of the hostilities, when the principal investigators published an autobiographical review of their researches (1098). The chemical studies were published separately (1099). The best organic chemists and x-ray crystallographers combined their efforts to elucidate the unexpected β-lactam structure and to attempt syntheses of the several penicillins found in the natural mixture.

The secret of the superiority of penicillin to all previously known antibacterials had been kept so well that an Allied scientific team interrogating German investigators in the field of chemotherapy received an astonishingly naive reply. The Germans said that they had heard rumors about penicillin but firmly believed that this new agent had been announced merely for commercial reasons to compete with the sulfonamide drugs.

The manufacture of penicillin had to overcome some unexpected hurdles; these had to be met by a combination of mycologists, fermentation experts, chemists, and chemical engineers. The original *Penicillium notatum* gave only low yields of penicillin; it was replaced by *P. chrysogenum* (1100–1102), which had been cultured from a mold growing on a grapefruit in a market in Peoria, Illinois. The original culture medium, yeast extract, was replaced by cornsteep liquor, with trace metal salts and side-chain precursors added to improve the yields. The earlier surface cultures of the organism in hundreds of Erlenmeyer flasks worked well, but when production was stepped up to big vat containers, yields of penicillin dwindled. It was recognized that the small-scale units permitted proper aeration of the fungus, whereas oxygen access to lower layers of the mycelium was inadequate in the large vessels. The mycelia therefore, had to be broken up by stirring, but the tough and sticky mass resisted all such attempts. Finally, a stirrer shaped like a ship's propeller was tried ("Queen Mary stirrer"), and this provided proper aeration in the fermentation process. Coghill has pointed out that these operations, quite logical in retrospect, were revolutionary in pharmaceutical manufacturing.

The correct structure of penicillin was elucidated in 1943 by groups around Sir Robert Robinson at Oxford and Karl Folkers at Merck (1098, page 667; 1099, page 5) and was corroborated by x-ray diffraction spectra. The difficulty in executing these studies was aggravated by the need to separate the more than seven naturally occurring penicillins and by the lability of the compounds in chemical manipulations. Several total syntheses have been described, but none of them, even the most rewarding (1103), have been able to compete commercially with fermentation methods.

The addition of appropriate side chain precursors to the fermentation mixtures led to a number of biosynthetic penicillins (1104), but success here was limited. The only material of this group that is still prescribed is phenoxymethylpenicillin (penicillin V), which was described in 1948 (1104) and marketed in 1953. It has an improved oral absorption because of its greater stability to acids.

It had been hoped for a long time that the penicillin "nucleus," called 6-aminopenicillanic acid, could be isolated on a large scale and become the starting material for molecular modification by acylation. The first preparation of this substance was by enzymatic deacylation of benzylpenicillin

6-Aminopenicillanic acid   Penicillin

(1105), but it became more widely known through observations of Batchelor (1106) that the "nucleus" occurs in fermentation mixtures to which no side chain precursor had been added. Best results were obtained by enzymatic cleavage of the side chain using penicillin acylases (1107) or by alkaline hydrolysis under special conditions (1108). More than 10,000 amides, ureas, imides and many other types of acyl derivatives now became available for exploration. Among these were compounds with increased acid stability and resistance to penicillinase, and broad-spectrum semisynthetic penicillins.

They have been reviewed by Hoover and Dunn (1109). A typical example is ampicillin (D-α-aminobenzylpenicillin), the second most widely used penicillin in clinical practice (1110, 1111).

The mechanism of action of the penicillins involves interference with the biosynthesis of bacterial cell walls, which leads to lysis of the microbial cell. Cells of the mammalian hosts have no cell walls that can be damaged in this manner. The constituents of bacterial cell walls include a network of lipopolysaccharide, teichoic acid, and peptidoglycan (1112). The last is composed of uridine nucleotides containing muramic acid and variable polypeptides that serve as cell wall precursors (1113, 1114). The β-lactam antibiotics interrupt the action of a membrane-bound D-alanine transpeptidase which attaches the pentaglycine bridge of the polypeptide network to the penultimate D-alanine of a neighboring pentapeptide chain with elimination of a terminal D-alanine (1115).

3.18.2b. Cephalosporins. An almost identical course of events as in the penicillin series led to the introduction and molecular modification of the cephalosporins, with the advantage that the pattern for these studies had been set in the work on penicillins. A Sardinian bacteriologist, G. Brotzu (1116), observed the production of antibiotics with broad-spectrum activity by a *Celhalosporium* species found in the Cagliari sewage. A similar material (cephalosporin C) was obtained in minute amounts from acid degradation experiments of penicillin (1117). The chemistry of these antibiotics, produced in adequate quantities from Brotzu's cultures by the Oxford penicillin team (1098), showed a considerable similarity to that of penicillin; a β-lactam ring was fused, however, to a 1,3-thiazine ring instead of a thiazole system, and the substituents in the thiazine moiety were unlike those encountered in the preceding penicillin studies. Cephalosporin C was one of the first found to occur in nature; it has a D-α-aminoadipoyl side chain attached as an amide to 7-aminocephalosporanic acid (1118). The thiazine ring fusion renders the β-lactam system more resistant, but not immune, to opening by both acids and penicillinase. For chemotherapeutic purposes, side chain modification seemed imperative, and the production of 7-aminocephalosporanic acid became a key step in this sequence. The earliest practical procedure, although cumbersome, was reported in 1962 (1119). The most widely adopted method converts the side chain amide groups to an imino chloride after protecting the carboxyl optimally as the trimethylsilyl ester (1120). The imino chloride reacts with alcohols to form imino ethers which hydrolyze in water to 7-aminocephalosporanic acid; the silyl ester group is cleaved during the workup (1121). For the reacylation of the 7-amino group in the preparation of semisynthetic cephamycins, additional steps of group protection and deprotection have been elaborated (1122, 1123).

Since the cephalosporins and cephamycins contain a functional ester group in position 3 (except where this has been replaced by CH_3 or SCH_3), these side chains have also been varied by molecular modification. For example, displacements of acetate have been carried out by pyridinium (1124) and sulfur nucleophiles.

The greater stability of cephalosporins to (gastric) acid has presented the puzzling problem why most of these compounds are orally inactive. Therefore, the efficient oral absorption of cephalexin and cephradine was unexpected. The apparent structural requirements for oral absorption have been reviewed (1109). Cephalotin, still popular with physicians, is regarded as "cephalosporin" in the same way benzylpenicillin is "penicillin."

Another example of the many molecular modifications tried in this series could be chosen from the cephamycins, which carry a 7α-methoxyl group; cefotixin (1122) is a member of this structural group under clinical investigation.

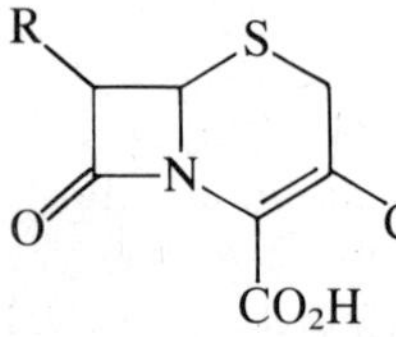

Cephalexin (1125, 1126): $R = C_6H_5\overset{(D)}{\underset{NH_2}{C}}HCONH-$

Cephadrin (1127, 1128): $R = \text{(phenyl)}-\underset{NH_2}{C}HCONH-$

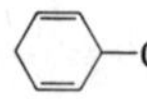

Cephalotin (1129, 1130): $R = \text{(thienyl)}-CH_2CONH-$

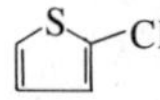

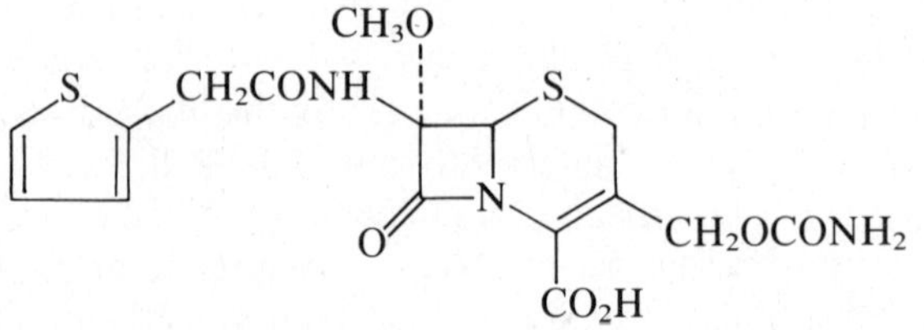

Cefotixin

The tremendous effort of modifying extensively and ingeniously the side chains and the nuclei of the fused-ring β-lactam antibiotics has been supported by the stunning market for such drugs. Without this commercial incentive, most of the work might never have been done. Physicians can now choose from more than a dozen specific compounds, with oral or parenteral activity, greater resistance to chemical and enzymatic attack, longer duration of action, negligible toxicity and allergenicity, and an activity spectrum that encompasses gram-negative organisms previously impervious to chemotherapy. From a medicinal–chemical point of view, several of these innovations, though of some practical pharmacologic advantage, constitute only extensions of previous fundamental discoveries. Therefore, only two additional structural types will be listed here. None of these compounds has reached clinical utility but, as in the earlier case of the not very potent cephalosporin C, may serve as starting points for more meaningful molecular modification.

ANTIBIOTICS

3.18.2c. Other β-Lactam Antibiotics. The first group of these compounds is represented by thienamycin and clavulanic acid, in which the β-lactam ring is condensed with different rings (pyrrolideine and oxazolidine, respectively) and the side chains are either absent or totally different from those standard in penicillins and cephalosporins.

Thienamycin (1131) is a broad-spectrum antibacterial agent but with a short half-life although resistant to β-lactamases. Clavulanic acid (1132), produced by *Streptomyces clavuligerus,* is a substance of low potency but it is a powerful inhibitor of β-lactamases; in combination with β-lactamase-sensitive antibiotics, it protects the latter from enzymatic destruction.

The second type worth mentioning includes β-lactam antibiotics in which the β-lactam ring no longer is condensed with other rings. Examples are the nocardicins (1133) and a low-toxicity compound, az-threonam, which can be synthesized from threonine. The sulfate group does not distort the β-lactam ring but is sufficiently electronegative to activate it (1134).

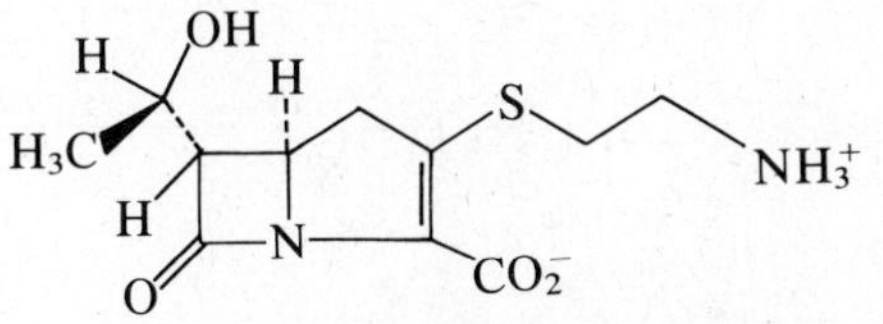

Thienamycin

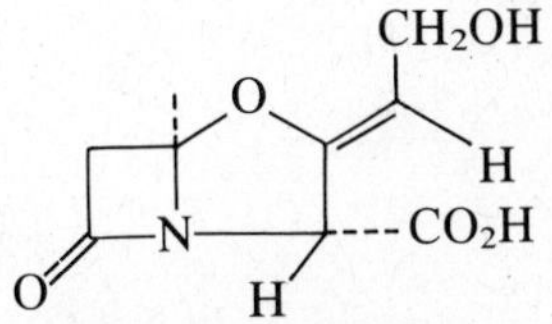

Clavulanic acid

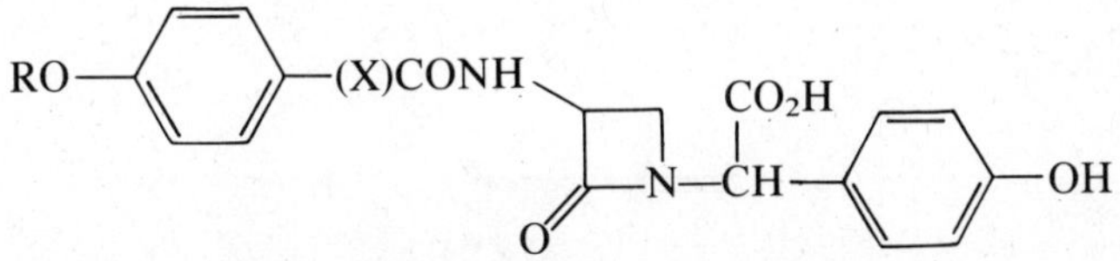

Nocardicins

X = C=NOH, CHNH$_2$, C=O

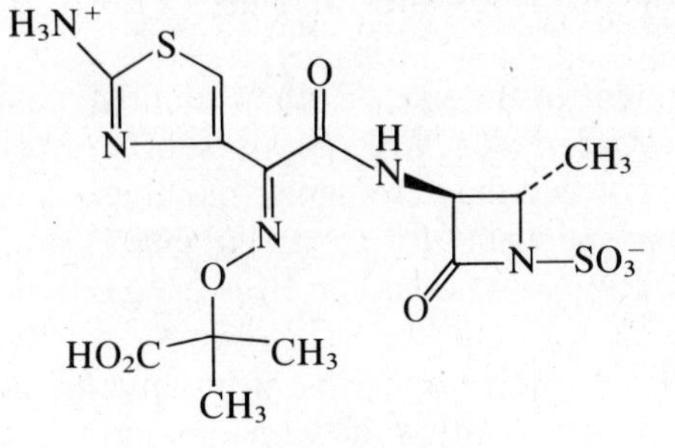

Az-threonam

3.18.3. Aminoglycoside Antibiotics

After penicillin, the next chemotherapeutically important antibiotic to be discovered was streptomycin, isolated in 1944 from the fermentation of *Streptomyces griseus* (179). Chemically, it is an aminoglycoside containing an aldehyde group; it is composed of three glycosidically linked units: streptidine, streptose, and *N*-methylglucosamine (220). It has activity against a variety of gram-positive and gram-negative organisms but its most important property is its inhibition of the life processes of *Mycobacterium tuberculosis* H37 Rv. It was the first really effective drug for human tuberculosis, a property also shared by its dihydro derivative (CH_2OH instead of CHO). Both compounds unfortunately damage branches of the eighth cranial nerve, and this toxicity has severely limited the chemotherapeutic use of these antibiotics.

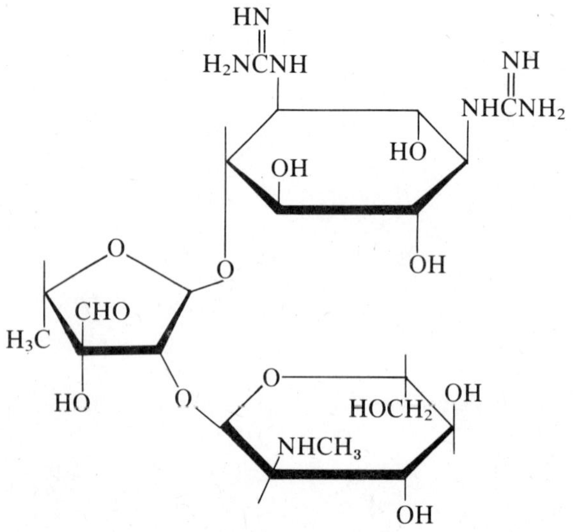

Streptomycin

Streptomycin inhibits protein biosynthesis in intact bacteria by acting on a protein called P10 that occurs in the 30S ribosomes. It causes a misreading of the genetic code (1135).

Ototoxicity has been an impediment in the use of other aminoglycoside antibiotics as well. *Streptomyces fradiae* was Waksman's source of neomycin (1136). This material consists of two diastereomers, neomycin B and C (1137). Although their antibacterial spectrum *in vitro* and *in vivo* is similar to that of streptomycin, the complex (mostly neomycin B) is used only topically.

A related group of antibacterial antibiotics are the paromomycins, discovered later (1138). Their activity resembles that of the neomycins, but in addition, *Entamoeba histolytica* is affected. They are used in amebic dys-

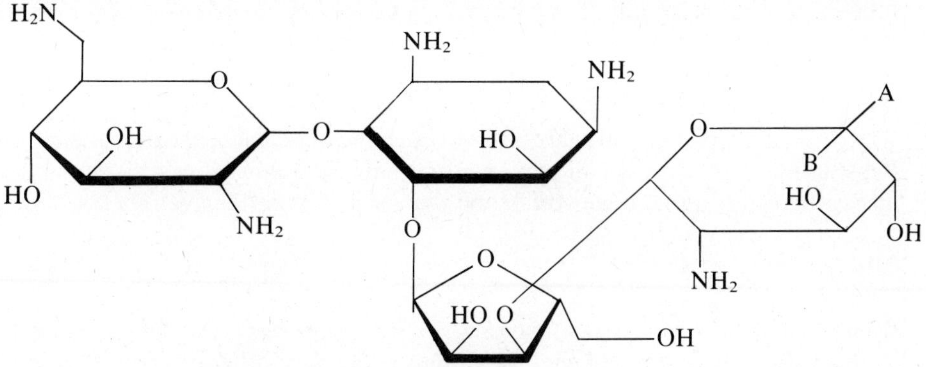

ANTIBIOTICS

Neomycin B: A = H, B = CH$_2$NH$_2$
Neomycin C: A = CH$_2$NH$_2$, B = H

entery, shigellosis, and salmonellosis. Another related group are the kanamycins (1139–1141) which are useful against staphylococcal and resistant urinary infections. The gentamicins, also with aminoglycoside structures (1142, 1143), are particularly valuable against *Pseudomonas* and *Proteus* infections. Aminoglycoside antibiotics such as gentamycin, tobramycin, and kanamycin are detoxified by sensitive bacterial enzymes which transfer adenyl residues from ATP to the antibiotic molecules. These enzymes are inhibited by 7-hydroxytropolone, and this not-yet perfect "lead" compound thereby decreases bacterial resistance to the aminoglycosides (1144). This cycloheptatriene derivative was not discovered by logic but was found to potentiate aminoglycosides during a systematic screening program.

3.18.4. Peptide Antibiotics

In 1939–1941, Dubos and Hotchkiss (1145, 1146) isolated a peptide antibiotic from *Bacillus brevis* and named it tyrothricin. It was a mixture of gramicidin, in itself a mixture of three related peptides (1147), and a cyclic peptide, tyrocidin. Gramicidin, a linear peptide, has been used as a topical agent against gram-negative cocci. It is not to be confused with gramicidin S, isolated in 1944 from a culture of *B. brevis* in the USSR (1148). Gramicidin S (S stands for Soviet) is a cyclodecapeptide studied by the British team that developed column chromatography (1149). It has been synthesized (1150). Gramicidin and gramicidin S were among the first peptides and conjugated peptides with antibiotic activity. Others with early clinically useful bacteriostatic properties are bacitracin (1151) and the polymyxins (1152), which counteract gram-negative bacterial infections.

A fascinating aspect of the chemistry of cyclic peptide antibiotics is the mode of action of these compounds as illustrated for valinomycin. This compound, obtained from *S. fulvissimus,* is a cyclododecadepsipeptide whose 36-membered ring also contains four nonpeptide lactone groups (1153, 1154).

It chelates potassium ions in a cavity of its molecule, changing its own conformation in the process. Since energy is released in these reactions, valinomycin constitutes an "electrogene pump." Its mode of action has been reviewed (1155).

Not all microbial peptides are antibiotics in the classical sense of the term. Pepstatin (1156), a microbial pentapeptide, inhibits renin, pepsin, isorenins and cathepsin D and reduces the blood pressure of hypertensive rats (1157).

3.18.5. Tetracyclines

Two other types of products from *Streptomyces* species were among those in the 1940's that heralded the age of antibiotic chemotherapy. Both were isolated in 1947. *S. aureofaciens* yielded chlorotetracycline (1158), soon to be followed by 5-hydroxytetracycline from *S. rimosus* (1159). Tetracycline, the parent antibiotic of this series, was first obtained by hydrogenolysis of 7-chlorotetracycline, and soon thereafter by fermentation in low-chloride media (1160). Tetracycline is now one of the most widely prescribed antibiotics; it is active against gram-positive bacteria, spirochetes, rickettsias, and some large viruses. Gram-negative bacteria are somewhat less susceptible and *Proteus* and *Pseudomonas aeruginosa* do not usually respond to tetracyclines. Another five chemically modified tetracyclines are also available commercially.

The virtually simultaneous isolation and structure elucidation of several tetracyclines in a number of industrial laboratories, aided by eminent consultants in universities, led to complex and acrimonious patent litigations that preoccupied courts of law for many years.

3.18.6. Chloramphenicol

The isolation of another broad-spectrum antibiotic, chloramphenicol, was also accomplished in 1947 (222, 223). This antibiotic has a simple, unorthodox structure that harbored several surprises, an aromatic nitro group and a dichloroacetyl group. Its synthesis (223) was an early triumph of natural-products chemistry in the antibiotics field. The commercial drug is the product of improved chemical syntheses and is no longer manufactured by bio-

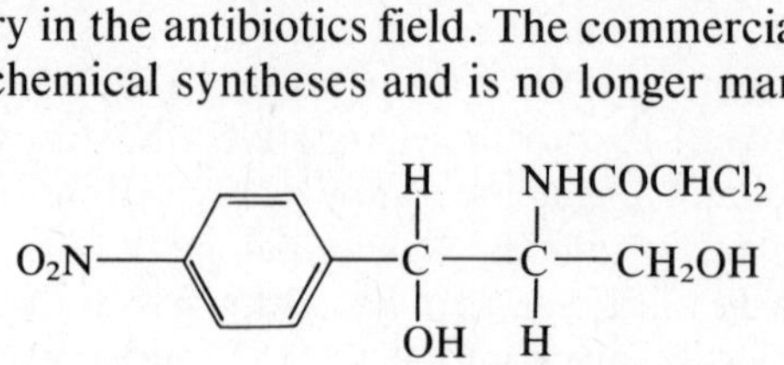

$$O_2N-\underset{}{\bigcirc}-\underset{\underset{OH}{|}}{\overset{\overset{H}{|}}{C}}-\underset{\underset{H}{|}}{\overset{\overset{NHCOCHCl_2}{|}}{C}}-CH_2OH$$

Chloramphenicol

synthesis from *Streptomyces*. The palmitate ester is the common, relatively tasteless, form for oral administration. Chloramphenicol is used most widely for rickettsial infections, typhus, and typhoid fever. It is safe when treatment is monitored for bone-marrow depression which, when unchecked, may lead to dose-related aplastic anemia. Chloramphenicol inhibits protein synthesis by interfering with the continuation of amino acid incorporation. This reaction is stereospecific.

3.18.7. Other Antibiotics

By the early 1950's, antibiotics research had settled into a standardized and predictable pattern. The flush of excitement that had accompanied the development of the early penicillins, the tetracyclines, chloramphenicol, and streptomycin gave way to an intense but stylized search for new antibiotics with broader activity spectra, special activities—especially toward hard-to-treat gram-negative microbes—and ever lower toxicity. The general stages of these searches were very similar in most cases.

Randomly collected soil samples from one's backyard or from exotic sources (offering the opportunity of travel to laboratory-weary scientists) were extracted with buffered aqueous solutions, which were then cleaned and added to the center of circular culture plates that contained wedges of representative pathogens. As the soil extract diffused, organisms sensitive to the extract would be inhibited, presumably by its antibiotic content. Without knowing anything about the chemistry or biology of the extracted soil inhabitants, one could soon judge whether the particular soil sample deserved further investigation.

If these inhibition tests looked promising, the soil sample was cultured and the antibiotic-producing soil organism isolated, characterized, and fermented. The antibiotic (or antibiotics) was purified by chromatography and subjected to structural analysis, aided by spectroscopic methods. If the structure could be ascertained and was not too complicated, it was confirmed by synthesis, often calling into play an arsenal of ingenious methods of organic chemistry. If synthesis appeared unfeasible, adequate supplies of the antibiotic were made by stepped-up fermentation. These fermentation procedures could be improved in many cases to give better yields of pure products by addition of precursor compounds, trace metal salts, and other chemicals useful in the biosynthesis of the antibiotic.

With pure material in hand, the pharmacological, microbiological and clinical study of the antibiotic followed the same prescribed pattern as that of any other natural or synthetic agent. From a historical point of view, much of the glamour of earlier antibiotics studies was dimmed by the sameness of approach and execution. This does not mean that unexpected structural or biological problems did not occur to tax the imagination of the research personnel.

Only a few of the hundreds of antibiotics researched have reached clinical utility. The vast majority are too toxic to be of medicinal value. Among the antibiotics discovered and developed by these schemes are the macrocyclic antibiotics such as erythromycin from *S. erythreus* (1161–1164), oleandomycin from *S. antibioticus* (1165–1167), and carbomycin from *S. halstedii* (1168–1170); the antifungal nystatin from *S. noursei* (1171), amphotericin B from *S. nodosus* (1172), cycloserine (1173, 1174); and griseofulvin from *Penicillium griseofulvum* (1175, 1176); rifamicin (1177, 1178); lincomycin from *S. lincolnensis* (1179); butirosin (1180); lividomycin (1181, 1182) and other newer aminoglycosides; the many peptide antibiotics—and other chemically interesting compounds which, because of toxic side effects, have remained laboratory curiosities. For a review, see ref. 1183.

In cases of highly complex structures such as that of erythromycin the total synthesis has opened avenues to analogs not otherwise obtainable (1184). In such cases, the chemistry of natural products and medicinal chemistry melt into one.

3.18.8. Antitumor Antibiotics

A number of antibiotics not only possess antimicrobial activity but also inhibit the reproduction and proliferation of neoplasms by interfering with their nuclear events. Like most anticancer drugs, these antibiotics have to be applied in high concentration to be effective and therefore inevitably are toxic to normal body functions, especially on prolonged administration.

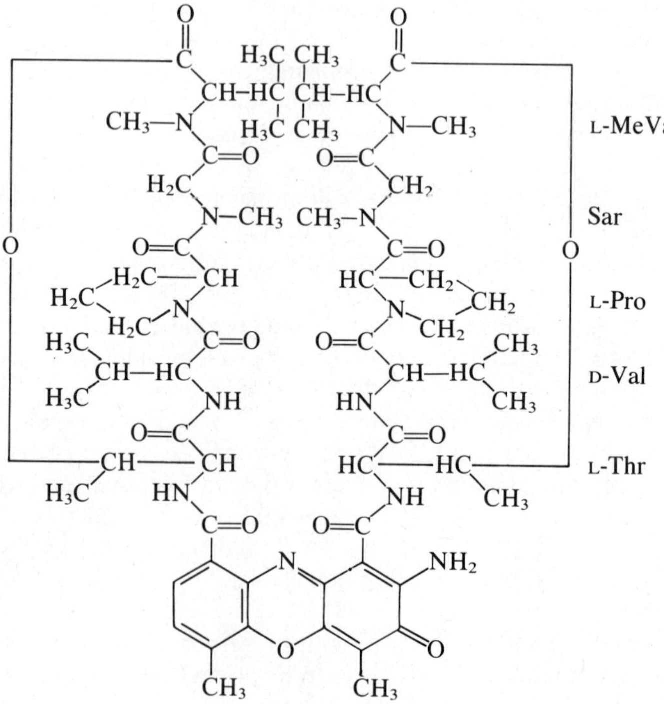

Actinomycins

ANTIBIOTICS

Nevertheless, some of them are of great value in prolonging life and making tumors regress in certain types of neoplastic diseases.

The earliest of these substances were the actinomycins (1185), deeply colored chromopeptides isolated from various species of *Streptomyces*. Indeed, they have turned up time and again in the screening of soil samples and have disappointed investigators who thought they had found a new antibiotic. Their structure contains a chromophore, actinocin (3-amino-1,8-dimethyl-2-phenoxazone-4,5-dicarboxylic acid), carrying two pentapeptides as amides of its two carboxyl groups (1186–1188). Actinomycins are highly effective in the treatment of Wilms' tumor, trophoplastic tumors, and rhabdonyosarcoma. It seemed hard to believe that Nature would have created the members with the best chemotherapeutic index, that is, the lowest toxicity, in this group of highly toxic substances, but extensive molecular modification of the chromophore nucleus and the peptide chains has failed to turn up therapeutically more suitable antitumor drugs. The actinomycins complex strongly with DNA, and several features of this binding mechanism have been elucidated (1189, 1190).

Other antineoplastic antibiotics have reached clinical trial and/or utility. Daunomycin (daunorubicin) from *S. peucitus* (1191, 1192) has been synthesized (1193). Adriamycin (doxorubicin), fermented from *S. peucitus* var. *caesius* (1194), has also been synthesized (1195) and has been introduced for the chemotherapy of leukemias and solid tumors (1196). Carcinomycin I, from *Actinomadura carminata,* is another anthracycline antibiotic with antileukemia activity and less severe cardiotoxicity (1197). Carzinophilin, from *S. sahachiroi* (1198) is rather toxic but is used in Japan for various malignancies.

From a chemotherapeutic point of view, the bleomycins, produced by *S. verticillatus,* are the most promising antitumor antibiotics now under study.

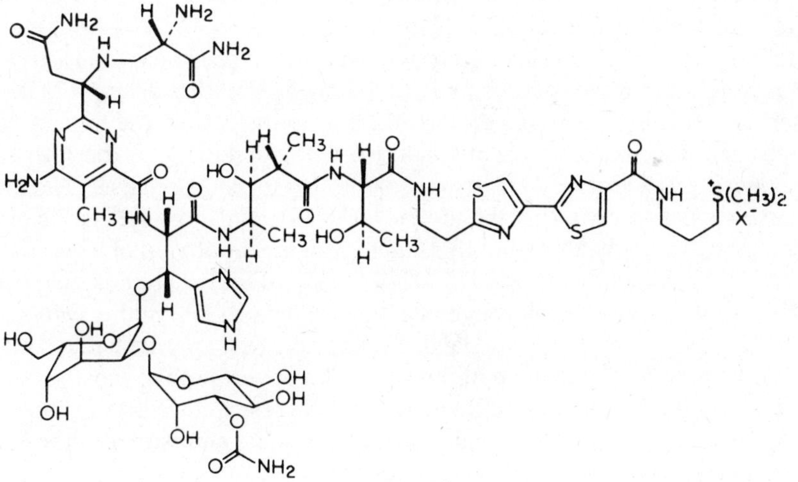

Bleomycin A_2

Elucidation of their complex chemical structure has undergone a number of revisions since 1964 but appears well established now (1199). They act against a variety of human tumors, especially squamous cell carcinomas, and have minimal myelosuppressive and immunosuppressive properties. Nine naturally occurring bleomycins have been isolated, and the addition of various amines to the fermentation beers has produced over 100 artificial bleomycins. They cause chain scission (nicking) of DNA molecules (1200). Bleomycin A_2 has been synthesized (1200 a).

As mentioned at the beginning of section 3.1.8, not all antibiotics are produced from microbes. For example, the African plant *Maytenus ovatus* Loes in the Celastraceae (staff-tree) family contains an *ansa* macrolide antibiotic, maytansine (1201), which is structurally related to the classical antibiotics rifamycins and streptovaricins. Maytansine is a promising antitumor agent (see Section 3.21).

3.19. ANTIPROTOZOAL AGENTS

3.19.1. Antitrypanosomal Drugs

Trypanosomiasis is a group of diseases of humans and animals caused by infection with protozoa of the genus *Trypanosoma*. It is transmitted by various blood-sucking insects, especially tsetse flies (genus *Glossina*). In Africa, the most pernicious types of trypanosomiasis are human sleeping sickness, nagana (Zulu for weakness), and surra disease, and other forms affecting wild and domestic animals. In South America, the human infection is called Chagas' disease, caused by *T. cruzi* and transmitted by reduviid bugs. The many forms of trypanosomiasis, their causative trypanosomes and vectors, the clinical manifestations and the enormous economic dislocations they cause have been reviewed (1202).

Most of the "wild" strains of trypanosomes have been adapted to greater virulence in laboratory rodents and can be used in testing drugs (1203).

Because of the devastating effects in the tropics on both the human populations and their domestic farm stock, the colonial nations of the 19th century concentrated research efforts on combatting trypanosomiasis. The most effective way of interrupting the life cycles of the trypanosomes is the eradication of the insect vectors, but this had to wait for the advent of insecticides such as chlorophenothane (DDT) (1204, 1205), dieldrin (1206), chlordane (1207), and gamma-hexachlorocyclohexane (lindane, 1208) in the 1940's and thereafter. Therefore, trypanocidal drugs were the first approach to the problem and have largely remained in this role. The poverty of the most severely affected regions has contributed to the hesitant research efforts in the field of tropical diseases because the large pharmaceutical companies in the West and in Japan could not expect to recover their investment in such ventures.

3.19.1a. Dyestuffs and Analogs. The history of trypanocidal drugs goes back to the early experiments of Ehrlich, who extended the selective staining of plasmodia with methylene blue to the staining of other fairly large microscopic organisms (56, 57). Ehrlich examined more than 100 cotton-substantive dyes related to Congo red in mice infected with *Trypanosoma equinum* and found a little activity in benzopurpurin ("nagana red"); introduction of a fifth sulfonate group (trypan red) produced a curative as well

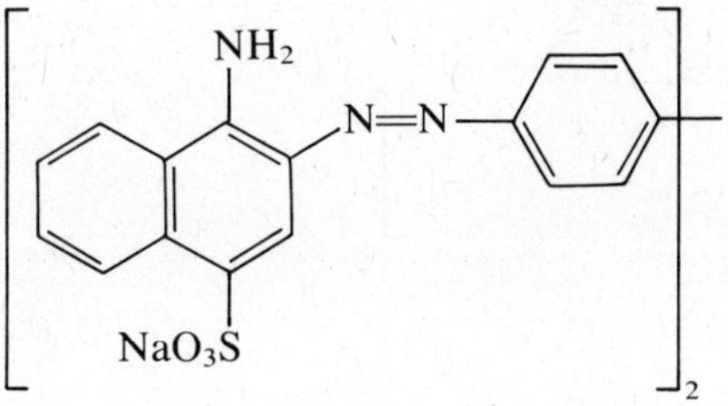

Congo red

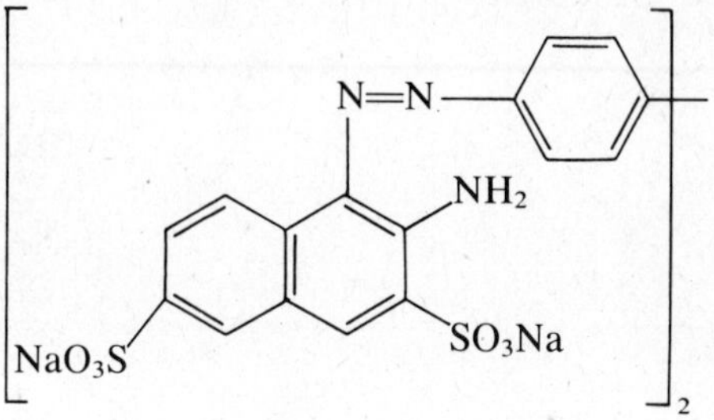

Benzopurpurin

as prophylactic dye (57). Keeping intact the 3,6-disulfonate arrangement in the naphthalene nuclei, trypan blue (62) and afridol violet (60) were developed; these dyes showed greater trypanocidal potency *in vivo*. Their syntheses in the Bayer works in Germany proceeded by standard azo coupling reactions but were not recorded until many years later, by others (1209, 1210). Following Ehrlich's line of thought that the trypanocidal action was

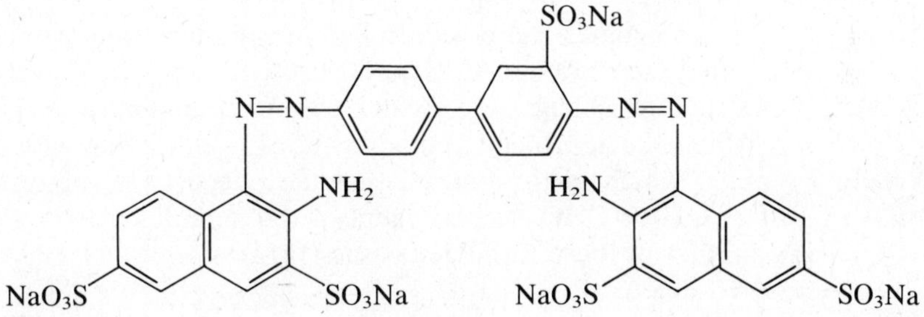

Trypan red

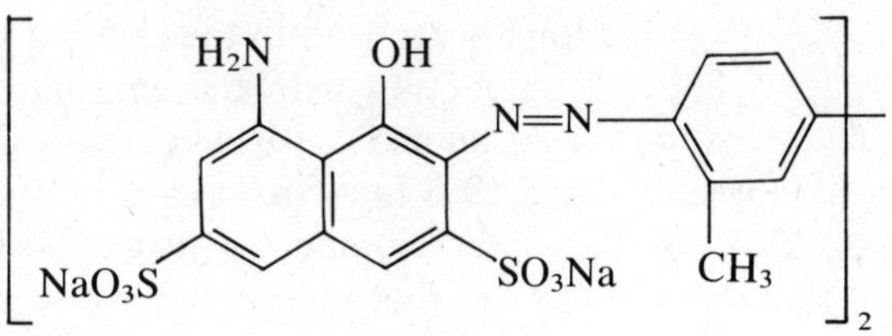

Trypan blue

caused by a direct binding of the dyes to biochemical receptors of the parasite by way of the functional groups (haptens) of the dye molecules, it appeared appropriate to lessen the staining properties by deletion or replacement of the chromophoric azo linkages. The haptenic properties should still cause similar biological activities. Indeed the colorless urea analog of trypan red in which the azo groups had been replaced by (we would say, bioisosteric) amide groups was active, and this activity was enhanced in *m*-aminobenzoyl in lieu of *p*-aminobenzoyl derivatives. The pharmacologically most suitable

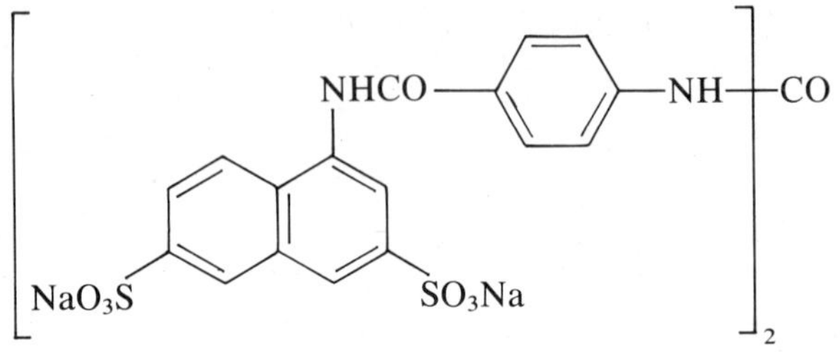

Colorless analog of afridol violet

agent in this series was suramin (Bayer 205) synthesized in 1917 by O. Dresel and R. Kothe (907). It is effective in acute cases of African sleeping sickness (1211). The structure of this compound was not revealed by the Bayer Company, but Ernest Fourneau deduced it from the patent literature and by independent synthesis (Compound 309F) (1212). A very similar instance occurred 10 years later when Pasteur Institute chemists again cracked an unpublished chemotherapeutic structure of the Bayer works, namely, Prontosil.

A number of trypanosomes can be controlled by suramin. Interestingly, in a molecule as large as suramin ($C_{51}H_{34}N_4Na_6O_{23}S_6$, molecular weight 1429.21), minor structural changes greatly decrease or even abolish trypanocidal activity. Among such prohibited modifications is tampering with the two methyl groups, which might indicate that they need to be activated oxidatively to CH_2OH or CHO. Replacement of the naphthalene rings by other nuclei is another of these forbidden cases (1213).

3.19.1b. Arsenicals. The use of inorganic compounds of arsenic dates back to Hippocrates; medieval alchemists worked with them, and Paracelsus (ca. A.D. 1520) described them in his writings. Fowler's solution (potassium arsenate) and orpiment (As_2S_3) had long been used in the tanning industry, as insecticides on flypaper, and in conjunction with the earliest dyestuffs for the treatment of trypanosomiasis in animals. Tartar emetic (antimony potassium tartrate) has been used since 1908 in African cattle trypanosomiasis and leishmaniasis (1214). With the development of organic chemistry and interest in chemotherapeutic dyestuffs that contain aromatic sulfonate groups, the possibility was contemplated that toxic arsenic atoms might replace sulfur, and that the resulting arylarsonates might have increased potency. The first observation in this series was that of H. Wolferstan Thomas (1215) who introduced Atoxyl into the therapy of clinical trypanosomiasis. Thomas misinterpreted the structure of the drug, thinking that it was arsenanilide, $C_6H_5NHAsO(OH)_2$, but this view was corrected by Ehrlich and Bertheim (1216), who identified Atoxyl as p-aminobenzenearsonate, p-$H_2NC_6H_4AsO(OH)_2$. Its high toxicity caused Atoxyl to be abandoned early but its structure provided the "lead" for other arylarsonates with a better therapeutic profile. The two most important of these drugs are tryparsamide, p-$H_2NCOCH_2NHC_6H_4AsO(OH)_2$ (1217), and orsanine (sodium 2-hydroxy-4-acetamidobenzenearsonate) (1218). Tryparsamide is still used extensivly in Gambian sleeping sickness. Orsanine has lost its toehold in the clinic.

Ehrlich's genius made him the first to consider drug metabolism by the host as a pathway for the activation of compounds that are active *in vivo* but inactive *in vitro* (41). This is the case for arylarsonic acids, which have to be reduced to trivalent arsenic states (—AsO or —As=As—) to become active (1219). By a similar token, some arsines are oxidized metabolically to arsenoxides, which represent the most active although frequently also the more toxic form. By changing the substituent on the amino nitrogen of aminobenzenearsonates, a considerable reduction in toxicity can be achieved. There was no leading idea involved but rather the utilization of available reactive heterocyclic halo substances. This sequence led Friedheim (1220) to Melarsen. Reduction of the arsonic acid group to the arsenoxide and condensation with dimercaprol yields melarsoprol (Mel B) (1221), an open bid for detoxification with this agent (dimercaprol, BAL). These compounds, though still toxic, are useful in late and resistant stages of human African trypanosomiases where penetration of the CNS is of the essence. In spite of the attractive working hypothesis that organic arsenicals act by inacti-

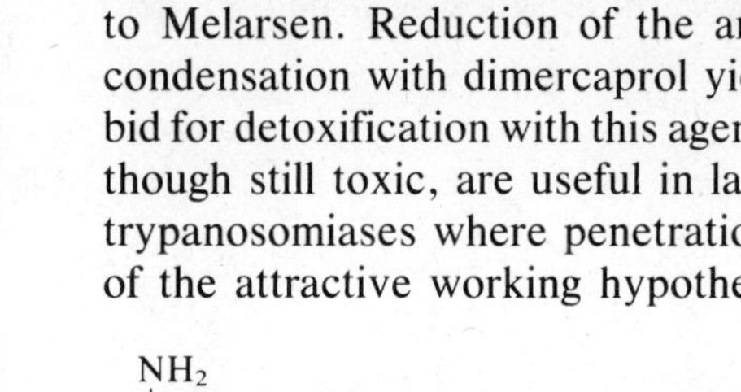

Melarsen sodium

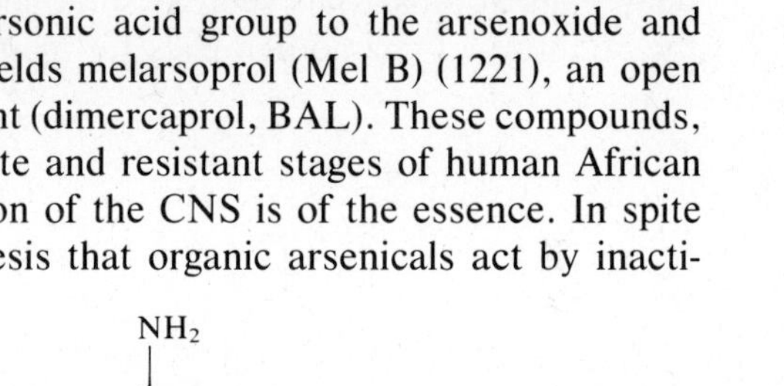

Melarsoprol

vating sulfhydryl enzymes of the parasites, the nonmetallic portion of the molecule appears to play a role in their mechanism of action, perhaps by contributing to proper pharmacokinetics of the drug.

3.19.1c. Aminoquinolines. In the course of his studies of trypanocidal dyes, Ehrlich tested acriflavine, a mixture of 3,6-diaminoacridine and its N^{10}-methochloride (1222); it is active against *T. rhodesiense* in mice (1219). Encouraged by these aminoacridine structures, Jensch (1223) tried out related ring structures, among them 4-aminoquinolines. Especially double mol-

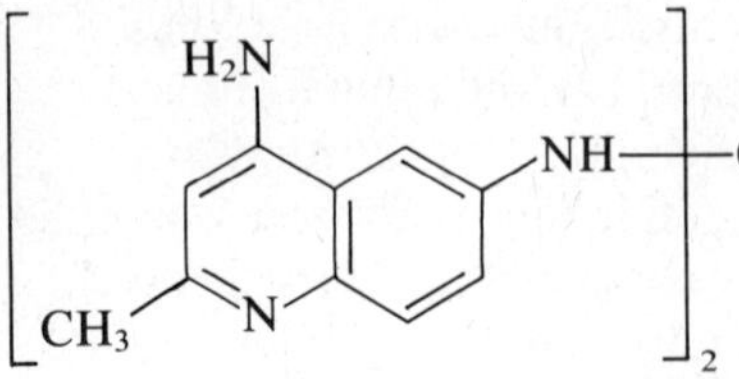

[Surfen structure]

Surfen

[Surfen C structure]

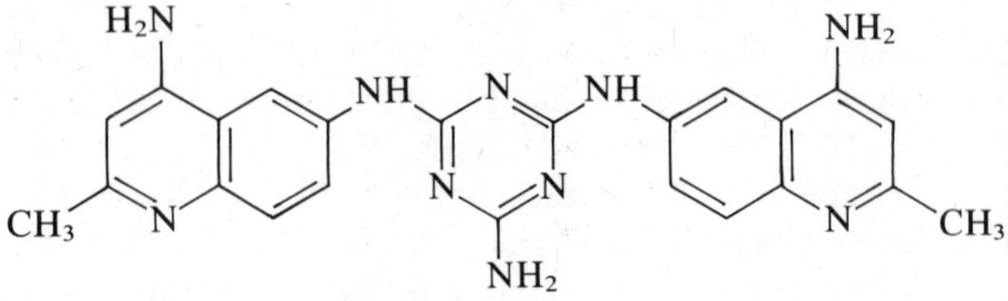

Surfen C

ecules such as surfen and surfen C had clinical and veterinary activity against *T. congolense*. When the structure of surfen C was modified systematically, both mono- and bimolecular derivatives of 4-aminoquinaldine were tested; quaternizaton of one of these rather weakly active compounds gave quinapyramine; it has become a potent standard therapeutic and prophylactic veterinary antitrypanosomal drug (1224). Note the similarity of these quinaldine derivatives with the compounds that had led to the 3-methyl-4-aminoquinoline antimalarials such as sontoquine; these types of structures held the attention of chemotherapists for many years.

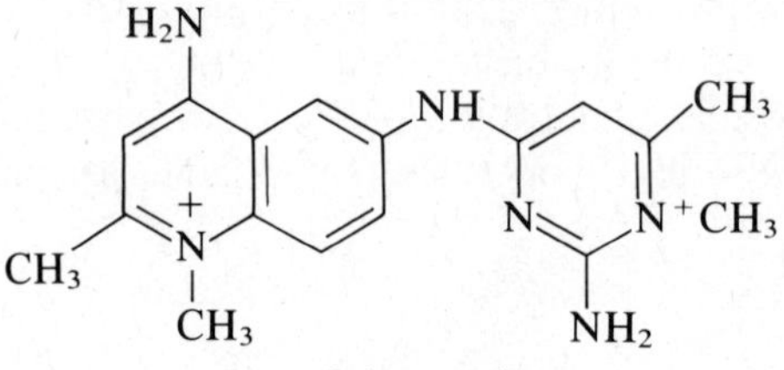

[Quinapyramine structure]

Quinapyramine

3.19.1d. Diamidines. Dissection of the formulas of bisaminoquinolines, such as surfen C, in which the aminoquinoline moieties are joined by a triazine ring also formed the basis for the choice of diminazene as the optimal molecular modification of the trypanocidal diamidines (1225, 1226). Diminazene [4,4'-diazoaminobenzamidine (diaceturate)] is a valuable veterinary agent in resistant cases of cattle trypanosomiasis (1227). Its structure

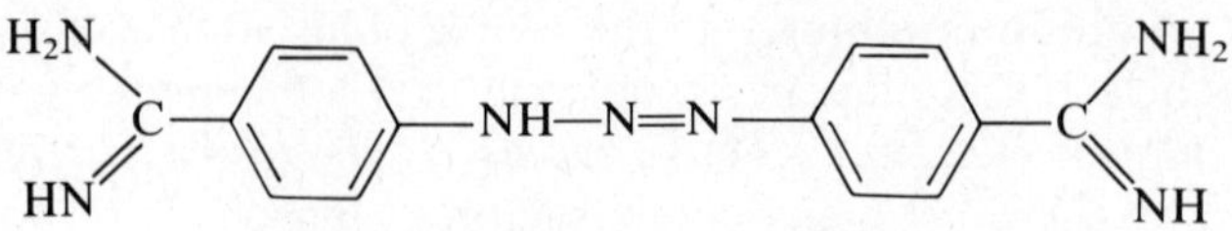

[Diminazene structure]

Diminazene

ANTIPROTOZOAL AGENTS

was also derived from an anti-*T. congolense* drug directly related to other diamidines (1226). The diamidines had been developed based on the hypothetical assumption that antitrypanosomal activity can be correlated to

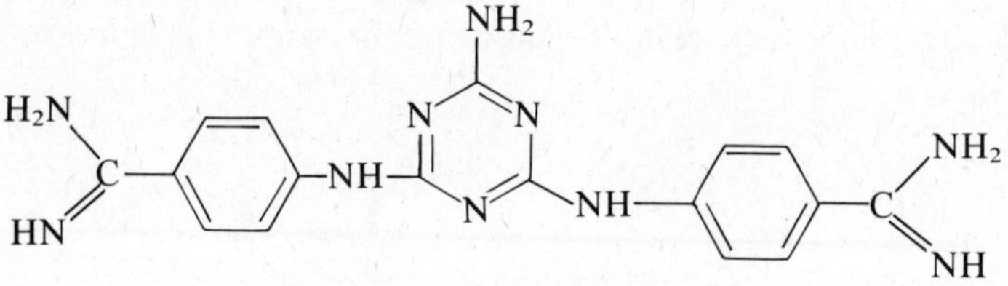

Diamidine-type antitrypanosomal drug

the high glucose requirement of the trypanosomes. More recently, blockade of glucose metabolism of trypanosomes *in vivo* has indeed been shown to lead to the temporary destruction of the parasites (1228). In any event, this doubtful "lead" was followed and numerous analogs were synthesized with

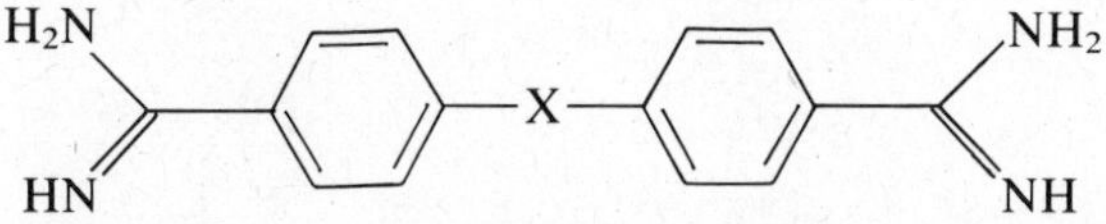

Diamidines

the general diamidine formula as shown, where X was CH_2, and so on (1229), and this actually led to diminazene described above.

3.19.1e. Quaternary Ions. The activity-enhancing effect of *N*-quaternization seen in quinapyramine offered an invitation to test the same procedure in other ring systems. Among the many examples tried, a number of

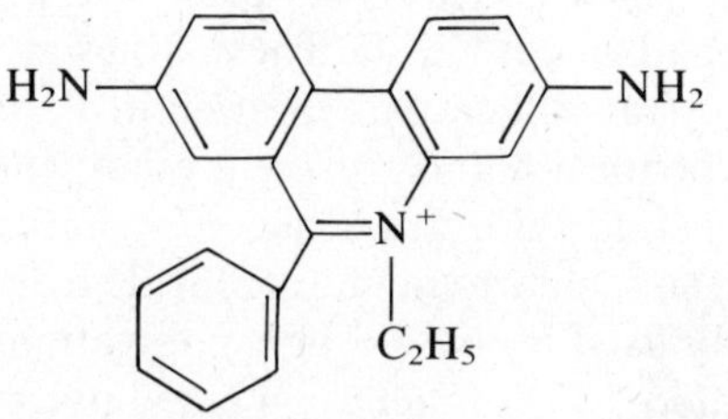

Ethidium

phenanthridinium salts were the most promising (1230). Especially 2,7-diamino derivatives were highly trypanocidal. The best-studied is ethidium (1231), with wide clinical applications.

3.19.1f. Nitroheterocyclics. The discovery of trypanocidal activity by testing compounds originally designed for other purposes, most commonly as antimicrobial agents for other infections, has usually been followed by molecular modification of these "leads" to enhance trypanocidal activity at the expense of the original specificity. Thus, nifurtimox, developed on the basis of the proven activity of nitrofurazole (1232), became the first drug of some use in Chagas' disease (1233). Similarly, a number of originally antibacterial nitroimidazoles exhibit significant trypanocidal activity. Several

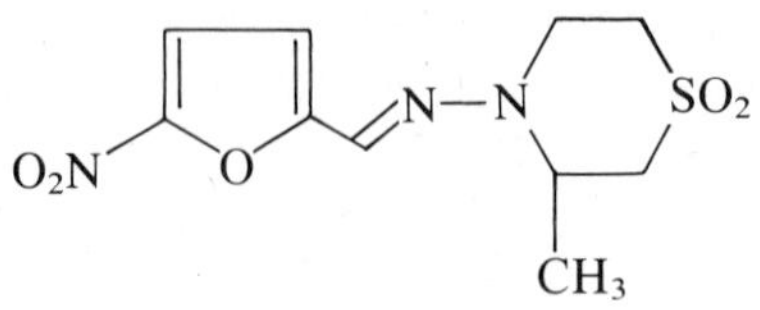

Nifurtimox 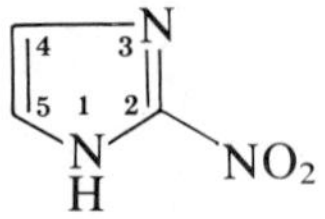   Azomycin

naturally occurring antibiotics such as aminonucleoside, nucleocidin, cordycepin, and amphotericin B also divulged trypanocidal actions in experimental infections, but they have not provided therapeutic or prophylactic usefulness under field conditions nor "leads" for successful molecular modification.

3.19.2. Drugs for Coccidiosis

Coccidiosis is a protozoan infection caused by microscopic parasites of the genera *Eimeria* and *Isospora*. It is prevalent in cattle and sheep and especially in poultry. *Eimeria* infections of chicken and turkeys are a serious threat to poultry growers.

Anticoccidial agents have been found by some rational methods but mostly by screening for "leads" and following up such structural information. Among the earliest anticoccidial drugs were *p*-aminobenzoate antagonists such as sulfanilamides (1234). These drugs are used in combination with other compounds that block participation of PAB-folic acid in metabolism at various biochemical points. As with other sulfonamide chemotherapeutics, the longer-acting sulfa drugs are now preferred, such as sulfadimethoxine (1235) and sulfachlorpyrazine (1236). The dihydrofolate reductase inhibitors most widely used in these drug combinations are pyrimethamine, diaverdine [2,4-diamino-5-(3,4-dimethoxybenzyl)pyrimidine], and ormetoprim [2,4-diamino-5-(3,4-dimethoxy-2-methylbenzyl)pyrimidine] (1237). Other 4-aminopyrimidine anticoccidials are conceived as thiamine antagonists, especially amprolium (1238) and beclotiamine (1239), the latter being preferred pharmacologically.

Of the many and structurally varied anticoccidials in the literature, only a few have been introduced as chemotherapeutics for poultry. Among them are a guanidine [(p-ClC$_6$H$_4$CH=NNH)$_2$C=NH] (1240), and several polyether ionophoric antibiotics, notably monensin, whose anticoccidial properties were discovered (1241) 16 years after its isolation.

3.19.3. Drugs for Trichomoniasis

Trichomoniasis is principally a disease of the genitourinary tract caused by pathogenic species of trichomonads, especially *Trichomonas vaginalis* in humans and *T. foetus* in cattle. *T. gallinae* affects the intestinal tract of fowl. These flagellate parasites can be cultured *in vitro* and can infect mice for purposes of screening drugs.

Although a variety of chemical structures have shown activity in such experimental trichomonal infections, the first compound that provided real chemotherapeutic promise was 2-amino-5-nitrothiazole (entramin) (1242, 1243). Its acetyl derivative (acinitrazole) was the most active of a series of analogs. These nitrothiazoles and many others tested for structure–activity relations were an outgrowth of the study of analogs of the antimicrobial nitrofurans and were selected from broad *in vitro* and *in vivo* screens. None of them became clinically useful but expansion to other nitro-substituted ring systems soon paid off in trichomonacidal compounds. When the antibiotic azomycin (1244, 1245) was found to be 2-nitroimidazole, it was natural to test it for antitrichomonal properties. Because 4(5)-nitroimidazoles are more readily accessible synthetically, such isomers were tested; this led to the selection of 1-hydroxyethyl-2-methyl-5-nitroimidazole (metronidazole) as a systemically active trichomonacide by a team of Rhone–Poulenc medicinal scientists (1246). This compound has become the drug of choice in human trichomoniasis. The competitive industrial research onslaught on this field which followed this "lead" produced many 1-substituted 5-nitroimidazoles but none was clinically superior to metronidazole. Interestingly, the 1-hydroxyethyl group of metronidazole is readily subject to metabolic oxidation—more so than the 2-methyl group—and efforts to replace it by less oxidizable groups resulted, as planned, in longer-lasting trichomonacides of which tinidazole (1247) and nitrimidazine (1248) have proved useful in medical practice.

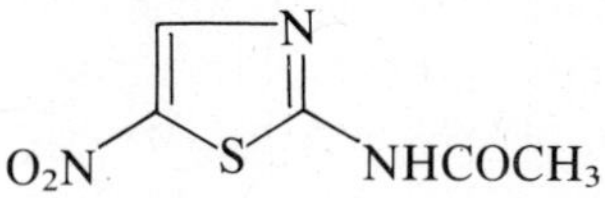

Acinitrazole

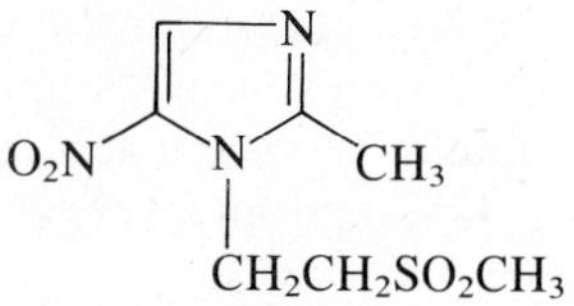

Tinidazole

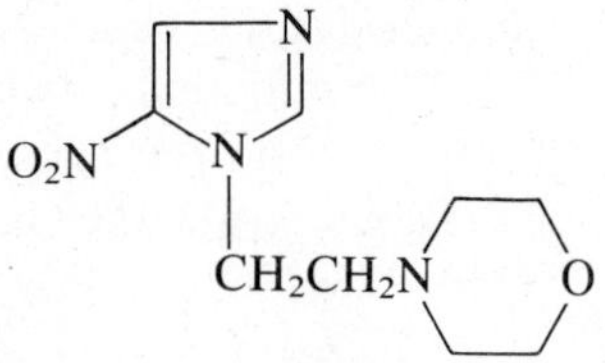

Nitrimidazine (nimorazole)

3.19.4. Drugs for Leishmaniasis and Histomoniasis

The hemoflagellate intracellular protozoa called leishmania cause visceral (Kala-azar) and monocutaneous diseases (1249). Before 1924 *Trypanosoma equiperdum* was used as an experimental model infection, until the Chinese hamster was found to be a suitable laboratory host for *Leishmania donovani* (1250). Screening in golden hamsters made it possible to expand an older finding that antimonials such as tartar emetic (1251), stibophen, and anthiomaline had activity in monocutaneous leishmaniasis. These early drugs have been superseded by sodium stibogluconate (1252) and meglumine (glucan-

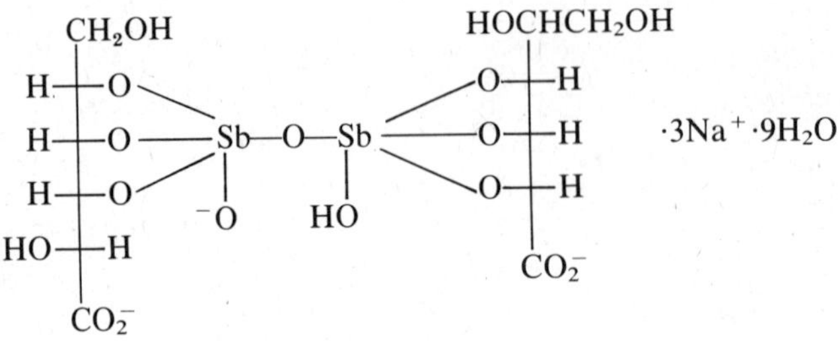

Sodium stibogluconate

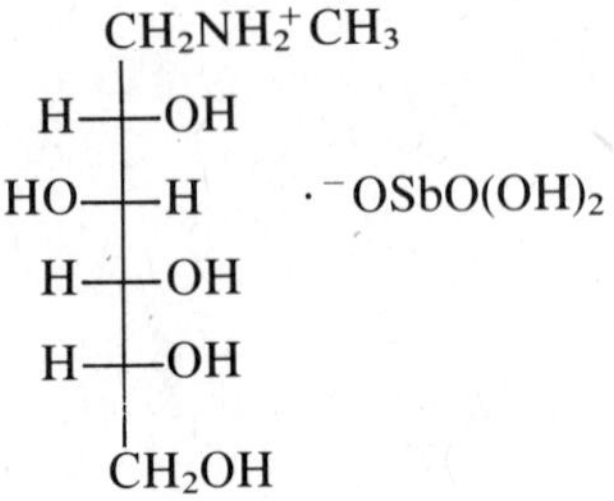

Meglumine

time) (1253), the antimonate of N-methylglutamine. These two substances are widely used in the clinical routine of the disease. Their chemical design was not prompted by rational approaches but rather by synthetic expediency in the field of organic antimonials (1249).

Nonantimonials have not made much of an inroad in the treatment of leishmaniasis. Pentamidine and the polyene antibiotic amphotericin B are used in cases resistant to antimonials, and the antimalarial cycloguanil generally performs well in *L. tropica* infections (so-called "oriental" sore). The input of drug design in these uses has been negligible.

The same lack of novel medicinal chemical ideas characterizes the chemotherapy of histomoniasis, a disease of turkeys and chicks caused by the protozoan *Histomonas meleagridis*. After an initial use of acetarsol (1254), a number of nitrothiazoles, nitroimidazoles, and nitrofural derivatives replaced the older drugs. In all cases, improved screening tests paced these advances, and standard molecular modifications of initially successful "lead" compounds unearthed by these tests led to acceptable veterinary drugs. Among the agents found by systematic modification are nithiazide (1255), dimetridazole (1256, 1257), ipronidazole (1258, 1259), and ronidazole (1260, 1261).

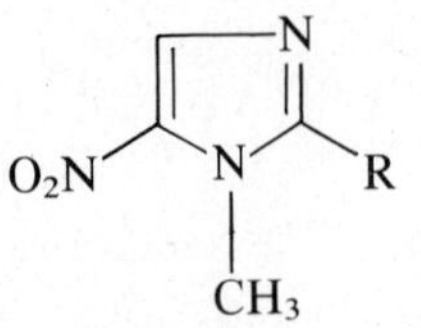

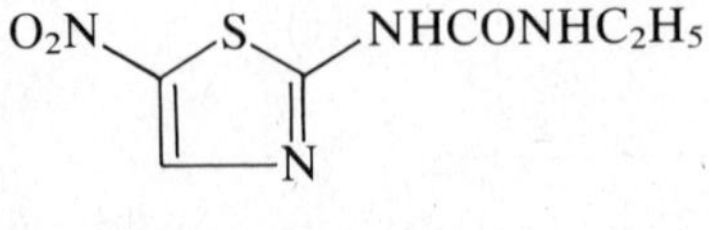

Nithiazide

Dimetridazole: R = CH$_3$
Ipronidazole: R = CH(CH$_3$)$_2$
Ronidazole: R = CH$_2$OCONH$_2$

3.19.5. Antiamebic Agents

Amebiasis is a widespread infection occurring generally in regions with poor economic and sanitary conditions but even found in as much as 6% of the U.S. population. While amebae may be present asymptomatically in some individuals, pathogenesis occurs either intra- or extraintestinally, giving rise to such symptoms as dysentery and intestinal and liver abscesses. Other organs may also be involved. The effectivenss of antiamebic agents will depend on the location of the infection and the ability of the drug to be transported there for action.

Amebiasis was recognized as a distinct disease about 1875 by Loesch (1262). The only ameba pathogenic to humans is *Entameba histolytica*. Drugs can be screened against it *in vitro*; *in vivo* methods make use of weanling rats for the assessment of activity against intestinal amebiasis and hamsters as models for hepatic amebiasis (1263).

3.19.5a. Emetine. Among the oldest antiamebic agents used in human chemotherapy are the ipecac alkaloids and certain arsenicals. The beneficial activity of the powdered root of the herb igpecaya had been known to Brazilian Indians for centuries (1264), and the root was exported to Europe by the conquistadores. It was first recorded in 1625 by Samuel Purchas in *Purchas his Pilgrimes* (1265) and reached greater popularity than cinchona, the universal antiparasitic drug, when the Dauphin, the son of Louis XIV, was cured from bloody flux by Helvetius towards the end of the 17th century. Ipecac was used increasingly in the second half of the 19th century (1266) but its administration was not placed on a secure basis before 1909 (1267–1269).

Extracts of *Cephaelis ipecacuanha* or *C. acuminata* yield a number of alkaloids, with emetine prevailing. Structural studies (1270) preceded the total synthesis (1271) of emetine and the determination of its stereochemistry

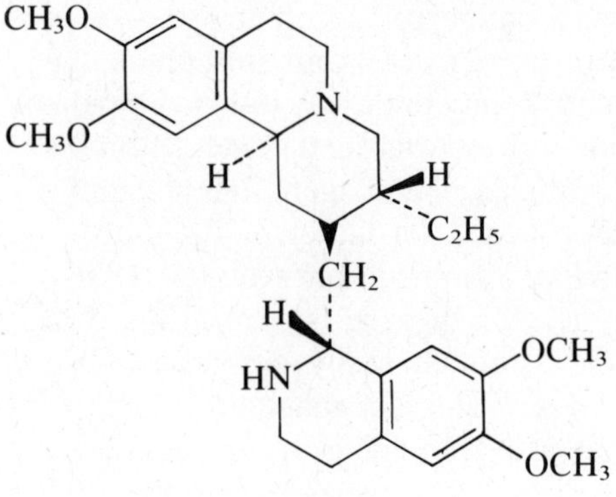

(−)-Emetine

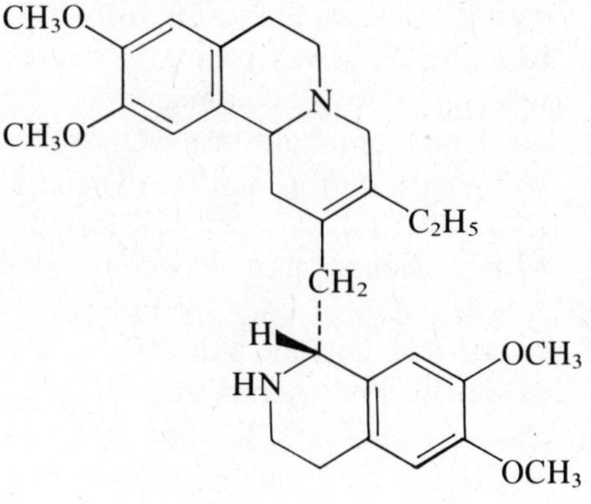

(±)2,3-Dehydroemetine

(1272, 1273). A stereospecific synthesis of (−)-emetine lends itself to commercial production (1274).

A large number of analogs of emetine have been made by every kind of molecular modification and simplification. Only one compound, 2,3-dehydroemetine (1275), has had success in clinical studies (1276, 1277). As far as the mechanism of action of emetine and its biologically active congeners is concerned, they all inhibit protein biosynthesis at the ribosomal level (1278).

3.19.5b. Other Natural Products. Various other natural plant products have amebicidal properties but have found either no clinical use or only very limited and controversial application. Among them is conessine, a steroidal alkaloid from *Holarrhena antidysenterica*, and berberine, which contains a condensed isoquinoline system and occurs in the plants *Berberis aristata* L., *Hydrastis canadensis* L., and other species. Compounds extractable from simarouba, especially glaucarubin, a pentacyclic non-nitrogenous material, and similar substances have been eyed as potential "leads" in this field. Nothing useful has come from these studies.

Broad screening of antibiotics has uncovered a number of such compounds that have antiamebic activity. The only antibiotic of clinical interest in this regard is paromomycin, an aminoglycoside metabolite of *Streptomyces rimosus* (1279). Paromomycin is clinically active in acute and chronic intestinal amebiasis (1280).

The hope that the natural products that have amebicidal activity might have provided "leads" toward other amebicides has not been fulfilled. The synthetic drugs that have found chemotherapeutic use in clinical amebiasis have all been discovered by screening, in some cases after observations of other biological properties. This was the sequence of events for the early synthetic arsenicals that had been synthesized before 1915.

3.19.5c. Arsenicals. When a compound is tested as an antibacterial agent, a representative array of pathogenic bacteria is exposed to the substance. Likewise, in antiprotozoal tests, a battery of different protozoa is used. Ehrlich was primarily interested in trypanocidal and spirocheticidal tests in rodents, but some of his compounds found their way into amebicidal tests in other laboratories. One of the first of these was acetarsone, Ehrlich's No. 594 test drug, not far ahead of the antispirochetal agent arsphenamine (No. 606). Acetarsone was tested by Fourneau, the father of French medicinal science, at the Pasteur Institute in Paris. Named Fourneau 190 (190F), it was renamed later by Fourneau (French for furnace, stove) in his own honor and became known as stovarsol. Although easily prepared (1281, 1282), its low therapeutic index in amebiasis did not maintain it long in clinical use (1283). It was replaced by the lower homolog, treparsol (1284), and this in turn yielded to a phenylurea derivative, carbarsone, which had also been prepared by Ehrlich's group (1285). It has been in use in amebic

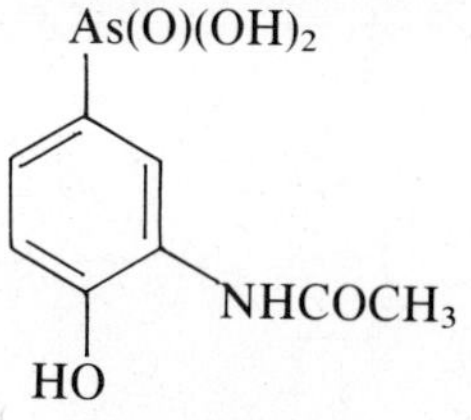

Acetarsone 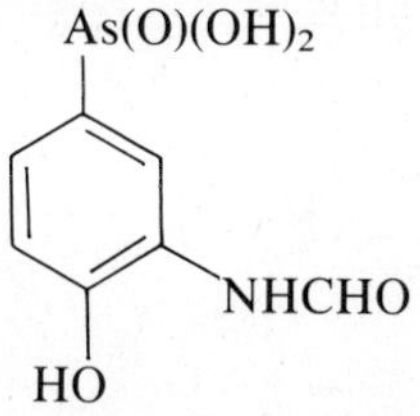   Treparsol 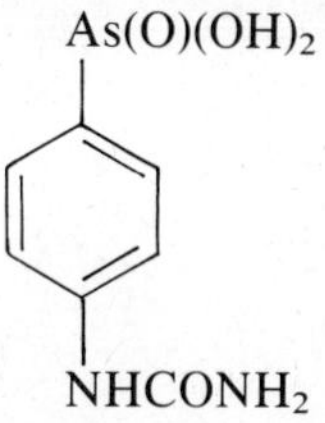   Carbarsone

dysentery since 1931 (1286). None of these arsonic acids can compete with more modern, less toxic antiamebic agents.

3.19.5d. Synthetic Amebicides. Blind screening of halogenated 8-quinolinols for antiseptic properties had culminated in chiniofon (7-iodo-8-quinolinol-5-sulfonate) and molecular modification had led to such analogs as iodochlorhydroxyquin (5-chloro-7-iodo-8-quinolinol) (1287) and 5,7-diiodo-8-quinolinol (1288). The amebicidal properties of these iodinated phenols were discovered subsequently (1289–1293). Aminoalkyl side chains were then introduced by a Mannich reaction on 5-halo-8-quinolinols, with 5-chloro-7-(diethylaminopropylaminomethyl)-8-quinolinol (clamoxyquin) (1294) being selected as a clinical amebicide. These compounds, like other 8-hydroxyquinolines, are chelating agents and specifically chelate Fe^{2+} needed for amebal life processes (1295).

It was only one step from 8-quinolinols to related ring systems that are capable of chelating metal atoms. A clinically interesting analog was encountered in phanquone (4,7-phenanthrolinequinone) (1296).

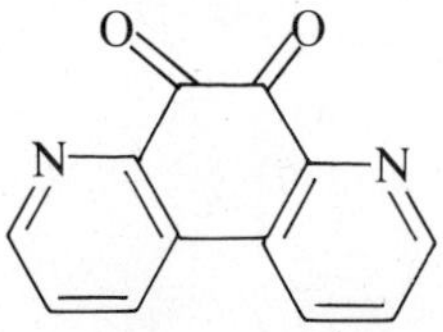

Phanquone 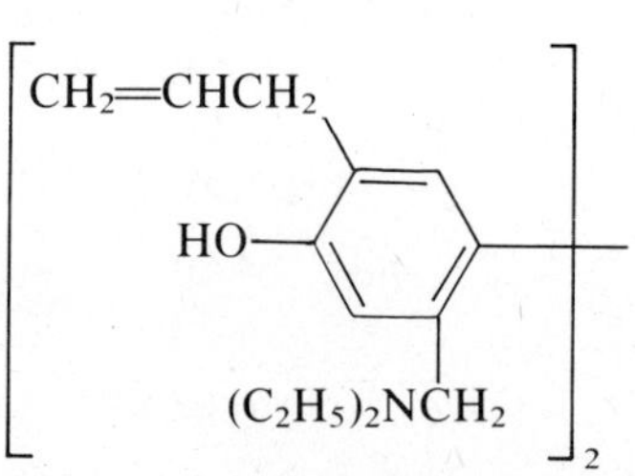   Bialamicol

A number of aminoquinoline antimalarials show antiamebic activity in hamsters (1297), and chloroquine has become an accepted drug for amebiasis in humans (1298).

Another related drug is quinacrine, which possesses some amebicidal activity. No acridine or benz(c)acridine congener of quinacrine has reached clinical utility. By contrast, molecular modification of the antimalarial Mannich base, amodiaquine, furnished 3,3′-diallyl-5,5′-bis(diethylaminomethyl)-4,4-dihydroxybiphenyl (bialamicol), with activity in intestinal and hepatic human amebiasis (1299).

Significant amebicidal activity in hamsters was discovered among certain 2-phenylthiazolidinone-1,1-dioxides, especially the 3,4-dichlorophenyl derivative (1300, 1301). These compounds were subjected to thorough molec-

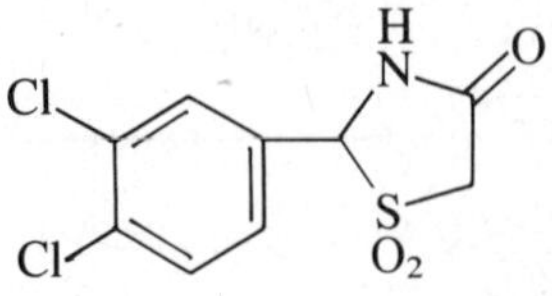

2-(3,4-Dichlorophenyl)-thiazolidone-1,1-dioxide

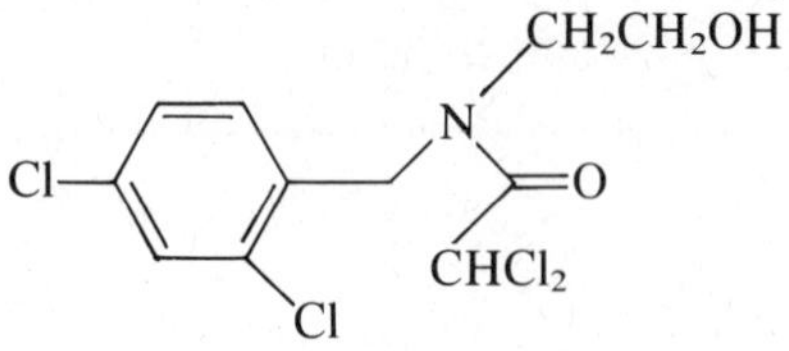

Chlorbetamide

ular dissection by Surrey (1302), including a deletion of the SO_2 group of the prototypes. The resulting benzylacetamides, $R_1R_2C_6H_3$-$CH_2NR_4COR_3$, were especially active when R_3 was $CHCl_2$. The most interesting member of this series was chlorbetamide; it excelled in experimental amebiases (1303) but remained equivocal in the clinic (1304). The dichloroacetamide group was suggested by that in chloramphenicol, which at that time dominated chemotherapeutic attention. Wide-ranging modification led to several analogs whose main common feature was the dichloroacetamide moiety. Among them were chlorphenoxamide [note the nitro group also found in chloramphenicol] (1304), teclozan (1305), and diloxanide furoate (1306).

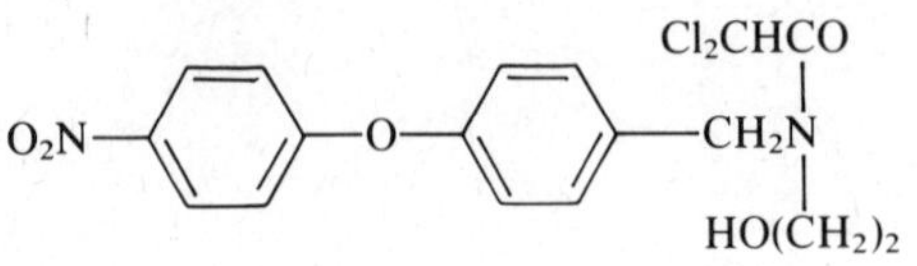

Chlorphenoxamide

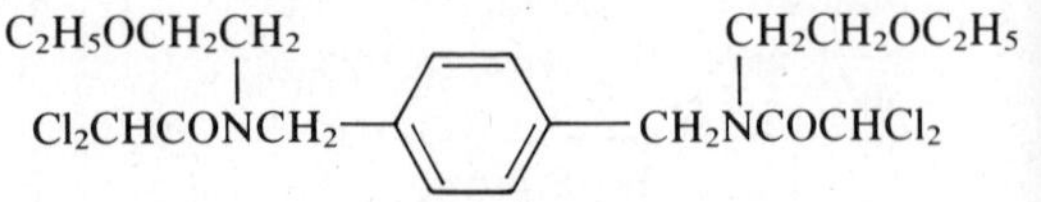

Teclozan

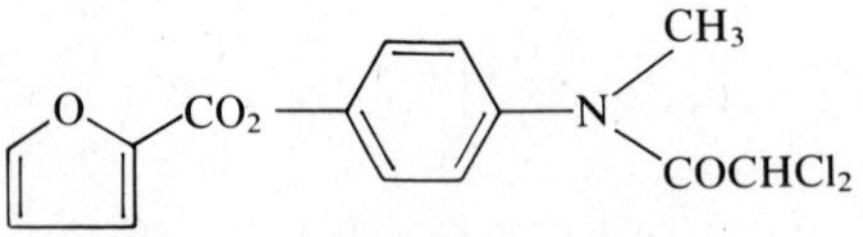

Diloxanide furoate

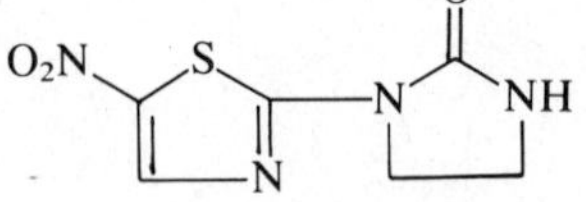

Niridazole

Chloramphenicol and chlorphenoxamide were by no means the first nitro compounds recognized as potent and relatively nontoxic antimicrobial agents. In 1944, Dodd (929) had observed the antibacterial properties of several nitrofurans, and this was followed by the introduction of 2-amino-5-nitrothiazole (entramin) as a veterinary systemic trichomonacide (1307, 1308) and antihistomonad in domestic fowl. Its N-acetyl derivative (aminitrozole) was found to be of value in the intestinal amebiasis of dogs. Another nitrothiazole, 1-(5-nitro-2-thiazolyl)-2-imidazolidone (niridazole) (1309) was first used against schistosomiasis but turned out to be a clinically useful

ANTIPROTOZOAL AGENTS

amebicide (1310) although still beset with CNS side effects. Many other 2-amino-5-nitrothiazoles have been tested in experimental amebiasis.

The seminal discovery of the antitrichomonal antibiotic azomycin (1244, 1245) (see Section 3.19.3) was the "lead" in a search for other nitroimidazoles. In this series, metronidazole appeared as the breakthrough drug for human trichomoniasis (1246). The amebicidal properties of metronidazole were also found by the original discoverers of the drug (1311) and confirmed later clinically (1312). These curative activities led to a flood of molecular modifications all over the world. Examples are tinidazole (1313) and nimorazole (1248; see Section 3.19.3). The subject has been reviewed (1314).

3.19.6. Other Protozoan Infections

Nitroimidazoles are also active against giardiasis, which is caused by *Giardia lamblia* and is manifested by intestinal disturbances (1315). In Eastern Europe, it is quite commonly caused by contaminated drinking water. Metronidazole (1316), furazolidone (1317), and nitrimidazine (1318) are useful drugs for this condition. Niridazole and furazolidone are of use in balantidiasis, another intestinal disease caused by *Balantidium coli*. In this infection, however, oxytetracycline is the drug of choice (1319).

Many trypanocides have been screened against babesiasis (piroplasmosis), a tick-borne disease mostly of domestic animals caused by *Babesia* species. Trypan blue (1320) and acriflavine were used early to a limited extent but were displaced by quaternary quinolinium compounds such as quinuronium methosulfate (1321), then by aromatic diamidines such as dim-

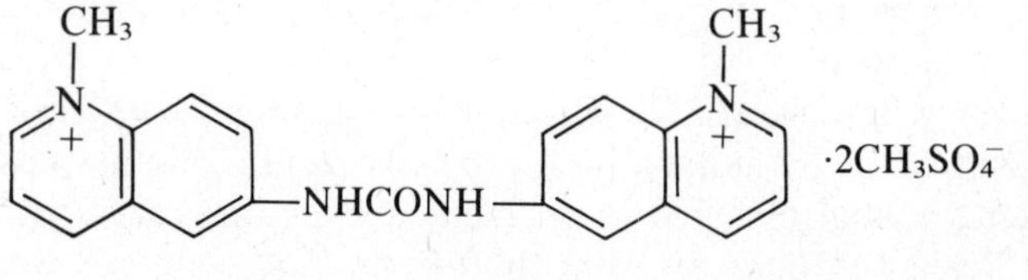

Quinuronium methosulfate

inazene (4,4'-diamidinodiazoaminobenzene) (1322), in various hosts. When the urea group present in the quinolinium salts was incorporated as the bridging moiety into diamidines, a number of compounds were obtained of which amicarbalide has become an effective veterinary antibabesiasis drug (1323, 1324).

Amicarbalide

Another urea babesicide, imidocarb, in which the amidine groups have become part of an imidazoline ring, exhibits additional activity against anaplasmosis, a tick-borne infection of cattle and sheep caused by *Anaplasma* species. This infection responds as well to tetracycline antibiotics and α-dithiosemicarbazones, which were tested after showing babesicidal and anticoccidial activity. The most acceptable of the rather toxic dithiosemicarbazone babesicides was gloxazone (1325).

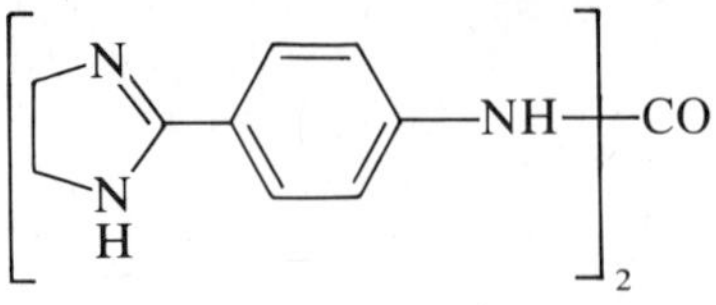

$$\left[\underset{H}{\overset{N}{\underset{N}{\bigvee}}}\!\!\!\!\!\bigcirc\!\!-NH-CO \right]_2 \qquad \underset{CH_3\ \ CH=NNHCSNH_2}{C_2H_5OCH-C=NNHCSNH_2}$$

Imidocarb 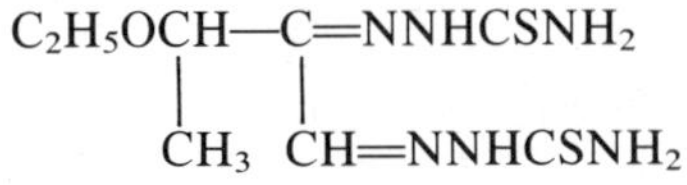   Gloxazone

Finally, toxoplasmosis, which is caused by the coccidian *Toxoplasma gondii* from cats and which infects humans, can be treated best with PAB-folic acid antagonists, especially sulfapyrazine (1326).

3.19.7. Antimalarials

3.19.7a. Introduction. The treatment of malaria can be divided into several sectors all involving interlocking scientific disciplines. They include the eradication of the insect vector (*Anopheles* mosquitoes) with insecticides; the resistance of the vector to insecticides; the study of plasmodial life cycles; the biochemial problems of drug-resistant plasmodial strains; the elaboration of laboratory and clinical methods for the evaluation of antimalarial drugs; the clinical use of such drugs in the various forms and stages of human malarias and in drug-resistant cases; and the discovery and development of antimalarial drugs.

The design of antimalarial drugs could not have progressed if it had not been complemented by suitable test methods. As in all other cases of drug development, animal test methods changed over the years. In the period 1926–1935, canaries parasitized with *Plasmodium relictum* were used in modifications of the method of Roehl (1327) based on the finding of Kopanaris (1328) that quinine had an effect on this avian infection. The years 1935–1948 made use of chicks, ducklings, and turkeys with a test on *Plasmodium gallinaceum* (1329). So great was the need for turkey chicks for this test during the World War II antimalarial crash program that a nationwide shortage of turkeys resulted during the Thanksgiving–Christmas season of 1943. Since 1948, *P. berghei* has become the standard for screening programs. Originally observed in wild rodents of the Belgian Congo (1330), as worked out by Leo Rane it can be established conveniently in laboratory albino mice (1331). Details of many of the test methods have been reviewed (1332, 1333).

Drugs have been used for the treatment of malaria for more than 2000 years. Ch'ang shan was prepared from powdered roots of *Dichroa febrifuga* Lour around 200 B.C., or from hydrangea leaves. An active ingredient, febrifugine, was isolated (1334) during World War II and its structure confirmed by synthesis (1335). However, neither febrifugine nor any of its synthetic congeners have become clinically useful antimalarials.

3.19.7b. Quinine. The classical antimalarials are the cinchona alkaloids: quinine, cinchonidine, and their C-9 diastereomers. The source of these alkaloids is the bark of *Cinchona ledgeriana* Moens (*Cinchona officinalis* L.), which has been used directly or in the form of powders as a remedy for malaria ever since it was included in the 1677 edition of the *London Pharmacopoeia* as "Cortex Peruanus."

The first report of this bark was written by an Augustinian monk, Calancha, of Lima, Peru, in 1633. It states that "a tree grows which they [the Peruvian Indians] call the fever tree in the country of Loxa whose bark, the color of cinnamon, made into a powder amounting to the weight of two small silver coins and given as a beverage, cures the fevers and tertians; it has produced miraculous results in Lima." A tale that has persisted through the centuries tells us that the bark was given to cure the Countess del Chinchón, wife of the Spanish viceroy of Peru, in 1638. In gratitude for her cure, her husband introduced the bark in his homeland in 1639 for the treatment of ague. It is not certain that the countess ever used the Peruvian bark; nevertheless, the powdered bark became known as *los Polvos de la Condesa*. In fact, one historian would have us believe that Anna del Chinchon died *before* her husband went to Peru, and that the lady who was cured from malaria was someone else in the viceroy's personal retinue. Another version is that the countess died during her passage back to Spain. In any case, we know that the viceroy shipped a large quantity of the bark to Spain and that by 1640 the drug was being used for "fevers" in Europe. It was listed in a medical pamphlet of 1643 by a Belgian, Hermann Van der Heyden, as *Pulvus indicus*.

The botanist Linnaeus called the plant genus *Cinchona*, misspelling the name of the alleged countess. It is possible that the real source of the name is the Incan "Kinia" (bark). After the 1640's, cinchona was imported mainly by Jesuit priests, and this led to the designation *Jesuits' bark*. The Roman Cardinal de Lugo recommended it, and an alternate name, *Cardinal's bark*, arose from this connection. The European medical establishment looked with suspicion at anything related to the Jesuits because of their role in the Inquisition and also because the new bark did not conform to the teachings of Galen. The drug was therefore relegated to the black market and sold as a secret remedy, its price set by its counterweight in gold.

The Peruvian government recognized the value of the native wild cinchona trees and decreed stiff penalties for their export. However, around 1880, two Dutchmen succeeded in smuggling seedlings out of Peru and transporting them to Java, where they were carefully cultivated, grafted, improved upon, and planted in immense orderly plantations. Extraction facilities were added later, and most of the world's supply of quinine has since come from Indonesia. This lent political and economic overtones to antimalarial research, for in World War I the Germans were cut off from this source of quinine, which they needed in their African colonies. The experience provided the impetus to the early systematic searches for synthetic antimalarials. The same situation arose again in 1942 when the Allies were deprived of quinine by the Japanese invasion of the East Indies and Malay. An enormous antimalarial crash program in Britain and the United States followed this emergency.

Quinine was extracted from cinchona bark by Pelletier and Caventou in 1820 (25) and its structure elucidated 100 years later (1336). Its synthesis by Woodward and Doering (1337) heralded the age of sophisticated modern synthetic methodology but was too complicated to compete with the natural product for medicinal purposes. Natural quinine, combined with other cinchona alkaloids, remained the sole antimalarial drug to combat the intraerythrocytic stages of the parasites until the advent of quinacrine in 1932. This monopoly of quinine dominated the minds of medicinal chemists searching for a "lead" structure for molecular modification. All the early synthetic antimalarials were to contain the 6-methoxyquinoline moiety as it is seen and most easily spotted in quinine, although a quinine-like antimalarial could not be developed in this manner (1338). Quinine has also been used in optical lenses as a polarizing agent.

3.19.7c. Synthetic Antimalarials. The first synthetic compound tested as an antimalarial was methylene blue. This dyestuff is bound firmly by plasmodia *in vitro* in staining preparations and is relatively nontoxic. For this reason, Guttmann and Ehrlich (56) tested methylene blue in one malarial patient as no animal test had been worked out for experimental purposes. It had "a beneficial effect" (56). This rather flimsy observation was sup-

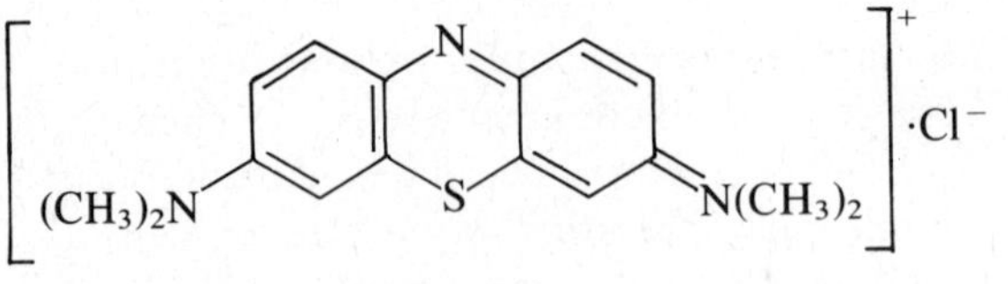

Methylene blue

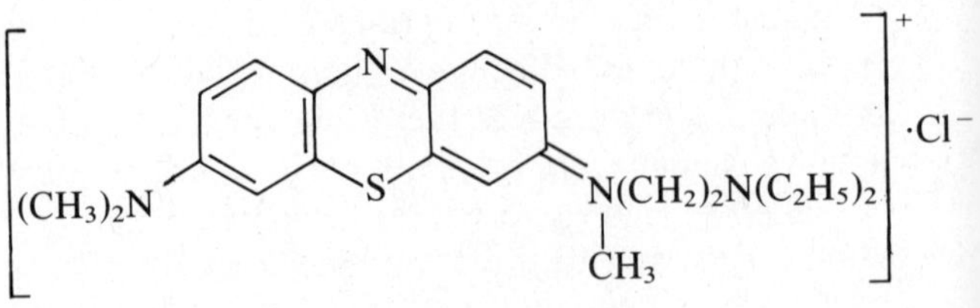

Diethylaminoethyl analog of methylene blue

ported later by Marks, who found that methylene blue is active against avian plasmodiases (1339). In the hope of making methylene blue "more basic" and thereby giving it even greater affinity to plasmodial cells, one of the *N*-methyl groups was replaced by diethylaminoethyl; the resulting derivative showed some enhancement of activity. This weak "lead" was combined with that provided by the 6-methoxyquinoline portion of the quinine molecule. Although most of these researches were not published by the German pharmaceutical industry, credit for them goes to the chemists F. Schönhöfer and W. Schulemann and to the malarialologist, W. Roehl (1340).

AMINOQUINOLINES. It seems likely that synthetic expediency played a role in choosing first 8- and then 4-aminoquinolines as points of departure for the attachment of the dialkylaminoalkyl side chains. Diethylaminoethyl was used first, then diethylaminopropyl; ultimately the 1-*N*,*N*-diethylamino-4-pentyl chain was chosen because the corresponding diethylamino-4-pentyl halide could be prepared in large batches from ethyl acetoacetate. The simple 8-(diethylaminopropyl)aminoquinoline was active in the Roehl test (1327) and provided the "lead" to pamaquine (Plasmochin) that emerged as the first clinically applicable, although still toxic, synthetic antimalarial in 1925 (1341). Because of its toxicity, it was discontinued during World War II (1342).

Approximately 3000 dialkylaminoalkylaminoquinolines are said to have been tested before 1932, which would make this series the most extensive in molecular modification before the sulfanilamides of the 1940's. These compounds are, on the whole, antirelapse drugs for vivax malaria. They are blood schizonticides. The clinically most important congeners, all with a 6-methoxy-8-aminoquinoline structure, carry the side chains at the 8-amino group as shown.

Pentaquine (1343–1345): $R = (CH_2)_5NHCH(CH_3)_2$
Isopentaquine (1343–1345): $R = CH(CH_3)(CH_2)_3NHCH(CH_3)_2$
Primaquine (1346): $R = CH(CH_3)(CH_2)_3NH_2$
Quinocide (1347): $R = (CH_2)_3CH(CH_3)NH_2$

Only primaquine has remained in clinical use in this class of drugs. Quinocide is used especially in Eastern Europe. All these compounds and thousands of others that did not make it to clinical trial were products of World War II antimalarial research in the United States, quinocide is the product of wartime research in the USSR.

Since the use of pamaquine was limited by its toxicity from the start, analogs were made not only in the aminoquinoline but also in related ring systems. Among these, 2-chloro-7-methoxy-9-(5-diethylamino-2-pentyl)aminoacridine (quinacrine, Atebrin) made its debut in 1932 (1348–1350) as one of many structural variants. This drug carried Allied troops engaged

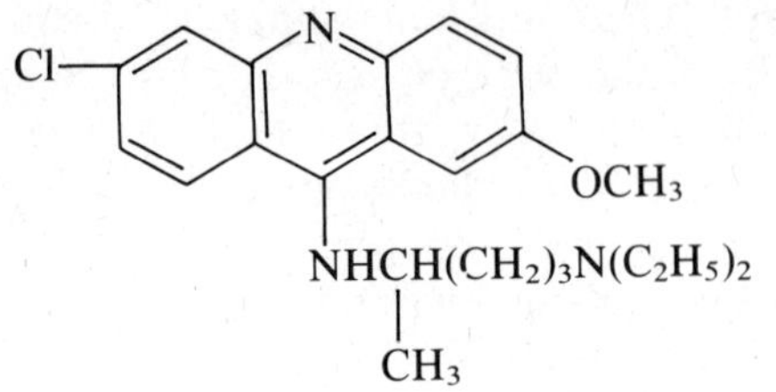

$$\text{Quinacrine}$$
Cl—[acridine ring with N]—OCH$_3$, with NHCH(CH$_2$)$_3$N(C$_2$H$_5$)$_2$ / CH$_3$ side chain

Quinacrine

in tropical regions through the first critical two years of World War II before better antimalarials could be developed. Its toxicity and dyestuff character, which stains the skin yellow, were its major drawbacks. It is not used any longer as an antimalarial but finds occasional application as an anthelmintic.

The basic side chain of quinacrine is attached *para* to the ring nitrogen. This corresponds to the 4-position in quinoline, and indeed some 4-substituted quinolines were tested in the German researches (1351) as well as in the USSR (1352). Schönhöfer (1351) postulated that 6- and 8-aminoquinolines might be bio-oxidized to quinones, 4-aminoquinolines to quinonimines (by prototropy, the so-called Schönhöfer tautomerism), and that quinonoid products were necessary for antimalarial potency. One of the 4-aminoquinolines with the traditional side chain, the 3-methyl-7-chloro derivative (sontoquine), was used by French troops during their North African campaign

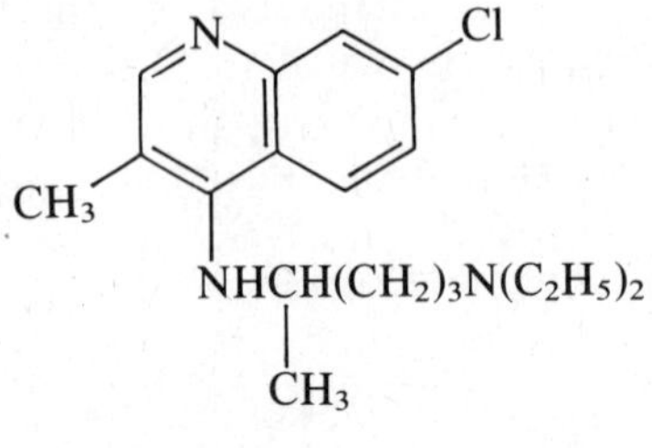

Sontoquine 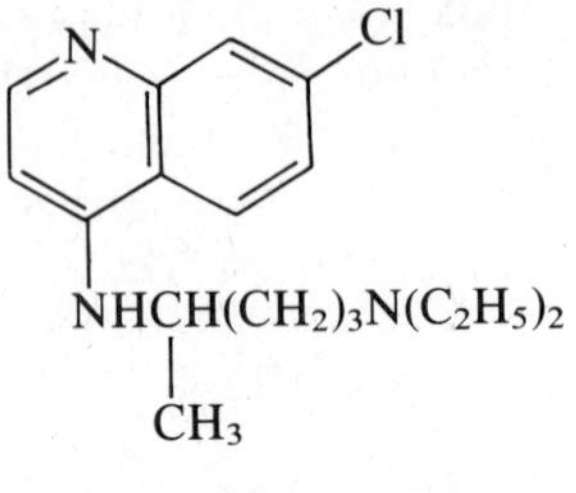   Chloroquine

(1353). Supplies of this drug were captured by American soldiers from the enemy and proof of its structure formed the basis for the synthesis of related analogs (1344, 1354–1356). The most effective drug of this series was chloroquine. It is ironic that this antimalarial had been studied in Germany in 1934 under the name of Resochin but had been considered too toxic (1333, pp. 347, 358; 1357).

Two analogs of chloroquine with alcoholic hydroxyl groups in the side chain have been found active clinically (1353). After phenolic biphenyl Mannich bases (ArCH$_2$NEt$_2$) had been found to exhibit antimalarial activity (1299), this functional group was combined with the 7-chloro-4-quinolyl nucleus—an additive practice not often crowned with success in medicinal chemistry. Several useful agents emerged from this work, especially amo-

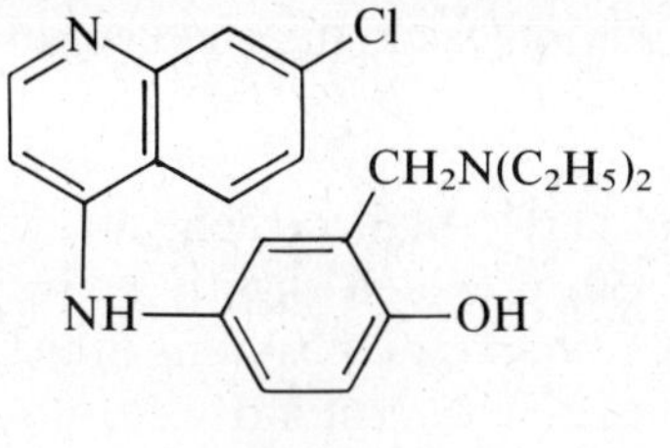

ANTIPROTOZOAL AGENTS

Amodiaquine

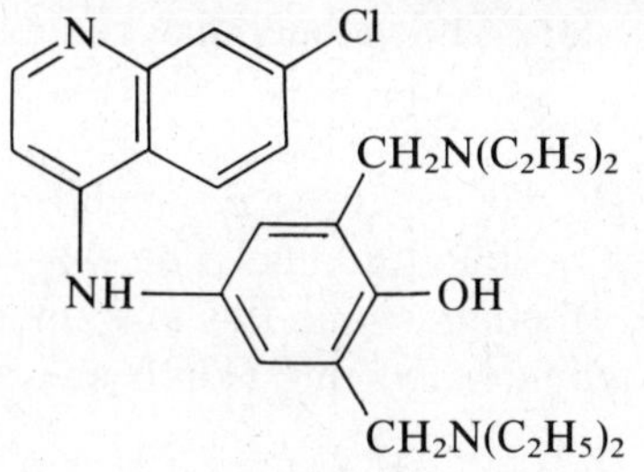

Cycloquine

diaquine. Russian workers later added even two Mannich base groups in cycloquine and observed high activity (1358).

Much effort has been expended on an understanding of the biochemical and morphological explanations of the mode of action of the aminoquinoline antimalarials as well as of the quinolinemethanol type of compounds, which act similarly in many respects. Observations bearing on these studies include those on the decomposition of hemoglobin by the plasmodia and the effect of antimalarials on the fusion of parasitic vacuoles; the accumulation of fast-acting blood schizonticides in parasitized erythrocytes (a specific distribution phenomenon); the intercalation of the drugs with nucleic acids, including details of this phenomenon; and the effects of the drugs on the biosynthesis and degradation of nucleic acids and the biosynthesis of proteins (1333, pp. 359–366). However, in the two types of antimalarials under discussion, these explanations have not been of help in the design of more effective, less toxic drugs with less tendency to cause resistance.

QUINOLINE AMINOALCOHOLS. From a historic point of view, one might expect the century-old cinchona alkaloids to have provided "leads" for molecular modification of more adequate antimalarials, but this was carried out only to a very limited extent. If the structure of the alkaloid is depicted schematically as

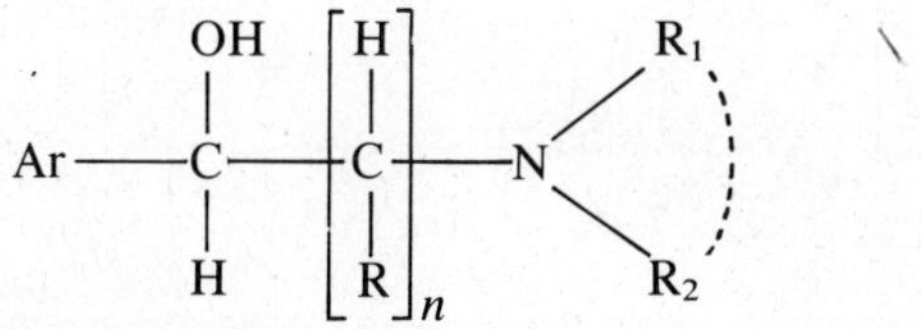

$$Ar-\underset{H}{\overset{OH}{\underset{|}{C}}}-\left[\underset{R}{\overset{H}{\underset{|}{C}}}\right]_n-N\begin{pmatrix}R_1\\ \\R_2\end{pmatrix}$$

where Ar is a substituted quinoline nucleus, $n = 1 - 2$, R and R_1 are H or an alkyl chain and R_1 and R_2 form the substituted quinuclidine ring, the following modifications suggest themselves:

1. Ar may be an aromatic or other heteroaromatic nucleus.
2. The stereochemistry of the alcoholic carbon could be altered or the CHOH oxidized to CO.

3. The quinuclidine could be replaced by other rings and the vinyl group could be saturated, altered, or removed.

The last of these possibilities was investigated early. Replacement of the substituted quinuclidine ring by 2-piperidyl (1359) provided slightly active compounds, and this also underscored that the asymmetric positions in the quinuclidine ring (1360) and at C-9 (the alcoholic carbon atom) were un-

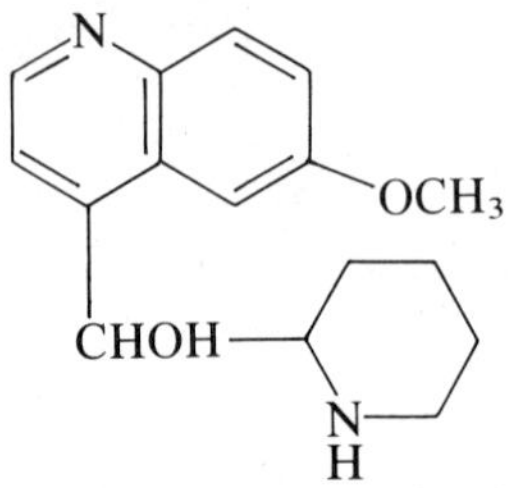

See ref. 1359

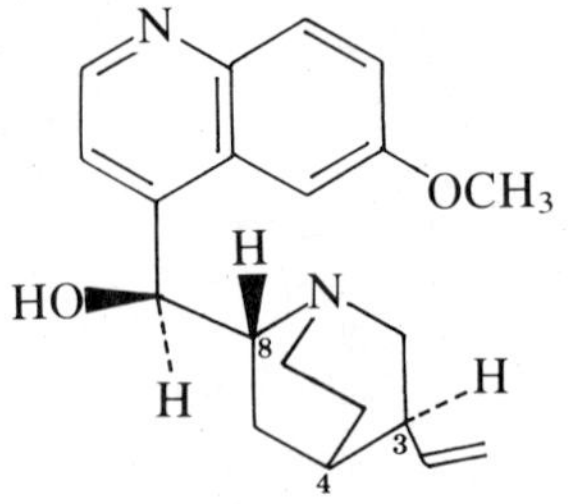

Quinine

important. Later (1361), only those derivatives of isomeric and related compounds were found to be active in which the distance between the oxygen and the nonaromatic nitrogen was about 3 Å. The 6-methoxy group was also soon found to be unnecessary although it had been staunchly adhered to in the 8-aminoquinoline antimalarials.

Metabolism of quinine yields 2-hydroxyquinine (1362); this metabolite is less active than the parent drug (1363). Blockade of the biological attack by synthetic 2-phenyl derivatives raised antiplasmodial activity in several avian test systems but caused phototoxicity in laboratory animals. Molecular mod-

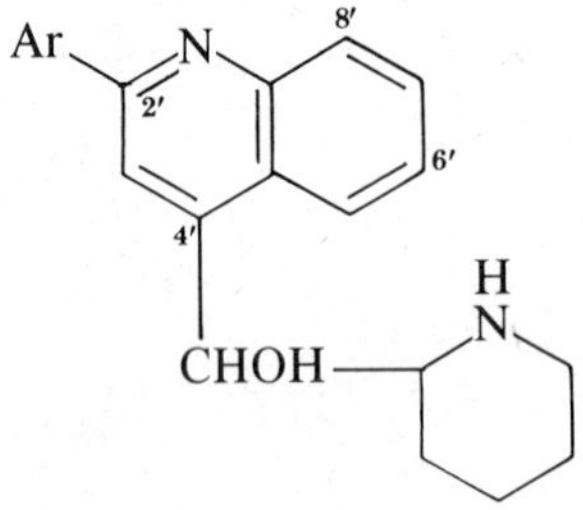

2-Aryl-substituted piperidinyl alcohols

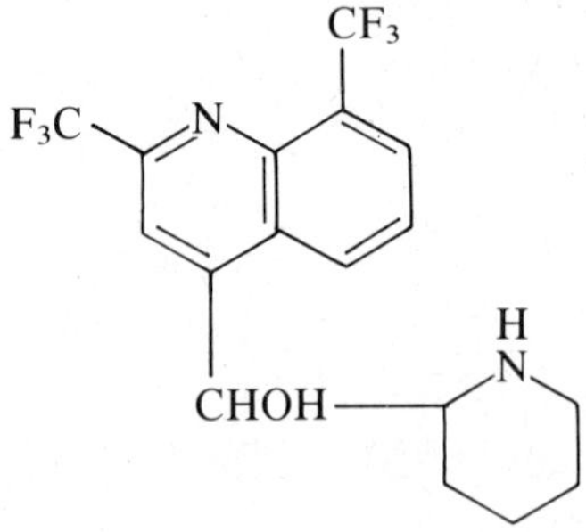

Mefloquine

ification of related 2'-aryl-substituted 2-piperidyl alcohols showed that 6',8'-dichlorination (plus chlorination of the phenyl group) diminished this objectionable side effect (1364). The Walter Reed Army Research Institute extended these observations in the late 1960's in the course of resuming antimalarial researches begun in wartime. It was found that the replacement of the 2′-aryl group by the electron-rich CF_3 group somewhat decreased photosensitizing properties (1365). Systematic molecular modification over an extended period of time led to mefloquine (1366), which cures clinical infections caused by chloroquine-resistant strains of *Plasmodium falciparum* (1367). Although mefloquine is the drug of choice for this condition, it may be difficult to introduce it commercially because the patents covering its synthesis have been assigned to an agency of the United States government.

3.19.7d. Aminopyrimidines. At the onset of World War II, with the demand for protection from and cure of malarias to which Allied troops were exposed, every kind of chemical was tested for antimalarial activity in avian systems (1332) and, if promising, in clinical cases in federal prisons and military hospitals. Among the earliest drugs of that period that were known to be systemic antibacterials—and therefore "better" candidates for other infections than randomly chosen compounds—were the sulfanilamides and bis(aminophenyl)sulfones. Gratifyingly, a number of sulfadiazines did show antimalarial activity, with sulfapyrimidines leading this type of drug. As very long-lasting members of this group were developed many years later, they corroborated and reemphasized the results of these early findings (1368). In the long run, these newer sulfapyrimidines are of greater value when used in two-drug or three-drug combinations with quinine, chloroquine, amodiaquine, or trimethoprim in a typical multifaceted attack on the dihydrofolate reductase system of the plasmodia.

With knowledge of the antimalarial activity of sulfapyrimidines at hand, British medicinal chemists under the leadership of F. H. S. Curd and F. L.

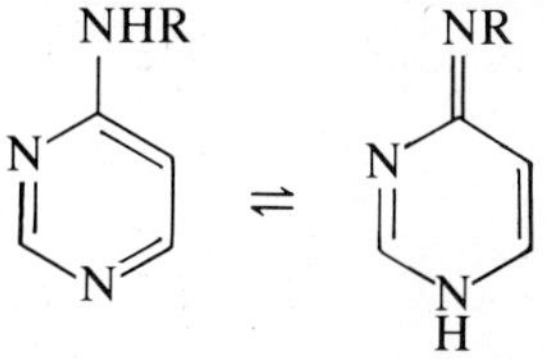

Proton tautomerism of aminopyrimidines

Rose set out to design improved antimalarials on a more rational basis. They had been impressed by Schönhöfer's hypothesis of tautomerism in the aminoquinoline series (1351) as a prerequisite for potent antimalarial activity, and they spotted an analogous proton shift in aminopyrimidines. The presence of aminopyrimidine moieties in some of the antiplasmodially active sulfanilamides reinforced their choice of pyrimidine structures as an entry to new researches. Patent considerations to circumvent the German hold on aminoquinolines may also have played a role in this decision, as may have the belief that the quinoline and acridine nuclei were foreign to human metabolism and less toxic agents might be encountered in the pyrimidine series since pyrimidines are components of nucleotides.

It should be remembered that the planar conformation of the acridine nucleus is optimal for antimicrobial activity but that separation of one or more planar aromatic rings from the whole tricyclic system still produces activity, whereas deletion of one of the three rings altogether abolishes it. Thus, 4-amino-6-phenylquinoline still retains much of the antibacterial activity of 10-aminoacridine whereas 4-aminoquinoline is inactive (1369). Such thoughts must have occupied Curd and Rose when they substituted their aminopyrimidines with *p*-chlorophenyl, thus retaining the overall flat area of their quinoline prototypes (1370). The importance of the dialkylaminoalkylamino side chain for antimalarial potency was reaffirmed when such chains were attached to the 2-(*p*-chlorophenyl)aminopyrimidines. After

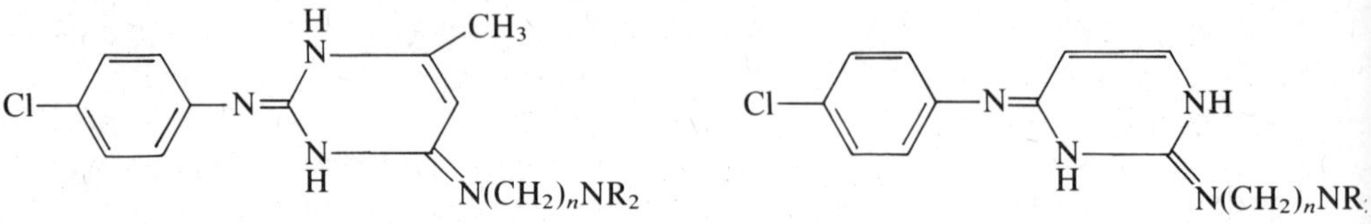

Prototropic changes in aminopyrimidines

sparring with some initial exploratory examples, the best "leads" were found in compounds capable of having both imino groups undergoing simultaneous prototropic change with a pyrimidine nitrogen. Both these structures contain guanidine moieties, one with the *p*-chloroanilino group participating, the other with the basic side chain furnishing the imino position (1371). In an effort to simplify the molecule, the pyrimidine could be regarded mainly as a source of tautomerizable nitrogens and could be opened to give biguanide derivatives (1372); disappointingly, such a model compound was inactive (1373).

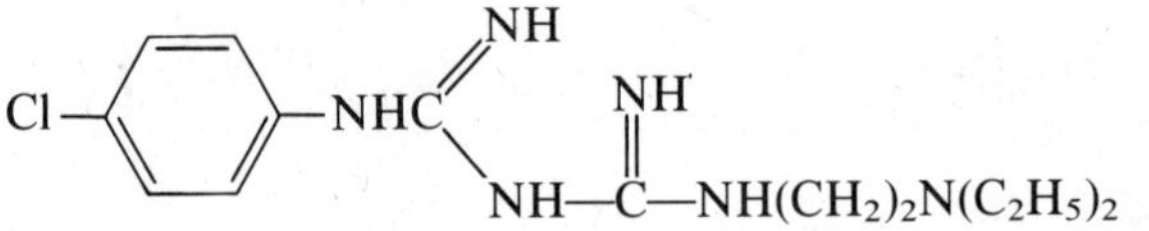

A biguanide derivative

What was the cause of this slump? Could it be the increased basicity of the compound? It contains not only a diethylamino but also two basic guanidine groups. This suggestion was followed up by substituting the terminal guanidino group with simple alkyls; the isopropyl derivative became a clinical drug (chlorguanide). In contrast with the aminoquinolines, chlorguanide does not destroy plasmodial gametocytes but is sporozonticidal, that is, it can be used prophylactically.

Chlorguanide is metabolized to several triazines depending on the animal species. Its human metabolite (1374) is more active than chlorguanide in avian but less active against simian and human malarias, probably because it is excreted too rapidly.

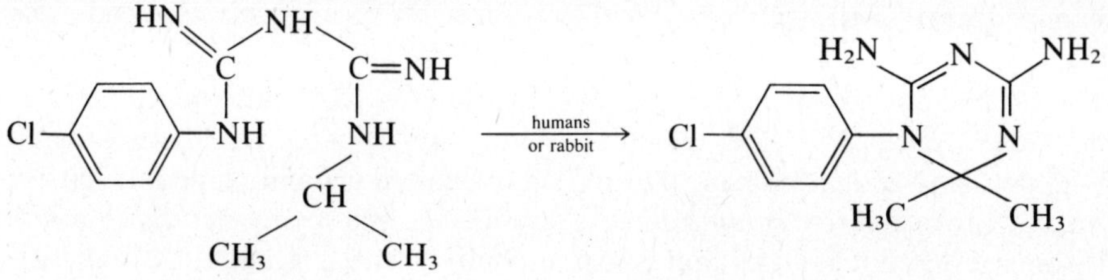

ANTIPROTOZOAL AGENTS

Chlorguanide

Human metabolite

If chlorguanide is viewed as an "open cyclic" form, it bears a formal resemblance to 2,4-diamino-5-(*p*-chlorophenoxy)pyrimidine (1375), which had been found to antagonize folic acid in *Lactobacillus casei in vitro* by an

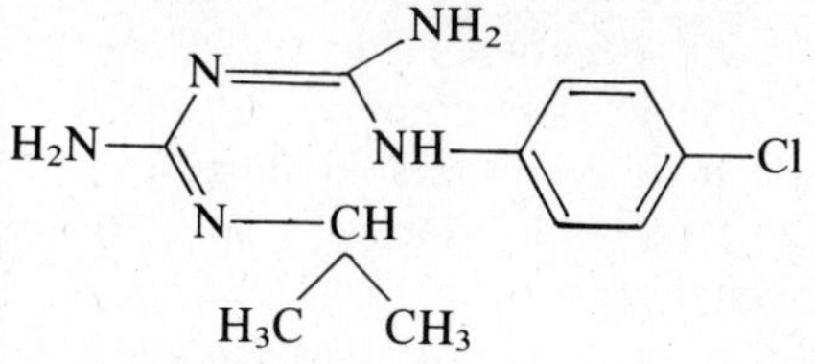

Chlorguanide

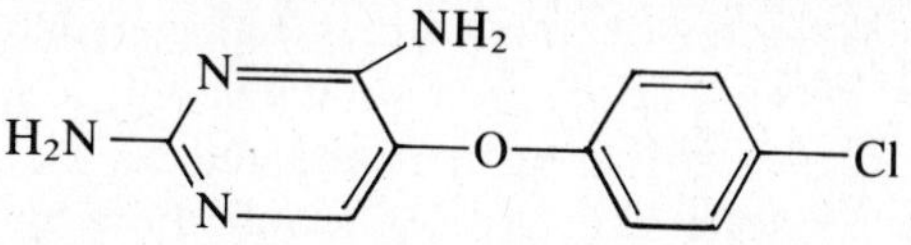

2,4-Diamino-5-(*p*-chlorophenoxy)pyrimidine

American research team (1376). Since chlorguanide has antifolate properties, the diaminopyrimidine derivatives and many of its congeners were screened as antiplasmodials and found to be potent in these test systems (1377). SAR variations included first the isosteric replacement of ether oxygen by CH_2 and later the deletion of any bridging atom between the two rings. The drug, pyrimethamine, was chosen as the most promising and active compound in this series.

The "malaria problem" had been proclaimed as solved several times, but on each occasion plasmodial resistance to available drugs has forced me-

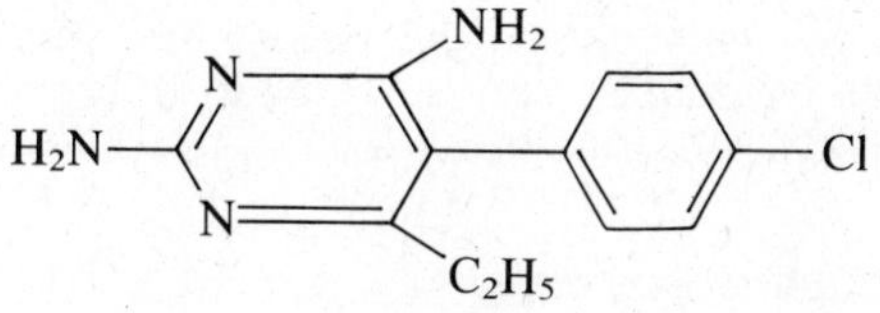

Pyrimethamine

dicinal scientists to approach the problem anew. With the better understanding of the isoenzymes on which host and parasite rely for their metabolism at widely different rates, it may become possible to surmount these difficulties in the future. Also, the development of a malaria vaccine may be in the offing based on improved culture methods of selected plasmodial strains (1377a).

3.20. ANTHELMINTICS

The human and commercial incentive to find effective anthelmintics is tremendous. About 20% of all the humans and animals in the world are infested with parasitic and pathogenic worms. An estimated one billion people harbor the roundworm *Ascaris lumbricoides,* up to 300 million are victims of schistosomiasis, about $100 million is lost annually in the United Kingdom alone by liver fluke infestations of cattle and sheep, tapeworms infest 150 million people worldwide and hookworms infest 750 million. The suffering due to secondary diseases and malnutrition cannot even be estimated. Yet, these figures in themselves tell the tale that the available drugs are inadequate to deal with the problems of worm-caused pathology in any part of the globe. Improved hygiene, control of the association of humans with worm-transmitting animals, and eradication of animal vectors such as snails in the case of schistosomiasis may help but have not been effective. That these conditions were prevalent when only a fraction of the present number of humans roamed the earth is attested by the discussion of symptoms of human nematode and trematode infections in the Egyptian Ebers Papyrus ca. 3500 years ago, and by the description of such infections in dogs, horses, and farm animals by Hippocrates and Aristotle. Lucretius mentioned the anemia accompanying hookworm infection in slaves used as miners by the Romans.

This chapter does not list the classes of worms, the test methods for experimental drugs, nor the pathological and clinical manifestations caused by helminthiases. They have been reviewed and referenced recently (1378). The number and structural variety of anthelmintics reads like a potpourri of organic chemicals. In some laboratories, one type of experimental helminthiasis is established by a suitable test procedure, in others broad anthelmintic testing is practiced—depending on the scientific evaluation of available test methodology, the apparent attractiveness of "lead" compounds, the experience of the staff in the chemistry of promising drugs, and the consent of market research or sponsoring public health agencies. In our discussion the evolution of the most important prototype drugs used in helminthiases is traced according to the classes of worms involved. Many types of these drugs overlap in their activity in different helminthiases, in different hosts, different species of worms, and also in unrelated diseases.

3.20.1. Drugs for Schistosomiasis

Schistosomiasis is caused by several trematodes (flukes), especially *Schistosoma haematobium, S. mansoni,* and *S. japonicum.* These worms have complicated life cycles, infecting aquatic snails as intermediate hosts, which then transmit swimming cercariae to humans.

The earliest chemotherapeutic attempts to control schistosomiasis used metallic antimony pills; they caused emesis, were recovered from the vomit, and used over again. After these experiments, credited to Theophrastus

ANTHELMINTICS

Bombastus von Hohenheim (Philipus Aureolus Paracelsus, 1493?–1541), the reputation of antimony declined. It was revived in 1918 in the form of tartar emetic (1379) and other trivalent antimonials, but their cardiotoxicity has impeded widespread use. Tying up the antimony atoms by chelation with, for example, penicillamine (1380) appears to improve the therapeutic index of these materials.

3.20.1a. Hycanthone. Continuing his successful synthesis of heteroaromatic—primarily tricyclic—compounds with dialkylaminoalkylamino side chains, which had led to quinacrine, Mauss (1381) attached diethylaminoethylamino groups to thioxanthene and xanthene systems. These compounds, trade-named Miracils, were found in the 1930's to exert activity

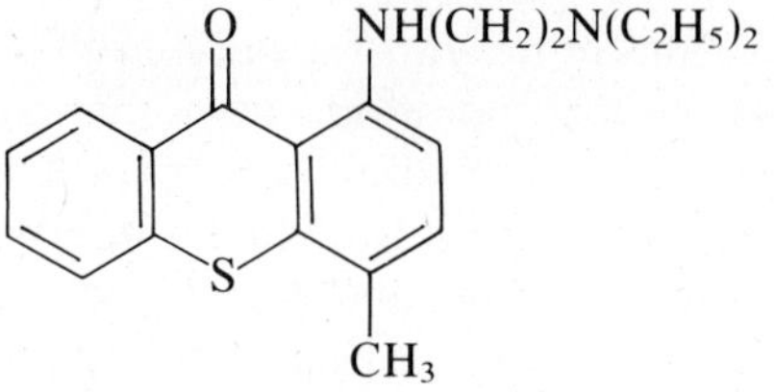

Lucanthone

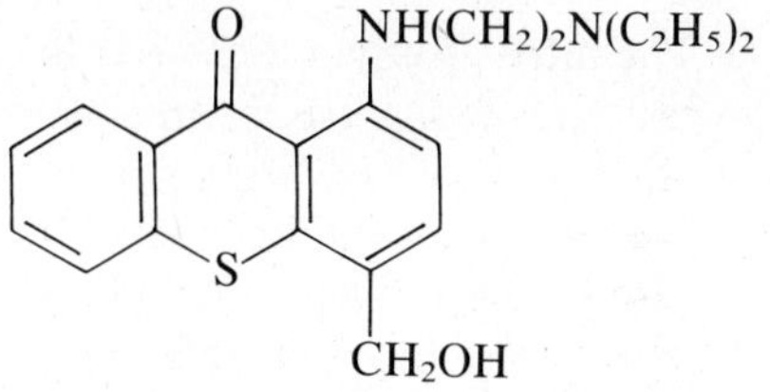

Hycanthone

against *S. mansoni* in mice (1382). One of these, Miracil D or lucanthone, was introduced as the first clinically active—though still very toxic—schistosomicide. Of the many analogs tested, all the active ones contained a *p*-toluidine system with a basic side chain at the amino group. Compounds without a *p*-methyl group were inactive. During a screening program of antifungal agents, Archer and colleagues (1383, 1384) isolated three metabolites from an exposure of lucanthone to *Aspergillus scleroticum*. These metabolites contained successive oxidative stages of the methyl group, namely CH_2OH, CHO and COOH. The drug hycanthone was synthesized and shown to be the active metabolite of lucanthone (1385).

Hycanthone (and lucanthone) are highly effective schistosomicides but they are carcinogenic in mice and mutagenic. Efforts to downplay these toxic effects resulted in benzothiopyrano[4,3,2-*cd*]indazoles (1386), but even though the acknowledged goal of separation of properties was achieved, some mutagenic activity still remained.

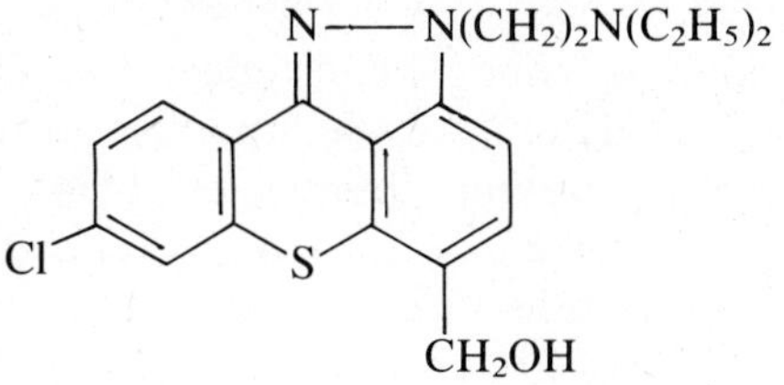

Chloro-hydroxymethyl-diethylaminoethylbenzothiopyrano[4,3,2-*cd*]indazole

3.20.1b. Nitroheterocyclics.

Another combination of the *p*-toluidine structure and the methanol group in lieu of methyl was attained in oxamniquine (1387, 1388), prepared by oxidative fermentation of the corresponding 6-methyl derivative (1389, 1390). Oxamniquine is clinically useful in *S. mansoni* infections.

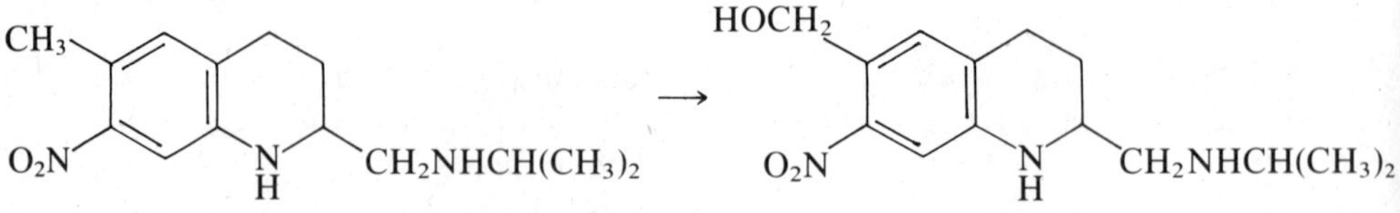

[Structure: 2-Isopropylaminomethyl-6-methyl-7-nitro-1,2,3,4-tetrahydroquinoline with CH$_3$ and O$_2$N substituents, N–H, CH$_2$NHCH(CH$_3$)$_2$] → [Structure: Oxamniquine with HOCH$_2$ and O$_2$N substituents, N–H, CH$_2$NHCH(CH$_3$)$_2$]

The nitro group in these tetrahydroquinolines represents the optimal electron-withdrawing group tried in this series. Its incorporation was prompted by the presence of NO$_2$ in several 5-membered heterocyclic compounds that exhibit schistosomicidal activity. Among these were 2-acetamido-5-nitrothiazole (1307), niridazole (1391) and other nitrothiazole and nitrothiazoline derivatives. Among nitrofurans, furapromidium is active against *S. japonicum*. *trans*-5-Amino-3-[2-(5-nitro-2-furyl)vinyl]-1,2,4-oxadiazole (SQ 18506)

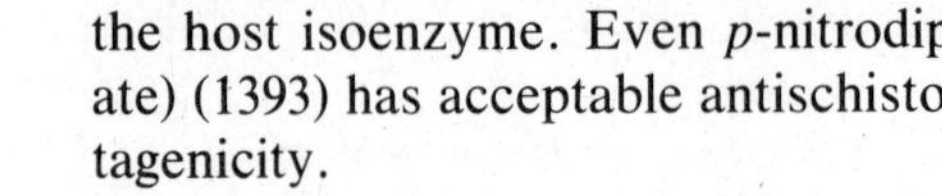

[Structure: Furapromidium — O$_2$N-furan-CH=CHCONHCH(CH$_3$)$_2$]   [Structure: SQ 18506 — O$_2$N-furan-CH=CH-1,2,4-oxadiazole-NH$_2$]

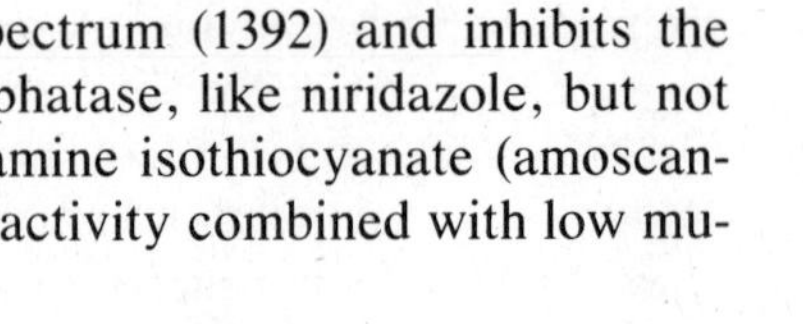

has a broader antischistosomal activity spectrum (1392) and inhibits the schistosomal glycogen phosphorylase phosphatase, like niridazole, but not the host isoenzyme. Even *p*-nitrodiphenylamine isothiocyanate (amoscanate) (1393) has acceptable antischistosomal activity combined with low mutagenicity.

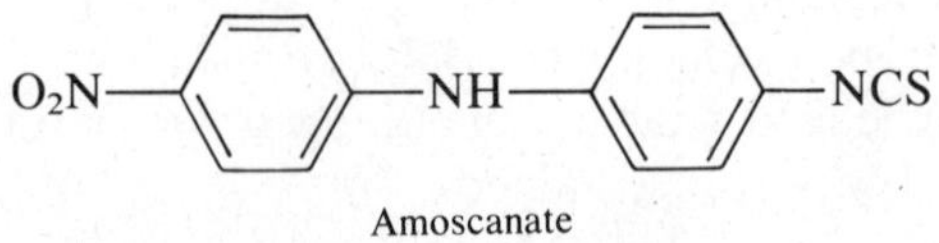

[Structure: Amoscanate — O$_2$N–C$_6$H$_4$–NH–C$_6$H$_4$–NCS]

The organophosphorus anticholinesterase insecticide metrifonate [Cl$_3$CCHOHP(O)(OCH$_3$)$_2$] is clinically highly active against *S. haemotobium* infections but lacks activity against *S. mansoni* (1394).

3.20.2. Other Trematode Infections

If sheep and man could be taught to avoid diets containing odd animals that serve as intermediate hosts for pathogenic trematode worms, much misery and agricultural loss could be avoided. As unlikely as it appears, sheep consume ants infected with metacercariae of the trematode, *Dicrocoelium dendriticum,* and thus get infected with this helminth. In the Far East, the liver flukes *Clonorchis sinensis* and two *Opisthorchis* species spend part of their life cycles in snails and freshwater fish; the predilection of people in that part of the world for raw fish often results in liver fluke infections. Metacercariae of *Fasciola hepatica* are encysted on wet grass and watercress, which respectively serves as food for sheep and humans and convey the infectious organisms into the body.

Carbon tetrachloride is a fasciolicide in sheep; it blocks cholesterol biosynthesis in the host (1395) and deprives the worms of this chemical since they depend on the host for it. Chloroform, the metabolite of CCl_4, probably contributes to the spasms caused by CCl_4; this chemical also leads to the destruction of the parasites (1396). By way of its spasmogenic metabolites, pentachloroethane and tetrachloroethylene, hexachloroethane [$(Cl_3C)_2$] also kills *F. hepatica.* Another halocarbon, p-$Cl_3CC_6H_4CCl_3$ [Hetol] (1397), is a broad-spectrum antitrematode agent as well as an insecticide. It uncouples oxidative phosphorylation (1398), but it causes a blood dyscrasia in dogs.

Just as 5-membered nitroheterocyclic derivatives have been screened in every kind of infection and infestation, phenolic compounds with their bactericidal record have been put through their paces, especially in veterinary applications where FDA regulations are more relaxed. Halogenated nitrophenols were found good candidates as flukicides, with 2,6-diiodo-4-nitrophenol (disophenol) (1399) providing a medium-good "lead." Its analog with a similar distribution of electron-withdrawing groups, 4-cyano-2-iodo-6-nitrophenol (nitroxynil), is more widely used. This theme has been varied further in bisphenolic compounds, led by the inexpensive hexachlorophene.

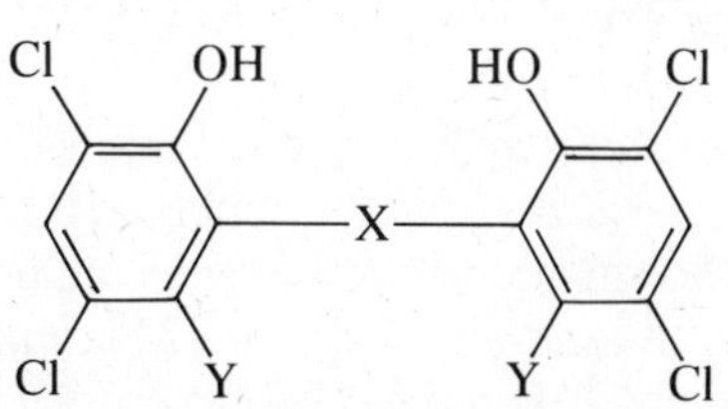

Hexachlorophene: X = CH_2, Y = Cl
Bithionol: X = S, Y = H
Bitin-S: X = SO, Y = H

Bithionol is particularly effective in human fascioliasis. Other flukicides of overall analogous structure contain CONH and CSNH bridges as well as bulky and different electron-withdrawing substituents such as CF_3. In the many examples of molecular variation in these series, Hansch analysis has pointed to the role of lipophilicity in substituents Y (1400). A particularly potent flukicide was found in 2,6-dihydroxy-3,4′,5-trichlorobenzanilide (1401). Most of these compounds uncouple oxidative phosphorylation; they have little effect on the development phase of *F. hepatica,* but this has been overcome largely in diamphenethide (1402, 1403), although this does not take into account the fine-tuning of these drugs in different stages of fascioliasis.

$$CH_3CONH-\langle\text{C}_6H_4\rangle-OCH_2CH_2OCH_2CH_2O-\langle\text{C}_6H_4\rangle-NHCOCH_3$$
Diamphenethide

The history of anthelmintics presents an example of chemotherapeutics developed so empirically that not much logic can be seen in it overall. Only in very recent years have biochemical processes in selected helminths been studied and suggestions made about potential metabolite analogs. Although the differences between the anatomies and life cycles of worms and mammalian hosts are vast, few exploitable variations have been encountered. In almost all institutions that are interested in anthelmintics, broad across-the-board screening of chemicals against worms representative of accepted zoological classes is being practiced. In some older pharmaceutical firms, emphasis has been placed on helminthiases of humans because of the interest of such companies in clinical medicine and their sales staff's training in visiting physicians' offices. With increasing diversification, the emphasis has frequently shifted to veterinary anthelmintics because the number of infested sheep, swine, cattle, dogs, cats, and even horses considerably exceeds that of human worm infestations. Moreover, regulatory agencies may relent in their condemnation of minor toxic side effects if they can be assured no human use of such drugs will be involved and if the drug is metabolized without leaving toxic residues in the animal carcasses.

3.20.3. Drugs for Filariasis

Living part of their complicated life cycles as microfilariae in insect vectors (mosquitoes for *Wuchereria bancrofti, Simulium* flies for *Onchocerca volvulus*), filarial nematodes invade various tissues including lymphatic vessels and the eyes, causing blindness and major economic dislocations. Of the large number of compounds screened in laboratory animals and humans, only three types of drugs have come into therapeutic use.

As in other chemotherapeutic areas, arsenicals initiated drug therapy in filarial infestations. Arsenamide (1404, 1405) is still used widely to treat dogs infected with heartworm (*Dirofilaria immitis*). Another arsenic disulfide derivative is F151 (1406, 1407), which is especially effective in this and similar filarial diseases when used in combination with a benzimidazolylbenzimidazole (Hoechst 33258) (1408).

Arsenicals have lost ground where nonmetallic organic drugs have surfaced, as in other therapeutic areas. Basing their conclusions on a rapid test in cotton rats naturally infested with *Litomosoides carinii* (1409), a group of biomedical scientists at the Lederle Laboratories of the American Cyanamid

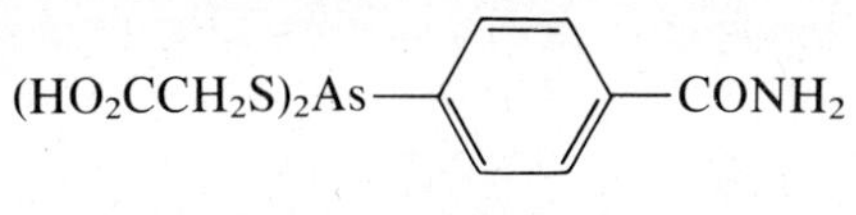

ANTHELMINTICS

$(HO_2CCH_2S)_2As$—⟨benzene⟩—$CONH_2$

Arsenamide

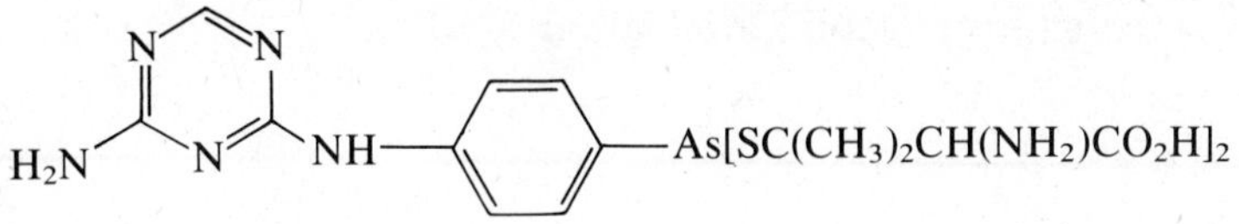

H_2N—⟨triazine⟩—NH—⟨benzene⟩—$As[SC(CH_3)_2CH(NH_2)CO_2H]_2$

F-151

CH_3N⟨piperazine⟩N—⟨benzimidazole⟩—⟨benzimidazole⟩—⟨benzene⟩—OH

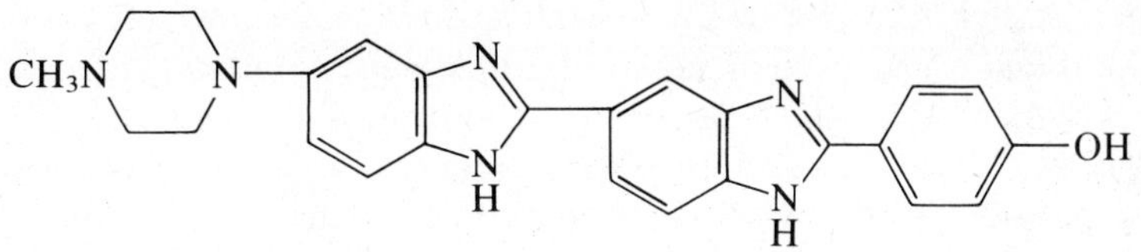

H-33258

Company screened a large number of compounds for antifilarial activity. Among them was a piperazine derivative, 1-carbethoxy-4-methylpiperazine, which had originally been prepared for comparison with meperidine, the structurally similar analgetic drug. This piperazine urethane had good selective properties. It showed considerable activity in the microfilarial test

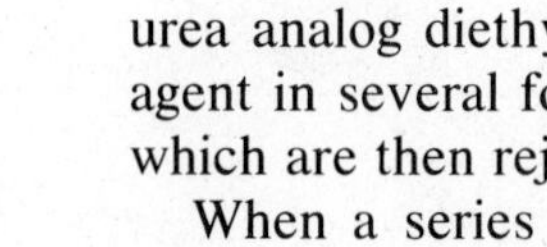

CH_3N⟨piperidine⟩$(C_6H_5)(CO_2C_2H_5)$ CH_3N⟨piperazine⟩$NCO_2C_2H_5$ CH_3N⟨piperazine⟩$NCON(C_2H_5)_2$

Meperidine 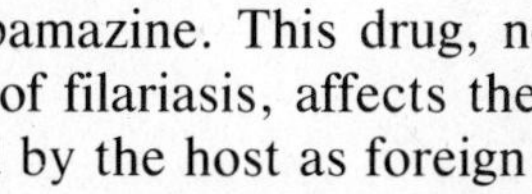   1-Carbethoxy-4-methylpiperazine 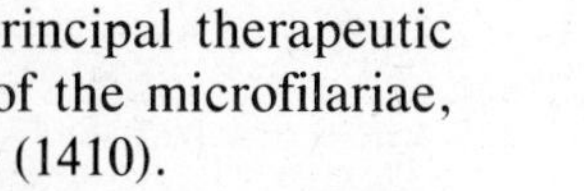   Diethylcarbamazine

and served as a "lead" for molecular modification, which culminated in the urea analog diethylcarbamazine. This drug, now the principal therapeutic agent in several forms of filariasis, affects the cuticle of the microfilariae, which are then rejected by the host as foreign particles (1410).

When a series of analogs of diethylcarbamazine had been established (1411), the basic nucleus, piperazine itself, was found to be of value as an ascaricide (1412) and as an oxyuricide (pinworm infections) (1413). Piperazine had been tried originally in rheumatism and found to be an anthelmintic accidentally (1414). It has replaced the older gentian violet and the tetracyclines in the treatment of human pinworm infections.

Onchocerciasis, caused by *Onchocerca volvulus,* responds well to suramin; other filariae remain unaffected by this drug. This was observed while searching for other activities of this primarily trypanocidal drug. The same process has been applied to other agents and has divulged a measure of filaricidal activity of haloxon, mainly an ascaricidal drug, the isothiocyanate drug amoscanate, and tetramisole. Metronidazole, a trichomonacide, has some clinical activity against *Dracunculus medinensis* (Guinea worm), which causes dracontiasis.

3.20.4. Drugs for Cestode Infestation

Cestode (tapeworm) infections in humans and animals have been treated historically with a variety of "remedies." These include extract of male fern *Dryopteris filix mas* rhizomes (oleoresin of aspidium), pharmaceutically still available. It contains filicin that has been fractionated to filixic acid (1415) and other phenolic ketones. Other older anthelmintics were the alkaloid arecoline (methyl 1,2,5,6-tetrahydro-1-methylnicotinate), which occurs in the betel nut and owes its taenicidal properties to its stimulation of intestinal peristalsis, which expels the tapeworms. Its indiscriminate cholinergic action restricts it to veterinary practice. The antimalarial quinacrine (mepacrine) has anthelmintic activity but its staining properties have limited its clinical use. A number of organometallic compounds such as arsenicals and lead and tin derivatives used to be advertised as taenicides but they are no longer used. Instead, more recent nonmetallic agents have taken their place. They were found during broad-spectrum anthelmintic screening programs; sometimes the same drugs proved of interest in different helminthiases, at other times further molecular modification was needed in one or the other infection in order to attain better overall performance or greater potency. Thus, in the series of halogenated bisphenols, dichlorophen (5,5'-dichloro-2,2'-dihydroxydiphenylmethane) had been a long-standing veterinary taenicide but found acceptance in clinical practice only in 1956. The best clinical agent in this series is niclosamide, a benzanilide derivative (1416). This drug also

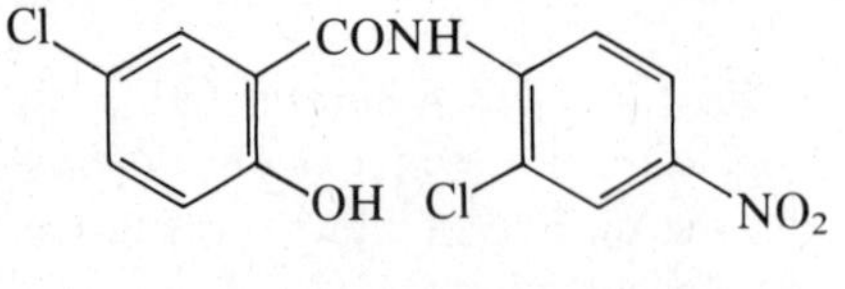

Niclosamide

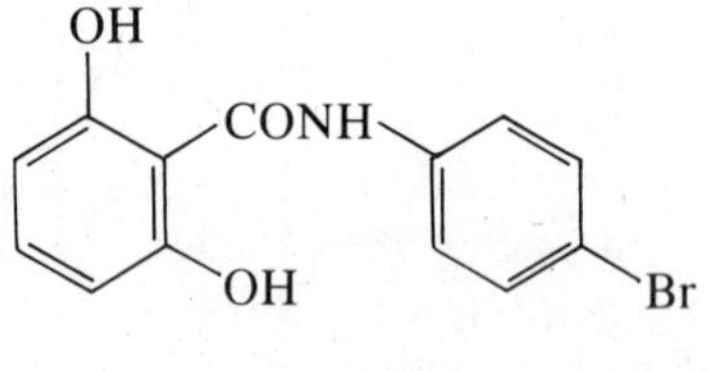

Resorantel

kills snails and can be of use against *Biomphalaria galbrata*, which transmits schistosomes. Similarly, resorantel (1417), a resorcinol derivative, shows high and useful cestocidal and trematocidal activity in animals.

In a series of nonhalogenated aromatic amidines of the type RO-naphthalene-C(=NH)NR$_2'$, bunamidine proved of value as a taenicide in dogs (1418). Bitoscanate (1419) and related diisothiocyanate compounds, and the organophosphorus compound vincofos (1420) are anticestode drugs. Vin-

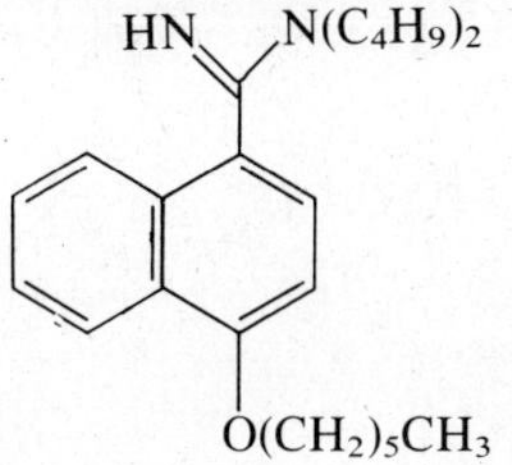

ANTHELMINTICS

Bunamidine

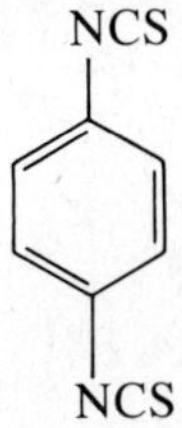

Bitoscanate

cofos appears to owe its anticestode activity to the increased lipophilicity induced by the *n*-octyl ester group; its parent prototype, dichlorvos (1421), is insecticidal and nematocidal but lacks cestocidal activity.

$$Cl_2C=CHOP(O)OCH_3$$
$$|$$
$$R$$

Dichlorvos: R = OCH$_3$
Vincofos: R = O-*n*-C$_8$H$_{17}$

3.20.5. Broad-Spectrum Anthelmintics

Broad-spectrum taenicidal (as well as schistosomicidal) activity has been observed for praziquantel (1422, 1423). It allows glucose to escape from the worm's covering tegument, and this has been suggested as a possible mode of action of the drug.

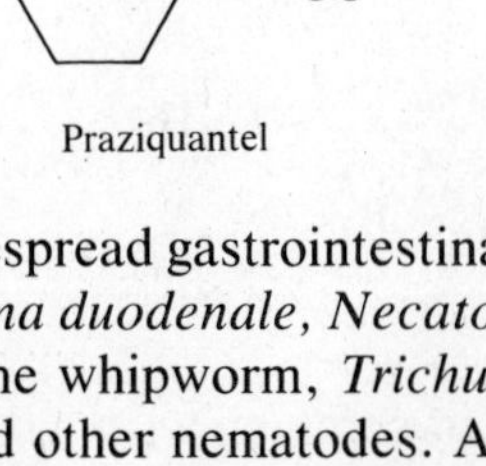

Praziquantel

The worms causing widespread gastrointestinal nematode infestations are the hookworms *Ancylostoma duodenale, Necator americanus*; the pinworm *Enterobius vermicularis*; the whipworm, *Trichuris trichiura*; the nematode *Haemonchus contortus* and other nematodes. As with other helminthiases, older and now abandoned drugs include arsenicals. Cyanine dyes such as dithiazanine iodide (1424) and pyrvinium pamoate (1425) have provided good activity for special helminthiases in selected hosts, evoking echoes of early chemotherapeutic dyestuffs.

A similar unsaturated chain joins the two cyclic moieties of pyrantel,

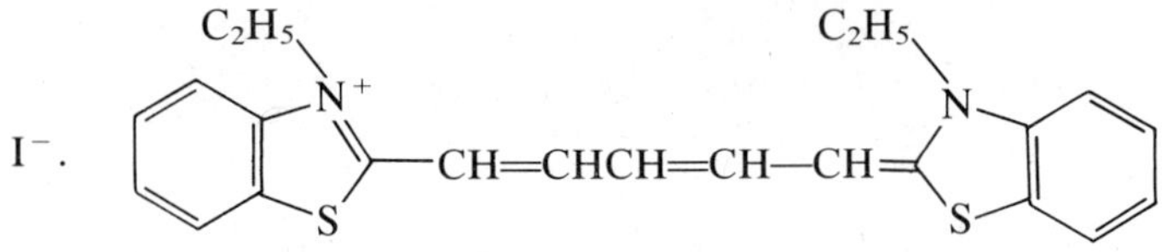

Dithiazanine

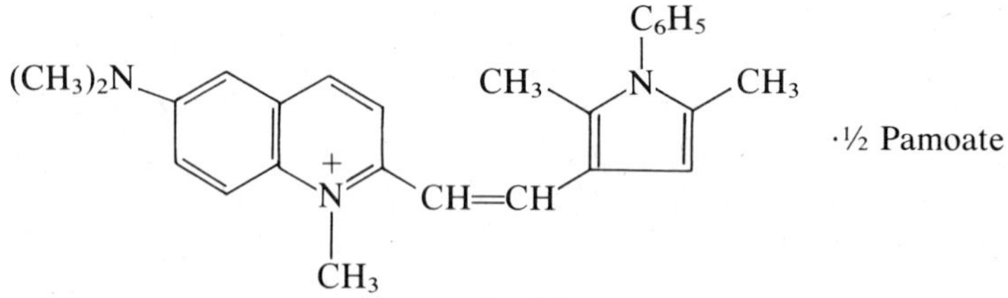

Pyrvinium pamoate

morantel, and oxantel. The first two (1426) are depolarizing blocking agents at the neuromuscular junctions of the worms and are broad-spectrum anthelmintics, especially for *Ascaris* infections. Oxantel (1427) is particularly suitable for the treatment of trichuriasis (1428). Their development was based

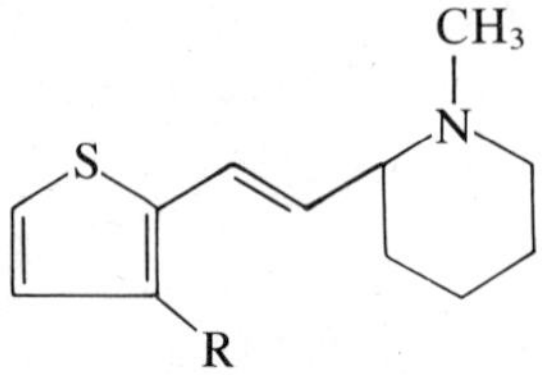

Pyrantel: R = H
Morantel: R = CH$_3$

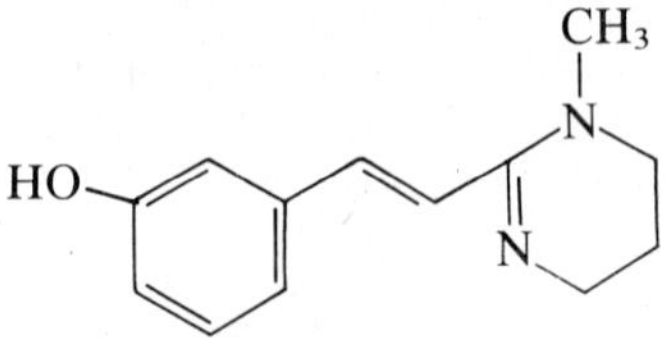

Oxantel

on a "lead" that isothiouronium salts had anthelmintic activity. Such salts have the structure

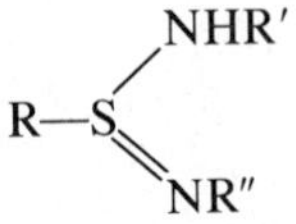

The well-known procedure of combining the substituents on nitrogen into a cycle led to the discovery of potent activity in 2-(4,5-dihydroimidazolyl) 2-thenyl sulfide (1429), and systematic molecular modification of the —CH$_2$S- group via the isosteric —CH$_2$CH$_2$— chain and the unsaturated analog with —CH=CH—, accompanied by expansion of the imidazoleine ring, pointed the way to pyrantel, which is useful in hookworm, pinworm and *Ascaris*

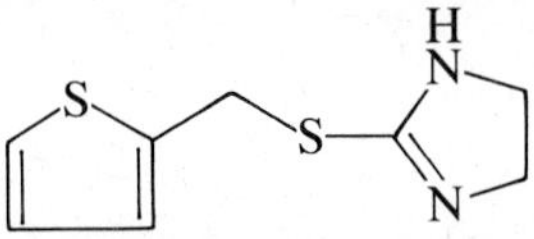

2-(4,5-Dihydroimidazolyl)
2-thenyl sulfide

diseases. The unsaturated bridge should be *trans* preferentially. Lipophilicity correlates well with activity. Of the many analogs tried, the 3-methylthienyl homolog (morantel) and the *m*-hydroxyphenyl analog (oxantel) (1427) were introduced into clinical practice.

Among other types of compounds active in intestinal nematode diseases are drugs previously used in other helminthiases such as bisisothiocyanates, several organophosphorus derivatives, and piperazine.

Broad-spectrum anthelmintics, both for veterinary and clinical purposes, were also initiated in 1961. The series concentrated on benzimidazole derivatives, with 2-(4-thiazolyl)benzimidazole (thiabendazole) as the first drug introduced to combat intestinal nematodes (except *Trichuris trichiura*)

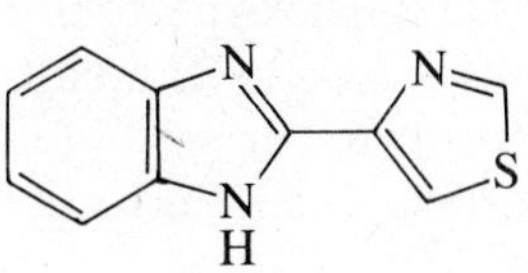

Thiabendazole

(1430). Its mode of action, at least *in vitro*, seems to involve inhibition of fumarate reductase in susceptible worms (1431), although such statements in the literature do not exclude other untested pathways. Thiabendazole is also antiinflammatory, and that may affect the clinical response to worm infections. Its action in sheep is rather brief, the drug being detoxified by hydroxylation in the 5-position (1432).

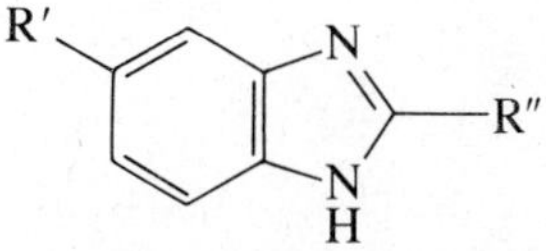

Cambendazole: $R' = (CH_3)_2CHOCONH-$, $R'' = $ [thiazolyl]
Parbendazole: $R' = C_4H_9-$, $R'' = NHCO_2CH_3$
Mebendazole: $R' = C_6H_5CO-$, $R'' = NHCO_2CH_3$

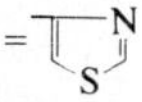

Wide SAR studies in the benzimidazole field led to a number of clinically interesting extensions of thiabendazole, especially cambendazole (1433), and going farther afield, carbamates like parbendazole (1434) and mebendazole (1435). Janssen Pharmaceutica, which had perfected mebendazole, also developed tetramisole, a mixture of S(−) (levamisole) and R(+) isomers (1436). These compounds had their origin in anthelmintic tests with thiazothienol, which had descended broadly from mebendazole. It turned out to be species-specific, giving rise to suspicions that the host's metabolism

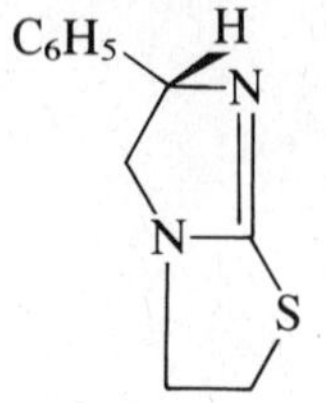

[Structures: Levamisole, Dextramisole (together Tetramisole), Thiazothienol, Thiazothielite]

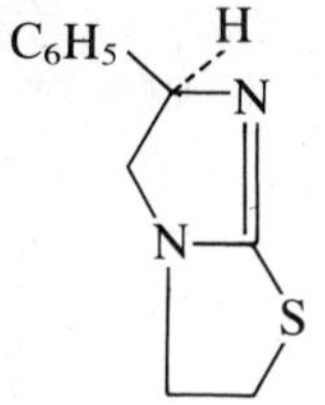

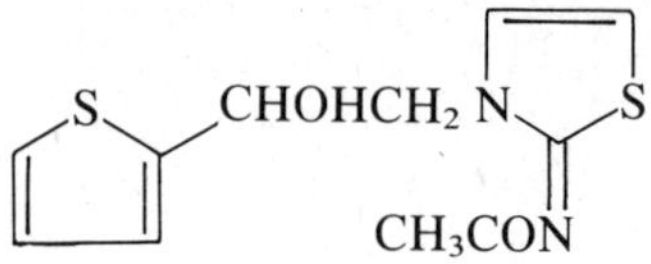

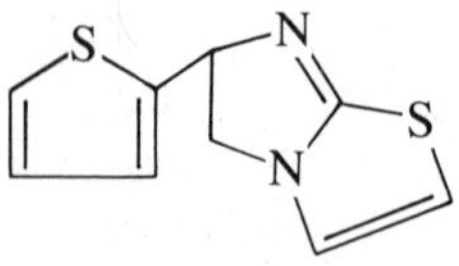

might be material in activating thiazothienol. Of the metabolites isolated, only thiazothielite had lost this species-specificity and was more potent than the parent compound. Extensive molecular modification, including replacing thienyl with the isosteric phenyl group, led ultimately to tetramisole. The broad activity of this substance was then located in the *levo*-isomer, levamisole.

Levamisole inhibits fumarate reductase even in thiabendazole-resistant nematodes, and it inhibits alkaline phosphatase in mammals. It has potent immunosuppressive properties in a number of clinical conditions. The R(+) isomer has even been tested as an antidepressant.

The history of anthelmintics provides a good insight into the progress of general medicinal science. All the early anthelmintics were found by empirical "blind" screening, whereas in recent years, biochemical reasoning has pinpointed the direction research for new drugs could take.

3.21. ANTIVIRAL AND ANTINEOPLASTIC AGENTS

Drugs for virus diseases are latecomers in therapy and prophylaxis. The target viruses were not well understood until the mid-1930's, when Stanley published his studies on the chemical composition of the tobacco mosaic disease virus. The recognition of viruses as nucleic acids encapsuled by proteins raised the question whether such organized macromolecules should be regarded as animate or inanimate. This point also posed the problem whether chemical reactions of viral macromolecules could be influenced by drugs. The answer to this was yes. Since the type—even if not the exact structure—of viral nucleic acids and proteins could be recognized, it appeared plausible to choose analogs of their monomer components as candidates for antiviral tests, as well as reagents that might interact with amino acids, nucleosides, and their polymers. This presented an opportunity for rational drug design, especially since both viral nucleic acids and coating proteins differ from the corresponding chemical components of the host cells. Indeed, in cell cultures many such inhibition experiments have shown promising results. In whole animals only few drugs could be pursued extensively. Apart from the ever possible effects of drug metabolism by the host, the fine-tuning of differential attack on similar biomacromolecules of host and virus has remained very difficult in the majority of experiments.

In the long run, it does not help the medicinal chemist's mission to fully understand the stages of action of antiviral drugs. The agents can interfere with the adsorption of a virus on host cells, they can prevent penetration of the host cell by the virus, or they can halt uncoating of the viral protein coat and thereby inhibit baring of the viral genome. It is important for the virologist to appreciate these mechanisms since they determine the antiviral testing systems to be employed. It is easier to get an overall idea of the antiviral potential of a compound *in vitro* than *in vivo* but the *in vivo* tests in infected animals remain indicators of the clinical potential of a drug. Infected embryonated eggs are the standard "*in vitro–in vivo*" systems in antiviral researches, but the few cases in which positive inhibition of viral multiplication in the embryonated egg was predictive for true *in vivo* experiments illustrate the difficulties of evaluating test results in these series. Viruses grown in tissue cultures are also used for *in vitro* tests. Hundreds of compounds found to be positive *in vitro* disappeared from further consideration when their activity could not be asserted in survival experiments in infected animals (1437). This is illustrated by the use of tests with enzymes found in viral life processes, such as neuraminidase, protein kinase, nucleotide phosphorylases, RNA- and DNA-dependent RNA polymerase, and reverse transcriptase. Few compounds that inhibit these enzymes are active clinically against the associated viral diseases.

Similar conditions prevail in testing antineoplastic drugs. Here, however, regression of the weight of about a half-dozen selected rodent tumors caused by a given compound indicates in 90–95% of all cases that the compound will be active in human cancers. Vice versa, 95% of all compounds found active in clinical malignant neoplasias will reduce those rodent tumors. In recent years it has become possible to clone biopsy specimens of human tumors and use them in human stem cell clonogenic assays. These sensitive tests and other *in vitro* and *in vivo* methods have been reviewed (1438).

3.21.1. Natural Compounds

As in all other fields of medicinal agents, natural products have been tried in antineoplastic and antiviral tests. In both types of tests, a few of these compounds have shown considerable activity. In tests against certain DNA viruses including *Orthopoxvirus* and *Herpesvirus,* the rifamycin antibiotics, especially the semisynthetic rifampin, have been found active. These macrocyclic compounds are effective inhibitors; they bind to the polymerase molecule in RNA leukemia viruses (1439–1441). Among other antibiotics, active *in vivo* against a variety of viruses (tumor viruses, influenza, vaccinia), are distamycin A, 9-methylstreptimidone, and bleomycin A2 (1442). The tetracyclines and a few other antibacterial antibiotics inhibit some large psittacosis-lymphogranuloma viruses.

Similarly, several antibiotics have cytostatic and particularly anticancer activity. Among the first were the bright-red, highly toxic actinomycins (1443). Others include the bleomycins (1444) and the anthracycline antibiotics duanomycin and doxorubicin (adriamycin) (1445, 1446). These latter

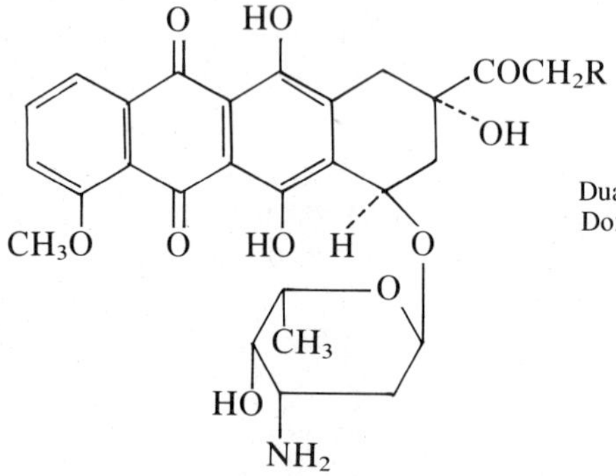

Duanomycin: R = H
Doxorubicin: R = OH

have lent themselves to molecular modification (1447). Not only microbial antibiotics like those just mentioned have been found active in cancer chemotherapy; a growing number of complex antitumor agents extracted from plants have joined their ranks. Among them are the mitotic inhibitors, the podophyllotoxins and some of their hydrazide derivatives, as well as the vinca alkaloids vinblastin and vincristine and some of their derivatives (1448, 1449). The vinca alkaloids are used against rapidly growing tumors (leukemias, lymphomas, choriocarcinoma, neuroblastoma, rhabdomyosarcoma, Wilms' tumor, and carcinoma of the testes). The formulas of a few other, newer antitumor agents isolated from botanical sources are pictured here (1438). Some of them, notably maytansine, are in clinical trial.

As can be seen, many functions could be modified in these compounds, but experiments in molecular modification have been hampered by limitations in the supply of the natural products.

3.21.2. Immunostimulants

Both malignancies and viral infections can fall prey to the organism's immune systems. Failure of the immuno-defenders of the body to rid the animal of the "foreign" tumor cells or the viruses is believed to be a component, if not the cause, of these invasive diseases. Of the many experiments de-

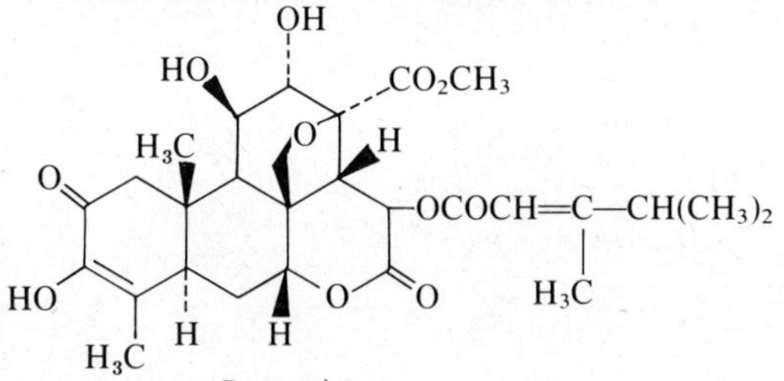

ANTIVIRAL AND ANTINEOPLASTIC AGENTS

Bruceantin

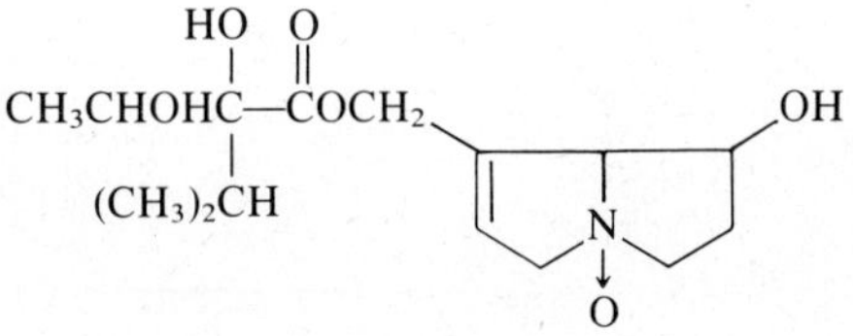

Indicine N-oxide

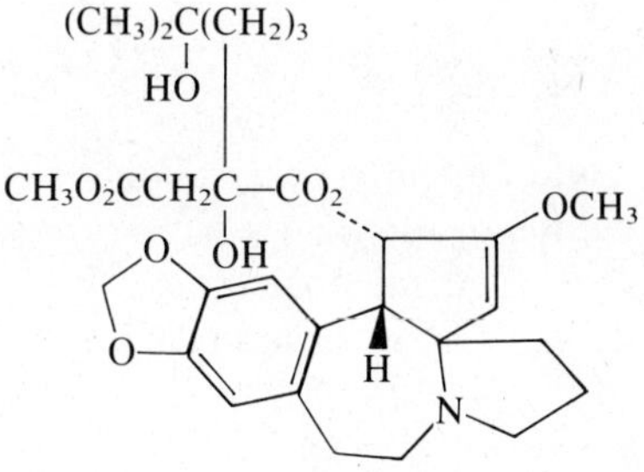

Homoharringtonine

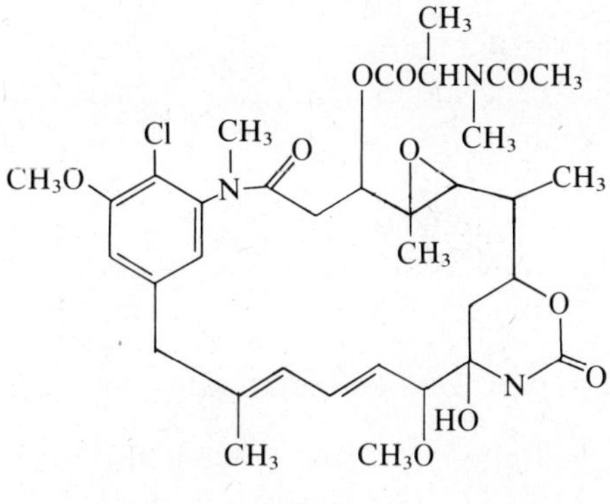

Maytansine

signed to strengthen the immune defenses, only those on interferon are touched upon as far as they permit chemistry to come into play (1450–1452).

Interferon is a species-specific glycoprotein made by all animal cells in minute amounts. After viral attack (1453, 1454) it is released by the cell and appears to regulate the immune system. In the intercellular fluid it signals neighboring cells to produce an antiviral protein. This chain reaction ultimately activates the natural phagocytes. In addition to its (indirect) antiviral activity, interferon also may have anticancer and possibly other activities. It has been cloned in bacteria, purified with monoclonal antibodies, and sequenced for amino acids. Leukocyte interferon is now being developed by recombinant DNA techniques; so far it has shown modest effects in cancer patients but is not free of toxicity (1455).

The extremely small quantities of animal interferons available have hampered experimentation to increase the biosynthesis of that protein. Nevertheless, a few inducers of interferon production have been found by random screening of compounds against interferon titers in animals. Examples are tilorone (1456–1458) and related bioisosteres (1442) and a number of 4-(3-dimethylaminopropylamino)-1,3-dimethyl-1*H*-pyrazolo[3,4-*b*] quinoline derivatives (1459). Apart from these more conventional, smaller molecules

$(C_2H_5)_2NCH_2CH_2O$ — [fluorenone structure] — $OCH_2CH_2N(C_2H_5)_2$

Tilorone

discovered by screening for a fashionable activity, the known sequence of amino acids of various interferons should encourage molecular modification of these proteins as a potential simplification of the unwieldy natural agents. In support of this contention, a lecture by Dausset (1460) may be quoted which points out that the products of human lymphocyte antigen (HLA) complex A, B, and C loci [Klein's class I] (1461) are present ubiquitously at the surface of almost all cells of the organism and appear to play a very general biological role. At least three variable regions of the heavy peptide chain of these Class I products are known (1462). The most distant α 1 domain has such a zone between amino acids 30 and 40; it interacts with the influenza virus. Other domains contain additional variable zones. It should, therefore, be possible to block the α 1 domain zone in the design of anti-influenza drugs.

3.21.3. Antimetabolites As Antineoplastic and Antiviral Agents

Antimetabolites prevent the normal biosynthesis or utilization of cellular metabolites. Usually enzymes involved in these processes are inhibited, but the metabolite analog may also serve as a faulty substrate and become incorporated into proteins or nucleic acids, which then cannot fulfill their normal biochemical functions. As well as inhibiting biosynthetic steps in viral metabolism, many antimetabolites inhibit a number of enzymes, not only those involved in the metabolism of neoplasms but also those regulating normal biochemical events. It is therefore not surprising that antimetabolites are not ideal therapeutic drugs. Pathogenic cells build up resistance to them, while the sensitivity of normal enzymes barely ever decreases. This limits the clinical utility of toxic antimetabolites severely, but as a logical approach to drug design they continue to occupy a prominent position. One or two bioisosteric changes in the structure of a metabolite known to be needed in nucleotide or polypeptide biosyntheses can be counted on to change metabolite catalysis to antimetabolite action. Both classical isosteric replacements and the more daring, newer bioisosteric changes of atoms, groups, and ring types, have been tried in hundreds of variations. Conjectures whether a given metabolite analog designed on these guidelines will be antineoplastic, antiviral, or interfere with different pathogenic enzymic reactions have remained mostly guesswork. In other words, only few avenues to real specificity have been found. Some of them, with a modicum of specificity, are listed here as an encouragement not to give up on the metabolite analog method.

Glutamine, the monoamide of glutamic acid, is a cofactor in transferring amino and amide groups to positions occupied by enolizable carbonyl functions in the biosynthesis of purine and pyrimidine nucleotides and intermediate steps (1463). Several of these biosyntheses are inhibited by O-diazoacetyl-L-serine (azaserine, $HO_2CCH(NH_2)CH_2OCOCHN_2$) or 6-diazo-5-oxo-L-norleucine (DON, $HO_2CCH(NH_2)CH_2CH_2COCHN_2$) which were first isolated from Streptomyces beers (1464, 1465) and then synthesized. In a way these diazocarbonyl compounds are (irreversible) alkylating agents with structures that are attracted to the catalytic sites ordinarily reserved for glutamine, which they resemble. As α-amino acids, these antagonists spill over to normal cells where they also compete with several natural amino acids. For details of the inhibition mechanisms, see reference 1463. This example has been chosen to illustrate a case of effective gradual replacement of one or two small groups, leaving the main molecular section of the prototype intact to act as a recognition region for the active site or receptor. In addition, the reactive diazocarbonyl group anchors the antagonist firmly in the vicinity of the active site.

With or without this feature of alkylation, which leads to the covalent binding of the antagonist, many other compounds that interfere with various biosynthetic processes leading to nucleotide and protein constituents have been "designed" on the same principle. "Design" has to be taken with a practical caveat. Most of the bioisosteres designed to be antagonists do interfere with enzymic reactions needed in the normal biosynthesis of intermediates of proteins and nucleotides. The few exceptions are explained on the basis of excess acidity or basicity, excess hydrophobicity or lack of this property, and other apparent discrepancies in physical properties. If properly considered in advance and combined with model calculations, as in the Free and Wilson method (475), the number of these failures can be minimized. Too few medicinal chemists take the trouble of reasoning out these factors on the chemical drawing board. The difficulties involved in finding that *in vitro* active compounds are inactive *in vivo* arise from similar differences in physical parameters plus biological transport complications and metabolic hurdles. Metabolic differences also account for failures to demonstrate antagonistic activity in disease processes in different species. All this results in a good deal of empiricism in the definitive choice of an antagonist as a useful drug although the intellectual choice of a structure designed as a metabolite antagonist is much more predictable.

Examples of alterations of metabolite structures that produce antagonists in such series are found in folic acid analogs and in analogs of preformed purines and pyrimidines from the intermediary metabolism cycles. Among the earliest antimetabolites were aminopterin and methotrexate in which the 4-OH of folic acid was replaced by 4-NH_2. In methotrexate, the 10-imino group was N-methylated (1464). The synthesis of other compounds in this series, which preceded these successful drugs, was motivated by expediency and by the usual desire to round out patent protection. After aminopterin and methotrexate had once been tested and found to be active *in vivo* and in the clinic in the treatment of leukemias, other diaminoheterocycles were prepared and tested (1465–1474). Some of them turned out to be of value in microbial infections but not in cancer chemotherapy. This may be due to their diffusion and mechanism of active transport in different cell lines, although on the basis of their *in vitro* antagonistic behavior they should have had a good chance of inhibiting biosynthetic processes in all kinds of cells.

The activity profile of folic acid antagonists underscores the need for broad screening against tumors and viruses whose nucleotide biosynthesis might be affected by suitable metabolite antagonists. This is seen again in analogs of natural pyrimidines and purines and their nucleosides in which (i) some of the C and N ring atoms have been moved to other positions or have been replaced (N → CH, CH → N, etc.); (ii) some of the substituents have been altered (NH_2 → OH, OH → NH_2, OH → SH, CH_3 → F or Cl, etc.); and (iii) the pentose moieties have been altered (e.g., ribose → hydroxyfurans; ribose → arabinose; deoxyribose → dihydroxycyclopentane; ribose → dihydroxypropane, etc.).

Examples of such antagonists are found among aza analogs of pyrimidines. Thus, 5-azacytidine is active against leukemia in AKR mice (1475) and 6-azauridine inhibits vaccinia and herpesvirus infections *in vivo* (1476, 1477), among other viruses, and in other hosts. Similarly, the pyridone nu-

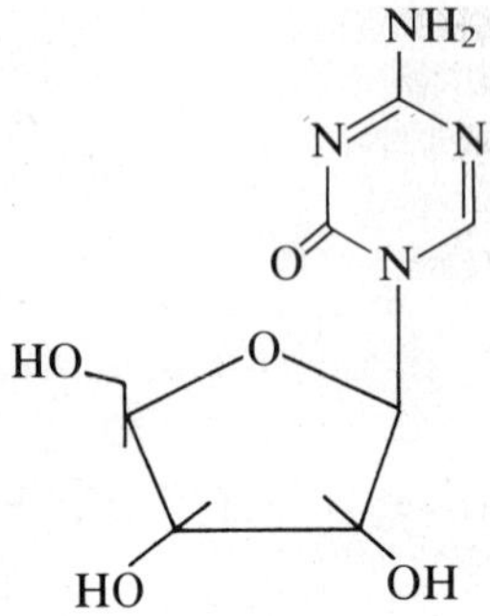

5-Azacytidine

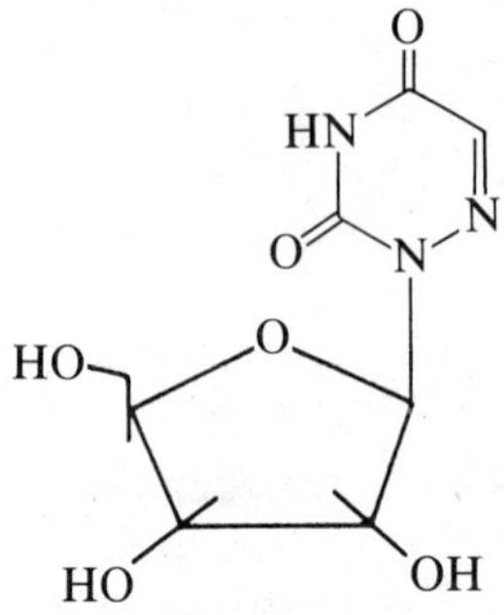

6-Azauridine

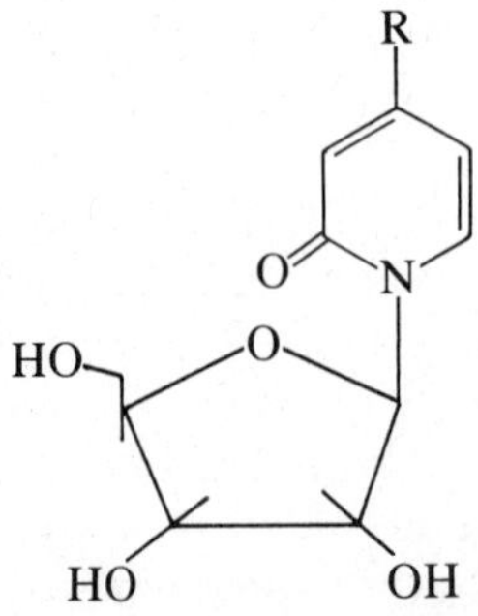

3-Deazauridine: R = OH
3-Deazacytidine: R = NH_2

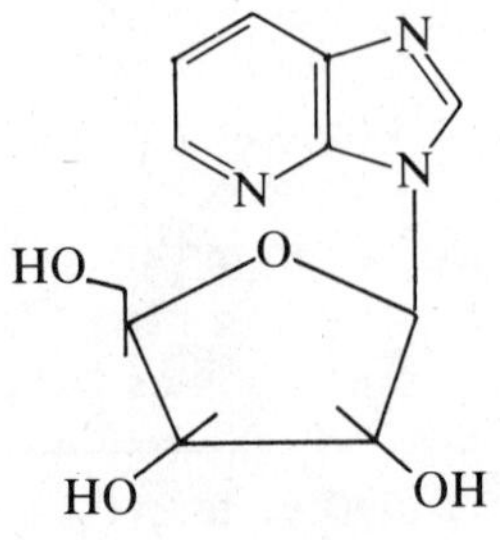

3-β-D-Ribofuranosylimidazole[4,5-*b*]pyridine

cleosides, 3-deazauridine and 3-deazacytidine inhibit the *in vitro* replication of several RNA viruses (1478–1480); in tumor cells they are converted to their 5'-triphosphates and these interfere with pyrimidine nucleotide biosynthesis (1481). Some deazapurine ribonucleosides, without substituents in the heteroring, are said to have activity against parainfluenza virus in the questionable embryonated egg test (1482).

Instead of replacing the imidazole portion, as was done in the aza- and deazanucleosides above, the pyrimidine portion of the purine nucleosides can be partly deleted, with further modification of the imidazole ring. Of the resulting azole nucleosides, ribavirin and especially its 5'-phosphate metabolite, are active against a variety of pathogenic viruses and tumor viruses (1480, 1483).

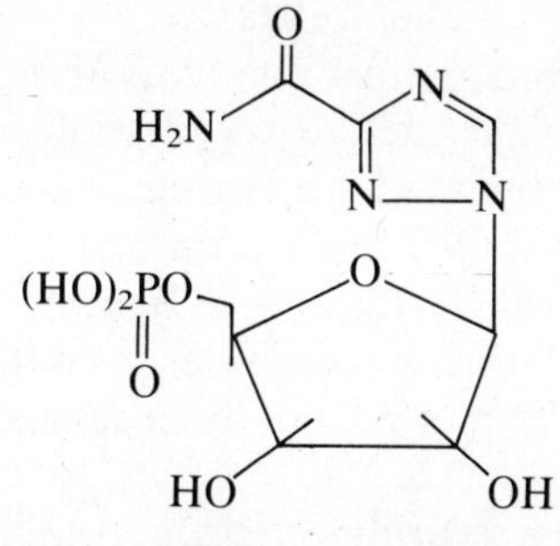

Ribavirin 5'-phosphate

One usually hesitates to draw any conclusions from differences in the biological activity of halogenated compounds in which different halogens (F, Cl, Br, I) or pseudohalogens (CHX_2, CF_3, CCl_3 etc.) occupy analogous structural positions. The van der Waals radii are 2.00 Å for CH_3, 1.95 Å for Br, and 2.15 Å for iodine. Thus these halogens, as well as the chloro substituent, are similar to methyl, and compounds that contain the halogens affect biochemical reactions of the parent methyl derivatives. In the case of 2'-deoxypyrimidine nucleosides, the prototypes are thymidine and its non-

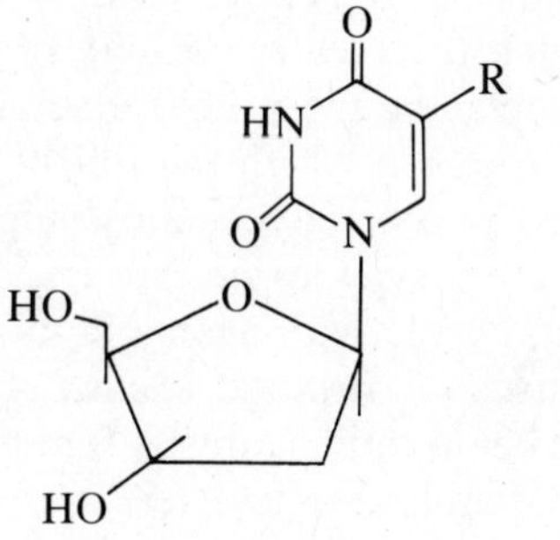

Thymidine: R = CH_3
5-Iodo-2'-deoxyuridine (IdU): R = I
5-Trifluorothymidine: R = CF_3

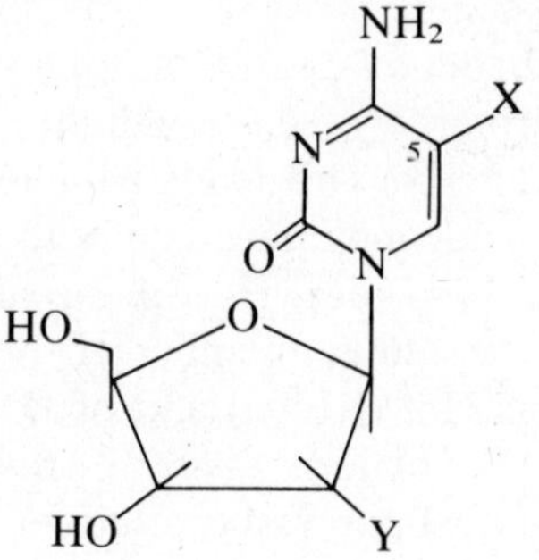

Cytidine: Y = OH, X = H
2'-Deoxycytidine: X = Y = H
5-Fluoro-2'-deoxyuridine: X = F, Y = H

methyl 6-amino analogs, cytidine and 2'-deoxycytidine. Every kind of halogen and pseudohalogen has been substituted in the pyrimidine ring, and those derivatives with halogens in position 5, where the halogen can mimic the methyl of thymine, have received the greatest attention in antiviral and anticancer chemotherapy. 5-Fluorouracil and its 2'-deoxyriboside are potent inhibitors of thymidylate synthetase (1484), the enzyme that ordinarily converts 2'-deoxyuridylic acid to thymidylic acid for DNA synthesis. Thus fluorouracil deprives neoplastic cells of thymine. This has been attributed to the relative similarity of H (radius 1.2 Å) and F (1.35 Å) and also to the 30 times greater acid strength of fluorouracil compared with uracil. The similar size of F makes the compound fit well at the active site of an enzyme that has 2'-deoxyuridylic acid as its traditional substrate. The 5-chloro and bromo derivatives do not fit as well and are less active; the 5-iodo derivative with its large halogen substituent is inactive (232) (but see the 5-iodo-2'-deoxyuridine structure). The higher activity of the fluoro compound must be due to the greater ease with which the drug forms a covalent linkage to a critical SH group of the enzyme (1485). 5-Fluorouracil is metabolized to aliphatic fragments by ring cleavage (—NH—//—CO—) but only by normal cells, not by animal cancer cells. This is regarded as a contribution to the selective therapeutic action of the compound, but other reactions also enter the picture (1463, p. 610).

5-Trifluoromethyl-2'-deoxyuridine (trifluorothymidine) (1486) inhibits thymidylate synthetase, is active against adenocarcinoma 755 in mice, and is useful in the topical therapy of herpes-infected eyes (1487). Its mechanism of action has been reviewed (1488).

5-Iodo-2'-deoxyuridine (IdU) is an effective antiviral agent, with clinical utility against herpes and vaccinia viruses in topical form. IdU and its other halogen analogs are phosphorylated, interact with thymidine kinase, thymidylate kinase, and DNA polymerase in infected and normal cells, and are incorporated into viral DNA which then no longer responds to normal DNA metabolism.

Similarly, 5-bromo- (BdC) and 5-iodo-2'-deoxycytidine (IdC) have activity against herpesviruses, and this extends to 5-halogeno-[and 5-trifluoromethyl]-2'-deoxy-5'-aminouridines in which the greater basicity of the cytidine derivatives has been imitated by replacing the primary alcoholic 5'-OH by 5'-NH_2. This leads over to nucleosides in which the riboside or 2'-deoxyriboside moieties have been replaced by other glycosidic groups, principally arabinosides. Here 1-β-D-arabinofuranosylcytosine (ara-C), the clinically most studied member of the group, has activity against a number of neoplasms (1489, 1490) and of DNA viruses *in vitro* and *in vivo* (1491). A large number of derivatives of Ara-C have been studied, mostly for antitumor and immunosuppressive activities. Several arabinosidic esters are resistant to cytidine deaminase, but as they release ara-C slowly they retain the activities of the parent compound. The enzymic reactions of ara-C have been reviewed (1463).

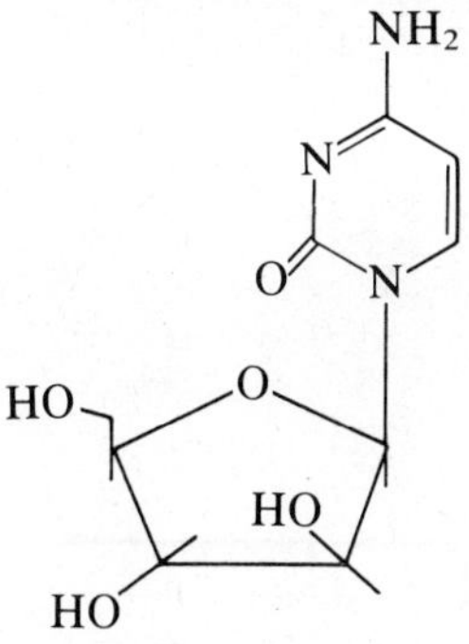

ANTIVIRAL AND ANTINEOPLASTIC AGENTS

Ara-C 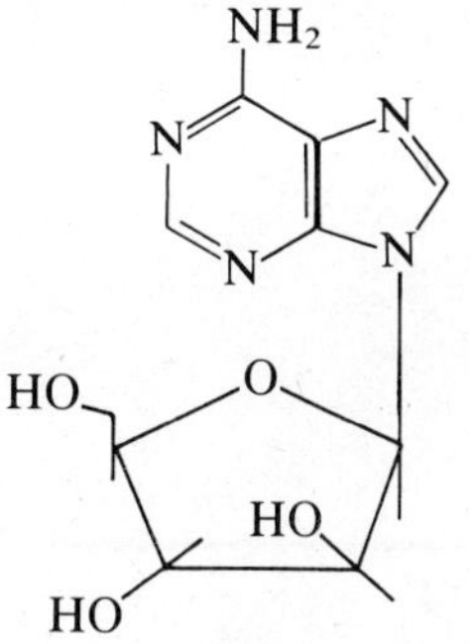   Ara-A 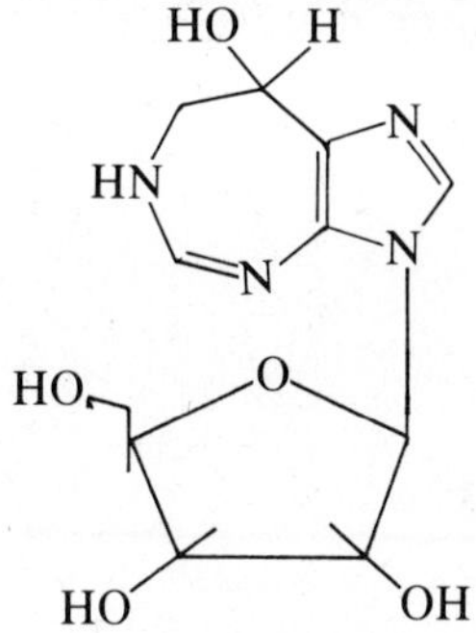   Coformycin

Purine nucleosides have been modified in analogous ways. The adenine arabinoside, ara-A (vidarabine), is effective against herpes infections, including brain infections. Its amino group is deaminated metabolically to 9-β-D-arabinofuranosylhypoxanthine (ara-Hx), which itself has some antiviral properties. This deamination can be inhibited by analogs having amino (or imino) as well as OH groups, such as coformycin and its 2'-deoxy derivative, which are regarded as transition-state analogs. Other glycosidic moieties can be substituted for arabinose without loss of antiviral activity. Likewise, the 2- and 6-substituents can be varied widely and C-8 can be replaced by nitrogen. Deaza derivatives such as deazaguanine and its riboside possess activity against DNA and RNA viruses as well as against solid tumors in animals.

It is not the purpose of this discussion to duplicate comprehensive reviews of antitumor (1463) and antiviral (1442) activities of the many modified pyrimidine and purine bases and their glycosides. Only the 6-SH-substituted purines are mentioned because 6-mercaptopurine (6-MP) (1492) and 6-thioguanine (1493) were among the earliest anticancer agents. The metabolism of these compounds and their ribosides plays a major role in maintaining their activity. The important antigout drug allopurinol, which inhibits xanthine oxidase, was originally studied as an adjunct to preventing the metabolic oxidation of 6-MP.

The design of antimetabolites in these series as listed under categories (i) to (iii) above presents good examples of an intimate interplay of biochemical thinking and bioisosteric molecular modification. It could not have succeeded if the isoenzymes of communal and invasive cells (tumor cells and viruses) had not shown enough differences that could be capitalized on in therapeutic procedures.

6-Mercaptopurine also suppresses the immune response to foreign serum albumin in rabbits (1494). In modifying the structure of this compound to improve the pharmacological profile of the drug, the thiol hydrogen was substituted by various groups that might alter the transport of the compound in the animal body and then be cleaved off metabolically to leave the parent

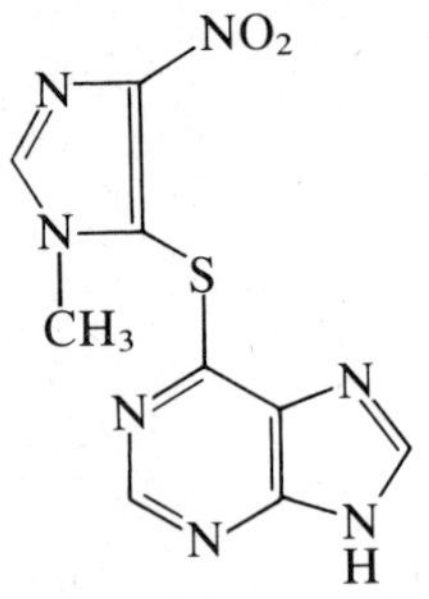

Azathioprine

6-MP in the most advantageous location. The drug of choice in this effort is azathioprine (1495, 1496).

Although azathioprine is metabolized to 6-MP, it is not certain whether it does not have immunosuppressive activity of its own (1497). It is used widely in kidney transplants; its effect is enhanced by corticosteroids.

3.21.4. Alkylating Agents

The alkylating agents are compounds that react with functionally important biomacromolecules, perhaps not by one specific type of reaction but probably by forming a covalent bond and thereby disorganizing the macromolecules. Some alkylating agents are active and clinically useful in inhibiting viruses and neoplasms and exerting immunosuppressive properties. The possibility that some chemicals could alkylate materials in cells and tissues was considered by Victor Meyer in 1887 (1498) and later by Ehrlich (1499) but was put to work only in 1946 when Gilman and Philips (226) observed that nitrogen mustard (HN2) could be used in the therapy of lymphomas. The mustards were also recognized early as being immunosuppressive (1500). The alkylating agents comprise chloroethylamines, aziridines, methanesulfonate esters, 1,2-epoxides, nitrosoureas, diazocarbonyl derivatives, and methylhydrazines. The mechanism of action has been studied most intensively for nitrogen mustards. Their biological activity depends on distribution characteristics, and these in turn are determined by the nature of the group that carries the mustard chain. Covalent linkages can be established with amino, carboxy, thiol, and phosphate groups of the macromolecules, with proteins or nucleic acids, or even with tertiary heterocyclic nitrogen atoms. Guanine groups are alkylated at position 7, which causes weakening of the glycosidic linkage. Cross linking of DNA may also occur, leading ultimately to errors in the direction of protein synthesis.

Clinically, the most effective and pharmacologically acceptable alkylating agent is cyclophosphamide (1501, 1502). This compound was conceived as a substrate–inhibitor of phosphoramidases, which were believed to be more abundant in neoplasms than in communal cells (1503). Nitrogen mustard

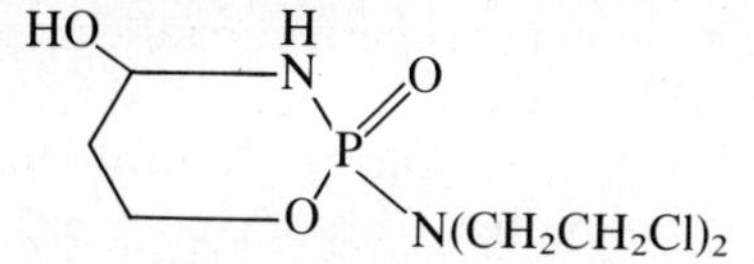

ANTIVIRAL AND ANTINEOPLASTIC AGENTS

Cyclophosphamide

phosphoramides can be regarded as latent forms of nornitrogen mustard, $HN(CH_2CH_2Cl)_2$, which might be liberated by cleavage of the phosphoramide linkage. When the metabolism of cyclophosphamide was studied, this concept turned out to be untenable. Nornitrogen mustard is apparently a fleeting intermediate in that scheme but is biotransformed to other products immediately. The active metabolite of cyclophosphamide seems to be as shown in the structure.

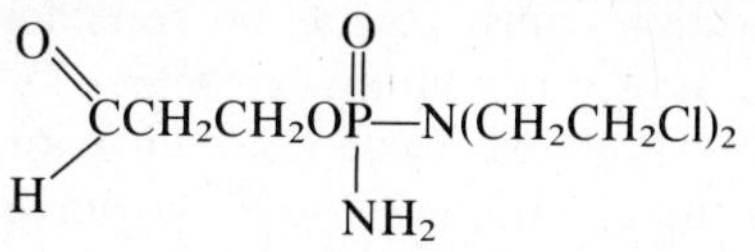

Metabolite of cyclophosphamide

3.21.5. Miscellaneous Drugs

There is apparently no uniform mechanism of action by which drugs inhibit neoplastic cell multiplication, but this can hardly be expected of hundreds of biochemically different types of cancer cells. Moreover, tumors consist of conglomerates of many types of cells, and during chemotherapy those affected least by the drug keep on metastasizing. This is a problem for the experimental oncologist. The medicinal biochemist must deal with one type of tumor cell at a time, with the characteristics of "growth" and metabolism of such cells and their susceptibility to a given type of drug structure. The metabolic analogs and the alkylating agents offer chemically rational sources of drugs in this field, but even here the fine tuning of potency and selectivity has to rely on screening. As understanding of cellular immunity will advance, some of the puzzling overlapping activities in the antinucleic acid field—

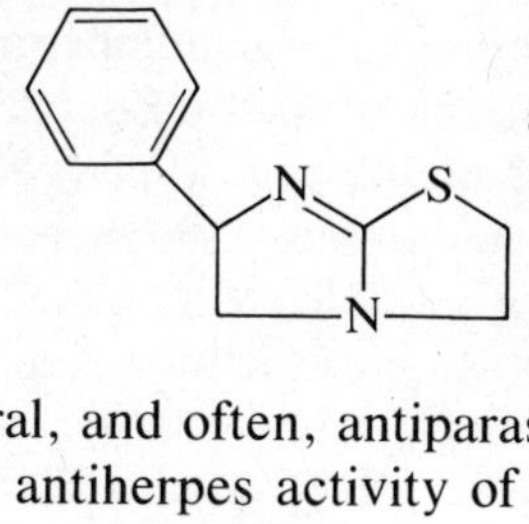

Tetramisole
Levamisole: *l*-isomer

antitumor, antiviral, and often, antiparasitic—may find a common denominator. Thus, the antiherpes activity of levamisole (1504), an anthelmintic drug, might find an explanation.

The huge screening programs of the National Cancer Institute and many other laboratories brought to light a variety of structural types that inhibit enzymes associated with the metabolism and proliferation of cell nuclei. In every case, overlapping inhibition of normal cellular enzymes and tumor or viral enzymes has given rise to toxicity, and these difficulties have not yet been resolved. It will take much more systematic molecular variation to separate activity and toxicity to host species and to reach more acceptable ratios. In the antiviral experiments, thiosemicarbazones are also beset with this drawback. They seem to owe their activity to the coordination of metal ions that are needed in some pertinent enzymic reactions.

It is difficult to say whether or not any antineoplastic agent will also exhibit antiviral activity. In some cases, failure to test a given compound in both areas may account for reports of greater specificity. The variety of viruses and neoplasms responding to different structural types of compounds and different test conditions must always be considered. Activity in either field and activity affecting the immune system may well be overlooked. However, some structures appear to be more virus-oriented. The cage type and intermediate-size alicyclic amines seem to have a specific action on several viruses.

The two clinically prophylactic compounds in this series are 1-adamantanamine (amantadine) and 1-adamantane-1-ethylamine (rimantadine). A

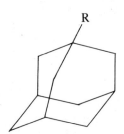

Amantadine: R = NH$_2$
Rimantadine: R = —CHNH$_2$
                    |
                    CH$_3$

diaminoadamantane had been tested as a curing agent for epoxy resins (1505); the prophylactic action of amantadine was subsequently disclosed (1506). Structure–activity relationships in a large number of derivatives (370) can be summarized as activity primarily against influenza viruses. This activity is also noted in other amines of the cage type (1507), such as bicyclo[2.2.2]octane and homoisotwistane amines (1508), but even cyclooctylamine (1509) exhibits similar *in vitro* and *in vivo* actions. This could indicate that the geometry of the alicyclic portion of the molecule is more important for antiviral activity than functionality.

Among agents that have a specific effect on the immune response are penicillamine [(CH$_3$)$_2$C(SH)CHNH$_2$COOH] and procarbazine [(CH$_3$)$_2$CH-

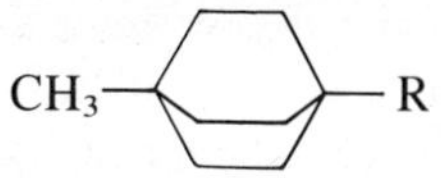

Bicyclo[2.2.2.]octane amines: R = NH$_2$ or CH$_2$NH$_2$

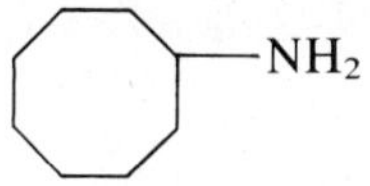

Cyclooctylamine

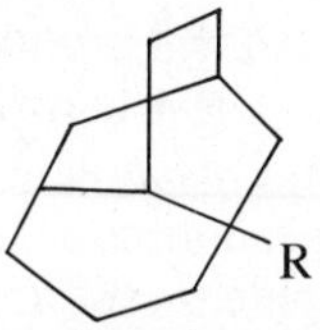

Homoisotwistane amines: R = amino groups

NHCOC$_6$H$_4$CH$_2$NHNHCH$_3$]. Penicillamine is a metal chelating agent that dissociates macroglobulins but this does not seem to be connected with its immunological properties. Procarbazine is an antitumor drug and was originally screened as one of the many hydrazines that inhibit monoamine oxidase (364). Its activity is due to the N-methylhydrazine moiety that is lost metabolically (1510). Other miscellaneous immunosuppressive agents suggestive of much additional research are listed in a recent review (1511).

3.22. ORPHAN DRUGS

Orphan drugs are agents that are used by relatively small patient populations and are therefore not profitable to pharmaceutical manufacturers. Nevertheless, the industry provides a limited number of these drugs by two routes. One is by preparing adequate supplies of an investigational new drug (IND) even if there are no plans to develop the agent through the required phases for FDA approval. The second route can be taken if a drug approved for a given disease entity turns out to be useful also for some rare or obscure disease. In this case the sales of the drug for such a secondary purpose can be regarded as a public service.

The medicinal chemist may find the study of orphan drugs fascinating for several reasons. Biochemical and metabolite prototypes for the disease syndrome may exist but will not be exploited in the industrial laboratories under a veto by the sales departments. Second, one of the ideal purposes of medicinal chemistry, to elaborate a drug that will be of real benefit to patients regardless of their number, can be achieved more comfortably by individual investigators or small cooperative teams in the field of orphan drugs.

The industry labels such compounds public service drugs if the patient population is not greater than 10,000, even if the same drug is sold to large numbers of patients for a different purpose. The examples given here are of drugs that are destined to be of help to truly small numbers of cancer patients.

The systematic study of antineoplastic agents has only very recently become a topic of interest to the pharmaceutical industry, after three decades of continuous research in governmental and cancer foundation laboratories. There is no dearth of "leads" since medicinal research on cancer drugs has relied heavily on metabolite antagonism. The multiplicity of neoplasms requires a multiplicity of chemotherapeutic agents, and a number of rare cancers have led to unusual drugs such as combination structures of the alkylating type and nitrosamines, and natural products. Examples are carmustine [ON—N(CH$_2$CH$_2$Cl)—CONHCH$_2$CH$_2$Cl], which is used in Hodgkin's disease and lung cancer; the enzyme asparaginase, which has been used in leukemia and melanoma; and the antibiotic mithramycin, which combats testicular cancer. The antimetabolite *p*-chlorophenylalanine was the earliest inhibitor of the biosynthesis of serotonin but is used only for the treatment of carcinoid, a rare cancer.

For additional remarks on orphan drugs, both in the anticancer–antiviral and other therapeutic areas, see (1512).

REFERENCES

1. H. Erlenmeyer and M. Leo, *Helv. Chim. Acta*, **15,** 117 (1932) and subsequent papers.
2. I. Langmuir, *J. Am. Chem. Soc.*, **41,** 1543 (1919).
3. O. Hinsberg, *J. Prakt. Chem.*, **93,** 302 (1916).
4. O. Schaumann, *Arch. Exp. Pathol. Pharmakol.*, **196,** 109 (1940).
5. D. D. Woods and P. Fildes, *J. Soc. Chem. Ind. (London)*, **59,** 133 (1940).
6. R. Buchheim, *Arch. Exp. Pathol. Pharmakol.*, **5,** 261 (1876).
7. *Burger's Medicinal Chemistry*, 4th ed., M. E. Wolff, Ed., Wiley, New York. (a) Part I: The Basis of Medicinal Chemistry (1980). (b) Part II (1979). (c) Part III (1981).
8. P. G. Waser, *TIPS*, **2,** 11 (1981).
9. H. M. Leicester, *Development of Biochemical Concepts from Ancient to Modern Times*, Harvard University Press, Cambridge, Mass. 1974, p. 80.
10. W. Pagel, *Bull. Hist. Med.*, Suppl. 2, 1944.
11. E. A. Underwood, *Endeavour*, **31,** 73 (1974).
12. A. Chatin, *Arch. Gen. Med. (Paris), 4th Series*, **28,** 245 (1852).
13. See L. Gershenfeld, in *Disinfection, Sterilization, and Preservation*, C. A. Lawrence and S. S. Block, Eds., Lea and Febiger, Philadelphia, 1968, pp. 329–347.
14. K. Schwarz and W. Mertz, *Arch. Biochem. Biophys.*, **85,** 292 (1959); W. Mertz, *Chemica Scripta*, **21,** 145 (1983).
15. D. Schneider, M. Boppré, J. Zweig, S. B. Horsley, T. W. Bell, J. Meinwald, K. Hansen, and E. W. Diehl, *Science*, **215,** 1264 (1982).
16. Quoted by T. L. Maugh II, *Science*, **216,** 722 (1982).
17. W. Withering, *An Account of the Foxglove and Some of Its Medicinal Uses: With Practical Remarks on Dropsy and Other Diseases*. C. G. J. and J. Robinson, London, 1785, reprinted in *Med. Class.*, **2,** 305 (1937).
18. A. Goris, *Centenaire de l'Internat en Pharmacie, des Hopitaux et Hospices Civils de Paris*, Imprimerie de la Cour d'Appel, Paris, 1920.
19. F. W. A. Sertürner, *J. Pharm.*, **13,** 234 (1805); *Ann. Phys. (Leipzig)* **55,** 56 (1817). See also Derosne, *Ann. Chim.*, **45,** 257 (1803).
20. P.-J. Robiquet, *Ann. Chim. Phys.*, **51,** 225 (1832); E. Grimaux, *Compt. Rend.*, **93,** 591 (1881).
21. C. R. A. Wright, *J. Chem. Soc.*, **27,** 1035 (1874).
22. O. Hesse, *Ann. Chem.*, **222,** 203 (1884).
23. P. L. Geiger and O. Hesse, *Ann. Chem.*, **5,** 43 (1833); **6,** 44 (1833); **7,** 269 (1833).
24. G. Merck, *Ann. Chem.*, **66,** 125 (1848).
25. J. Pelletier and E. Caventou, *Ann. Chim. Phys. [2]*, **15,** 291, 337 (1820).

26. A. Niemann, *Vierteljahrsschr. Prakt. Pharm.*, **9**, 489 (1859); *Arch. Pharm.*, **153**, 146, 291 (1860).
27. F. Wöhler, *Ann. Chem.*, **114**, 213 (1860).
28. S. Siddiqui and R. H. Siddiqui, *J. Indian Chem. Soc.*, **8**, 667 (1931); **9**, 539 (1932); **12**, 37 (1935).
29. J. M. Mueller, E. Schlittler, and H. J. Bein, *Experientia*, **8**, 338 (1952).
30. K. Bryn Thomas, *Curare, Its History and Usage*, Lippincott, Philadelphia, 1963.
31. C. Bernard, *Compt. Rend.*, **43**, 825 (1856).
32. A. Crum-Brown and T. R. Fraser, *Trans. Roy. Soc. Edinburgh*, **25**, 151, 693 (1868–1869); *Proc. Roy. Soc.*, **6**, 556 (1869); **7**, 663 (1872).
33. A. Burger, *Medicinal Chemistry*, Vol. 1, Interscience, New York, 1951, p. 370.
34. Richardson, *Med. Times Gaz.*, **18**, 703 (1869).
35. J. S. Fruton, *Science*, **192**, 327 (1976).
36. W. Kühne, *Untersuch. Physiol. Inst. Heidelberg*, **1**, 291 (1878).
37. E. Buchner, *Chem. Ber.*, **30**, 117 (1897).
38. J. N. Langley, *J. Physiol*, **1**, 339, (1878).
39. A. J. Clark, in *General Pharmacology, Vol. 4, Handbuch der experimentellen Pharmakologie*, A. Heffter, Ed., Springer-Verlag, Berlin, 1937, pp. 8, 63, 215.
40. P. Ehrlich, *Das Sauerstoff-Bedürfnis des Organisms, eine farbenanalytische Studie*, Hirschwald, Berlin, 1885.
41. P. Ehrlich, *Chem. Ber.*, **42**, 17 (1909).
42. H. G. Mautner, in ref. 7a, pp. 271–284; I. D. Kuntz, Jr., *ibid.*, pp. 285–312; P. A. Kollman, *ibid.*, pp. 313–330.
42a. L. F. Kuyper, B. Roth, D. P. Baccanari, R. Ferone, C. R. Beddell, J. N. Champness, D. K. Stammers, J. G. Dann, F. E. A. Norrington, D. J. Baker, and P. J. Goodford, *J. Med. Chem.*, **25**, 1122 (1982).
43. G. L. Olson, H.-C. Cheung, K. D. Morgan, J. F. Blount, L. Todaro, L. Berger, A. B. Davidson, and E. Boff, *J. Med. Chem.*, **24**, 1026 (1981).
44. A. Burger, "Paul Ehrlich," *Chem. Eng. News*, **32**, No. 42, 4172 (1954).
45. P. Ehrlich, *Z. Angew. Chem.*, **23**, 2 (1910); *Lancet* **2**, 445 (1913).
46. See P. J. Cohen, "History of Surgical Anesthesia," in *The Pharmacological Basis of Therapeutics*, 5th ed., L. S. Goodman and A. Gilman, Eds., Macmillan, New York, 1975, p. 53.
47. J. Lister, *Lancet*, **1**, 326, 352, 387, 501 (1867); **2**, 95, 353, 668 (1867).
48. C. D. Leake and M. Y. Chen, *Proc. Soc. Exp. Biol. Med.*, **28**, 151 (1930).
49. W. L. Ruigh and R. T. Major, *J. Am. Chem. Soc.*, **53**, 2662 (1931).
50. E. G. Klarmann, *J. Am. Chem. Soc.*, **48**, 2358 (1926).
51. W. C. Harden and E. E. Reid, *J. Am. Chem. Soc.*, **54**, 4325 (1932).
52. E. G. Klarmann, in *Medicinal Chemistry*, 2nd ed., A. Burger, Ed., Interscience, New York, 1960, pp. 1109–1142.
53. See T. S. Kuhn, *Science*, **136**, 760 (1962).
54. M. Nencki, *Arch. Exp. Pathol. Pharmakol.*, **20**, 396 (1886); **36**, 400 (1895); *Opera Omnia*, Vieweg, Braunschweig, 1904.
55. H. Dreser, *Arch. Ges. Physiol.*, **76**, 396 (1899); *Ther. Monatsh.*, **17**, 131 (1903).
56. P. Guttman and P. Ehrlich, *Klin. Wochenschr.*, **28**, 953 (1891).
57. P. Ehrlich and K. Shiga, *Klin. Wochenschr.*, **41**, 329, 362 (1904).
58. J. W. Churchman, *J. Exp. Med.*, **16**, 221 (1912); *J. Am. Med. Assoc.*, **85**, 1849 (1925).

REFERENCES

59. C. H. Browning and W. Gilmour, *J. Pathol. Bacteriol.*, **18**, 144 (1913).
60. M. Nicolle and F. Mesnil, *Ann. Inst. Pasteur*, **20**, 417 (1906).
61. P. Ehrlich, *Klin. Wochenschr.*, **44**, 233 (1907).
62. G. H. F. Nuttall and S. Hadwen, *Parasitology*, **2**, 156, 236 (1909).
63. M. Mayer and H. Zeiss, *Arch. Schiffs Tropen-Hyg.*, **24**, 257 (1920).
64. E. Fourneau, *Bull. Méd. Paris*, **91**, 458 (1924).
65. E. Fourneau, J. Tréfouël, and J. Valée, *Ann. Inst. Pasteur*, **38**, 81 (1924).
66. J. Tréfouël, Mme. J. Tréfouël, F. Nitti, and D. Bovet, *Compt. Rend. Soc. Biol.*, **120**, 756 (1935).
67. K. Alt, *Münch. Med. Wochenschr.*, **56**, 1457 (1909); **57**, 561, 1774 (1910).
68. P. Ehrlich and A. Bertheim, to Hoechst A. G., U.S. Pat. 986,148 (1911).
69. A. L. Tatum and G. A. Cooper, *Science*, **75**, 541 (1932).
70. A. Cahn and P. Hepp, *Zentralbl. Klin. Med.*, **7**, 561 (1886).
71. V. Hinsberg and A. Kast, *Zentralbl. Klin. Med.*, **25**, 145 (1887).
72. H. N. Morse, *Chem. Ber.*, **11**, 232 (1878).
73. W. Lossen, *Ann. Chem.*, **133**, 351 (1865).
74. Moréno and Maiz, *Recherches Chimiques et Physiologiques sur l'Erythroxylum Coca du Peru et la Cocaine, Paris*, 1868.
75. B. V. Anrep, *Arch. Ges. Physiol. Menschen Tiere*, **21**, 38 (1880).
76. K. Köller, *Wien. Med. Bl.*, **7**, 1224, 1352 (1884); *Wien. Med. Wochenschr.*, **43**, 46 (1884); *Lancet*, **2**, 127, 990 (1884).
77. J. M. Ritchie and P. J. Cohen, in *The Pharmacological Basis of Therapeutics*, 5th ed., L. S. Goodman and A. Gilman, Eds., Macmillan, New York, 1975, Chapter 20.
78. H. Salkowski, *Chem. Ber.*, **28**, 1917 (1895).
79. E. Ritsert, German Pat. 147,790 (1903).
80. Anonymous, *Pharm. Ztg.*, **37**, 427 (1892); **47**, 356 (1902).
81. A. Einhorn and B. Pfyl, German Pat. 97,333 (1898); *Ann. Chem.*, **311**, 34 (1900).
82. A. Einhorn, *Ann. Chem.*, **311**, 26 (1900).
83. Schering A. G., German Pat. 91,121 (1895).
84. E. Fourneau, German Pats. 169,746, 169,787 (1903); *Compt. Rend.*, **138**, 766 (1904).
85. A. Einhorn and E. Uhlfelder, German Pats. 179,627, 180,291, 180,292 (1904).
86. A. Einhorn and M. Oppenheimer, *Ann. Chem.*, **311**, 154 (1900).
87. H. von Euler and H. Erdtman, *Ann. Chem.*, **520**, 1 (1935).
88. H. Erdtman and N. Löfgren, *Svensk Kem. Tidskr.*, **49**, 163 (1937).
89. N. Löfgren, *Ark. Kem. Mineral. Geol.*, **22A**, 30 (1946).
90. N. Löfgren and G. Widmark, *Svensk Kem. Tidskr.*, **58**, 206, 323 (1946).
91. N. Löfgren and B. G. Lundquist, U.S. Pat. 2,441,498 (1948).
92. B. af Ekenstam, B. Egner, and G. Petterson, *Acta Chem. Scand.*, **11**, 1183 (1957).
93. J. W. Wilson, G. E. Ullyot, N. D. Dawson, and W. Brooks, *J. Am. Chem. Soc.*, **71**, 937 (1949).
94. G. Oliver and E. A. Schäfer, *J. Physiol. (London)*, **18**, 230 (1895).
95. J. J. Abel, *Z. Physiol. Chem.*, **29**, 318 (1899).
96. O. v. Fürth, *Z. Physiol. Chem.*, **29**, 105 (1899).
97. J. Takamine, *Am. J. Pharm. Sci. Support Publ. Health*, **73**, 523 (1901).
98. T. B. Aldrich, *Am. J. Physiol.*, **5**, 457 (1901); **7**, 359 (1902).
99. H. A. D. Jowett, *J. Chem. Soc.*, **85**, 192 (1904).

100. F. Stolz, German Pats. 152,814, 157,300 (1903); *Chem. Ber.*, **37,** 4149 (1904).
101. H. D. Dakin, *Proc. Roy. Soc.*, **76,** 491, 498 (1905).
102. F. Flächer, *Z. Physiol. Chem.*, **58,** 189 (1908).
103. F. Stolz and F. Flächer, *German Pat.* 155,632 (1903).
104. W. B. Cannon and J. E. Uridil, *Am. J. Physiol.*, **58,** 353 (1921).
105. U. S. von Euler and U. Hamberg, *Nature (London)*, **163,** 642 (1949).
106. M. Goldberg, M. Faber, E. J. Alston, and E. C. Chargaff, *Science,* **109,** 534 (1949).
107. B. F. Tuller, *Science,* **109,** 536 (1949).
108. U. S. von Euler, *Acta Physiol. Scand.*, **16,** 63 (1948).
109. A. v. Baeyer, *Ann. Chem.*, **142,** 322 (1867).
110. R. Hunt, *Am. J. Physiol.*, **3,** xviii (1899–1900).
111. H. H. Dale, *J. Pharmacol. Exp. Ther.*, **6,** 147 (1914).
112. O. Loewi, *Arch. Ges. Physiol.*, **189,** 239 (1921).
113. O. Loewi and E. Navratil, *Arch. Ges. Physiol.*, **214,** 678 (1926).
114. A. Windaus and W. Vogt, *Chem. Ber.*, **40,** 3691 (1907).
115. H. H. Dale and P. P. Laidlaw, *J. Physiol. (London)*, **41,** 318 (1910); **43,** 182 (1911); **52,** 355 (1919).
116. C. H. Best, H. H. Dale, H. W. Dudley, and W. V. Thorpe, *J. Physiol. (London)*, **62,** 397 (1927).
117. G. Barger and H. H. Dale, *J. Physiol. (London)*, **41,** 19 (1910).
118. N. Nagai, *Pharm. Ztg.*, **32,** 700 (1887).
119. K. K. Chen and C. F. Schmidt, *J. Pharmacol. Exp. Ther.*, **24,** 339 (1924).
120. L. Edeleanu, *Chem. Ber.*, **20,** 616 (1887).
121. H. Konzett, *Arch. Exp. Pathol. Pharmakol.*, **197,** 27 (1940).
122. J. Pohl, *Z. Exp. Pathol. Ther.*, **17,** 370 (1914).
123. J. Weijlard and A. E. Erickson, *J. Am. Chem. Soc.*, **64,** 869 (1942).
124. Sankyo Ltd., British Pat., 939,287 (1963).
125. M. J. Lewenstein and J. Fishman, U.S. Pat. 3,524,088 (1966).
126. R. L. Clark, A. A. Pessolano, J. Weijlard, and K. Pfister III, *J. Am. Chem. Soc.*, **75,** 4963 (1953).
127. H. H. Meyer, *Arch. Exp. Pathol. Pharmakol.*, **42,** 109 (1899); **46,** 338 (1901).
128. E. Overton, *Studien über die Narkose, zugleich ein Beitrag zur allgemeinen Pharmakologie.* G. Fischer, Jena, 1901.
129. C. Hansch and T. Fujita, *J. Am. Chem. Soc.*, **86,** 1616 (1964).
130. J. Traube, *Arch. Ges. Physiol.*, **105,** 541 (1904).
131. R. Höber, *Arch. Ges. Physiol.*, **120,** 492 (1907).
132. L. Pauling, *Science,* **134,** 15 (1961).
133. A. Burger, ref. 7a, pp. 7–22.
134. V. Vercello, *Arch. Exp. Pathol. Pharmakol.*, **16,** 265 (1883).
135. M. Conrad and M. Gutzeit, *Chem. Ber.*, **15,** 2844 (1882).
136. Farbenfabriken vorm. F. Bayer, German Pat. 247,952 (1911).
137. U. Hörlein, U.S. Pat. 1,025,872 (1912).
138. A. Hauptmann, *Münch. Med. Wochenschr.*, **59,** 1907 (1912).
139. H. A. Shonle and A. Moment, *J. Am. Chem. Soc.*, **45,** 243 (1923).
140. C. Funk, *J. State Med.*, **20,** 341 (1912).
141. C. R. Harington, *Biochem. J.*, **20,** 293 (1926).

REFERENCES

142. J. Gross and R. Pitt-Rivers, *Biochem. J.,* **53,** 645, 652 (1953).
143. J. Roche, S. Lissitzky, and R. Michel, *Compt. Rend.,* **234,** 997, 1228 (1952).
144. C. R. Harington and G. Barger, *Biochem. J.,* **21,** 169 (1927).
145. H. A. Selenkow and S. P. Asper, *Physiol. Rev.,* **35,** 426 (1955).
146. C. M. Greenberg, B. Blank, F. R. Pfeiffer, and J. F. Pauls, *Am. J. Physiol.,* **205,** 821 (1963).
147. E. C. Jorgensen, *Mayo Clinic Proc.,* **39,** 560 (1964).
148. C. L. Gemmill, J. J. Anderson, and A. Burger, *J. Am. Chem. Soc.,* **78,** 2434 (1956).
149. G. W. Anderson, I. F. Halverstadt, W. H. Miller, and R. O. Roblin, *J. Am. Chem. Soc.,* **67,** 2197 (1945).
150. C. Rimington, in *Advances in Thyroid Disease,* R. Pitt-Rivers, Ed., Pergamon Press, New York, 1961, p. 47.
151. A. M. Chesney, T. A. Clawson, and B. Webster, *Bull. Johns Hopkins Hosp.,* **43,** 261 (1928).
152. L. Loeb and R. B. Bassett, *Proc. Soc. Exp. Biol. Med.,* **26,** 860 (1929).
153. H. L. Fevold, F. L. Hisaw, A. Hellbaum, and A. Hertz, *Am. J. Physiol.,* **104,** 710 (1933).
154. O. Riddle, R. W. Bates, and S. W. Dykshorn, *Am. J. Physiol.,* **105,** 191 (1933).
155. A. Butenandt, *Naturwissenschaften,* **17,** 879 (1929).
156. E. A. Doisy, C. D. Veler, and S. A. Thayer, *Am. J. Physiol.,* **90,** 329 (1929); *J. Biol. Chem.,* **86,** 499 (1930).
157. W. M. Allen and O. Wintersteiner, *Science,* **80,** 190 (1934).
158. A. Butenandt, U. Westphal, and W. Hohlweg, *Z. Physiol. Chem.,* **227,** 84 (1934).
159. M. Hartmann and A. Wettstein, *Helv. Chim. Acta,* **17,** 878, 1365 (1934).
160. K. H. Slotta, H. Ruschig, and E. Fels, *Chem. Ber.,* **67,** 1270 (1934).
161. A. Butenandt, *Z. Angew. Chem.,* **44,** 905 (1931).
162. L. Ruzicka, M. W. Goldberg, J. Meyer, and H. Brüngger, *Helv. Chim. Acta,* **17,** 1395 (1934).
163. K. David, E. Dingelmanse, J. Freud, and E. Laqueur, *J. Physiol. Chem.,* **233,** 281 (1935).
164. E. Fernholz, *Ann. Chem.,* **507,** 128 (1933); *Chem. Ber.,* **67,** 1855, 2027 (1934).
165. A. Butenandt, U. Westphal, and H. Cobler, *Chem. Ber.,* **67,** 1611 (1934).
166. A. Butenandt and U. Westphal, *Chem. Ber.,* **67,** 2085 (1934).
167. A. Butenandt and G. Hanisch, *Chem. Ber.,* **68,** 1859 (1935); *J. Physiol. Chem.,* **237,** 89 (1935).
168. L. Ruzicka and A. Wettstein, *Helv. Chim. Acta,* **18,** 1264 (1935).
169. H. L. Mason, C. S. Myers, and E. C. Kendall, *J. Biol. Chem.,* **114,** 613 (1936); **116,** 267 (1936).
170. T. Reichstein, *Helv. Chim. Acta,* **19,** 29, 223, 402, 979, 1107 (1936); **20,** 953, 978 (1937); **21,** 1490 (1938).
171. M. Steiger and T. Reichstein, *Helv. Chim. Acta,* **20,** 817, 1040, 1164 (1937).
172. L. F. Fieser and M. Fieser, *Steroids,* Reinhold, New York, 1959, p. 634.
173. D. H. Peterson, H. C. Murray, S. H. Epstein, L. M. Reineke, A. Weintraub, P. D. Meister, and H. M. Leigh, *J. Am. Chem. Soc.,* **74,** 5933 (1952).
174. A. Fleming, *Brit. J. Exp. Pathol.,* **10,** 226 (1929).
175. F. Mietzsch and J. Klarer, German Pat. 607,537 (1935).
176. E. Fourneau, J. Tréfouël, Mme. J. Tréfouël, F. Nitti, and D. Bovet, *Compt. Rend. Soc. Biol.,* **122,** 652 (1936).

177. F. Y. Wiselogle, *A Survey of Antimalarial Drugs, 1941–1945*, Edwards, Ann Arbor, Mich., 1946.
178. F. H. S. Curd, D. G. Davey, and F. L. Rose, *Ann. Trop. Med. Parasitol.*, **39,** 157, 208 (1945).
179. A. Schatz, S. Bugie, and S. A. Waksman, *Proc. Soc. Exp. Biol. Med.*, **55,** 66 (1944); **57,** 244 (1944).
180. J. Lehmann, *Lancet,* **1,** 14, 15 (1946).
181. F. Bernheim, *Science,* **92,** 204 (1945).
182. V. Chorine, *Compt. Rend.,* **220,** 150 (1945).
183. S. Kushner, H. Dalalian, J. L. Sanjurjo, F. L. Bach, Jr., S. R. Safir, V. K. Smith, Jr., and J. H. Williams, *J. Am. Chem. Soc.,* **74,** 3617 (1952).
184. T. S. Gardner, E. Wenis, and J. Lee, *J. Org. Chem.,* **19,** 753 (1954).
185. R. I. Meltzer, A. D. Lewis, and J. A. King, *J. Am. Chem. Soc.,* **77,** 4062 (1955).
186. D. Liberman, N. Rist, and F. Grumbach, *Bull. Soc. Chim. Biol.,* **38,** 231 (1956).
187. H. H. Fox, *J. Org. Chem.,* **17,** 555, 1653 (1952); *Trans. N.Y. Acad. Sci.,* **15,** 234 (1953).
188. J. Bernstein, W. A. Lott, B. A. Steinberg, and H. L. Yale, *Am. Rev. Tuberc.,* **65,** 357 (1952).
189. H. A. Offe, W. Siekfen, and G. Domagk, *Z. Naturforsch.,* **7b,** 446, 462 (1952).
190. E. V. Cowdry and C. Ruangsiri, *Arch. Pathol.,* **32,** 632 (1941).
191. C. C. Shepard, L. Levy, and P. Fasal, *Am. J. Trop. Med. Hyg.,* **17,** 769 (1968).
192. A. Burger, *Medicinal Chemistry,* Interscience, New York, 1951, Chapter 4.
193. H. L. Friedman, in *Symposium on Chemical–Biological Correlation.*, National Academy of Science–National Research Council, No. 206, Washington, D. C., 1951, p. 295.
194. H. Erlenmeyer, *Bull. Soc. Chem. Biol.,* **30,** 792 (1948).
195. V. B. Schatz, in *Medicinal Chemistry,* 2nd ed., A. Burger, Ed., Interscience, New York, 1960, Chapter 8.
196. C. W. Thornber, *Chem. Soc. Rev.,* **18,** 564 (1979).
197. C. R. Ganellin, *J. Med. Chem.,* **24,** 1 (1981).
198. D. Bovet and A.-M. Staub, *Compt. Rend. Soc. Biol.,* **124,** 547 (1937).
199. B. N. Halpern, *Arch. Int. Pharmacodyn. Ther.,* **68,** 339 (1942).
200. G. R. Rieveschl, U.S. Pat. 2,421,714 (1947).
201. R. L. Mayer, C. P. Huttrer, and C. R. Scholz, *Science,* **102,** 93 (1945).
202. C. P. Huttrer, C. Djerassi, W. L. Beears, R. L. Mayer, and C. R. Scholz, *J. Am. Chem. Soc.,* **68,** 1999 (1946).
203. G. Daumezon and L. Cassan, *Ann. Méd.-Psychol.,* **101,** 432 (1943).
204. P. Guiraud and C. David, *Compt. Rend. 48th Congr. de Neurologie,* Masson, Paris, 1950.
205. H. Laborit, P. Huguenard, and R. Alluaume, *Presse Méd.,* **60,** 206 (1952).
206. T. G. Putnam and H. H. Merritt, *Science,* **85,** 525 (1937).
207. H. Biltz, *Chem. Ber.,* **41,** 1379 (p. 1391) (1908).
208. O. Eisleb and O. Schaumann, *Dtsch. Med. Wochenschr.,* **65,** 967 (1939).
209. O. Eisleb, *Chem. Ber.,* **74,** 1433 (1941).
210. A. Hofmann, *Die Geschichte des LSD-25.* Triangel Sandoz Z., *Med. Wiss.,* **2,** 117 (1955), as quoted by A. Hofmann, in *Drugs Affecting the Central Nervous System,* A. Burger, Ed., Marcel Dekker, New York, 1968, pp. 184–186.
211. H. Andersag, S. Breitner, and H. Jung, German Pat. 683,692 (1939).
212. F. Schönhöfer, *Z. Physiol. Chem.,* **274,** 1 (1942).

REFERENCES

213. N. L. Drake, J. Hook, J. Garman, S. Haywood, R. Peck, W. Walton, D. Draper, and H. Creech, *J. Am. Chem. Soc.*, **68**, 1214 (1946).
214. E. A. Steck, L. L. Hallock, and C. M. Suter, *J. Am. Chem. Soc.*, **70**, 4063 (1948).
215. J. H. Burckhalter, F. H. Tendick, E. M. Jones, P. A. Jones, W. F. Holcomb, and A. L. Rawlins, *J. Am. Chem. Soc.*, **70**, 1363 (1948).
216. R. C. Elderfield, L. Craig, W. Laner, R. Arnold, W. Gensler, J. Head, T. Bembry, H. Mighton, J. Tinker, J. Galbreath, A. Holley, L. Goldman, J. Maynard, and N. Picus, *J. Am. Chem. Soc.*, **68**, 1516, 1524 (1946).
217. F. H. S. Curd, D. G. Davey, and F. L. Rose, *Ann. Trop. Med. Parasitol.*, **39**, 157, 208 (1945).
218. E. A. Falco, L. G. Goodwin, G. H. Hitchings, I. M. Rollo, and P. B. Russell, *Brit. J. Pharmacol. Chemother.*, **6**, 185 (1951).
219. G. H. Hitchings, *Trans. N.Y. Acad. Sci.*, **23**, 700 (1961).
220. N. G. Brink and K. Folkers, in *Streptomycin*, S. A. Waksman, Ed., Williams & Wilkins, Baltimore, 1949, p. 55.
221. W. B. Stillman, A. B. Scott, and J. M. Clampit, U.S. Pat. 2,319,481 (1943).
222. Q. R. Bartz, *J. Biol. Chem.*, **172**, 445 (1948).
223. J. Controulis, M. C. Rebstock, and H. M. Crooks, Jr., *J. Am. Chem. Soc.*, **71**, 2463 (1949).
224. G. Domagk, R. Behnisch, F. Mietzsch, and H. Schmidt, *Naturwissenschaften*, **33**, 315 (1946).
225. E. B. Krumbhaar and H. D. Krumbhaar, *J. Med. Res.*, **40**, 497 (1919).
226. A. Gilman and F. S. Philips, *Science*, **103**, 409 (1946).
227. R. A. Peters, L. A. Stocken, and R. H. S. Thompson, *Nature (London)*, **156**, 616 (1945).
228. D. R. Seeger, D. B. Cosulich, J. M. Smith, Jr., and M. E. Hultquist, *J. Am. Chem. Soc.*, **71**, 1753 (1949).
229. S. Farber, L. K. Diamond, R. D. Mercer, R. F. Sylvester, and V. A. Wolff, *New Engl. J. Med.*, **238**, 787 (1948).
230. R. Duschinsky, E. Pleven, and C. Heidelberger, *J. Am. Chem. Soc.*, **79**, 4559 (1957).
231. S. S. Cohen, J. G. Flaks, H. D. Barner, M. R. Loeb, and J. Lichtenstein, *Proc. Natl. Acad. Sci. U.S.*, **44**, 1004 (1958).
232. K.-U. Hartmann and C. Heidelberger, *J. Biol. Chem.*, **236**, 3006 (1961).
233. F. M. Berger and W. Bradley, *Brit. J. Pharmacol. Chemother*, **1**, 265 (1946); *Lancet*, **257**, 97 (1947); *Nature*, **159**, 813 (1947).
234. F. M. Berger, *J. Pharmacol. Exp. Ther.*, **112**, 413 (1954).
235. F. M. Berger, C. D. Henley, B. J. Ludwig, and T. E. Lynes, *J. Pharmacol. Exp. Ther.*, **116**, 337 (1956).
236. R. P. Ahlquist, *Am. J. Physiol.*, **153**, 586 (1948).
237. W. Paton, quoted in *TIPS*, **2**, No. 10, October, 1981, p. IV.
238. G. A. Stein, M. Sletzinger, H. Arnold, D. Reinhold, W. Gaines, and K. Pfister, *J. Am. Chem. Soc.*, **78**, 1514 (1956).
239. W. E. Rosen, V. P. Toohey, and A. C. Shabica, *J. Am. Chem. Soc.*, **79**, 3167 (1957).
240. A. P. Gray, W. L. Archer, E. E. Spinner, and C. J. Cavallito, *J. Am. Chem. Soc.*, **79**, 3805 (1957).
241. G. E. Ullyot and J. F. Kerwin, in *Medicinal Chemistry*, Vol. 2, F. F. Blicke and C. M. Suter, Eds., Wiley, New York, 1956, pp. 234–307.
242. M. Nickerson and W. S. Gump, *J. Pharmacol. Exp. Ther.*, **97**, 25 (1949).
243. M. Nickerson and L. S. Goodman, *J. Pharmacol. Exp. Ther.*, **89**, 167 (1947).

244. M. Bockmühl and G. Ehrhart, German Pat. 711,069 (1943).
245. V. P. Dole and M. Nysvander, *J. Am. Med. Assoc.*, **193**, 646 (1965).
246. E. E. Williams, *J. Am. Med. Assoc.*, **109**, 1472 (1937).
247. J. Hald and E. Jacobsen, *Lancet*, **2**, 1001 (1948).
248. J. Hald, E. Jacobsen, and V. Larsen, *Acta Pharmacol. Toxicol.*, **8**, 329 (1952).
249. I. Fischer, *Svenska Läkartidn.*, **42**, 2513 (1945).
250. W. C. Clark and H. R. Hulpieu, *J. Pharmacol. Exp. Ther.*, **123**, 74 (1958).
251. J. F. J. Cade, *Med. J. Aust.*, **36**, 349 (1949).
252. H. Laborit, *Acta. Chim. Belg.*, **50**, 710 (1951).
253. J. Delay, P. Deniker, and J. M. Harl, *Ann. Méd.-Psychol.*, **110**, 112 (1952).
254. P. Charpentier, U.S. Pats., 2,519,886, 2,530,451 (1950).
255. A. Pletscher, P. A. Shore, and B. B. Brodie, *Science*, **122**, 374, 968 (1955).
256. P. A. J. Janssen, A. H. M. Jageneau, P. J. A. Demoen, C. van de Westeringh, J. H. M. de Cannière, A. H. M. Raeymaekers, M. S. J. Wouters, S. Sanczuk, and B. K. F. Hermans, *J. Med. Pharm. Chem.*, **1**, 309 (1959).
257. L. O. Randall, W. Schallek, G. A. Heise, E. P. Keith, and R. Bagdon, *J. Pharmacol. Exp. Ther.*, **129**, 163 (1960).
258. L. H. Sternbach and E. Reeder, *J. Org. Chem.*, **26**, 1111 (1961).
259. L. H. Sternbach and E. Reeder, *J. Org. Chem.*, **26**, 4936 (1961).
260. L. H. Sternbach, *J. Med. Chem.*, **23**, 1 (1980).
261. E. A. Zeller, J. Barsky, J. R. Fouts, W. F. Kirchheimer, and L. S. Van Orden, *Experientia*, **8**, 349 (1952).
262. A. Burger and W. L. Yost, *J. Am. Chem. Soc.*, **70**, 2198 (1948).
263. C. L. Zirkle, C. Kaiser, D. H. and R. E. Tedeschi, and A. Burger, *J. Med. Pharm. Chem.*, **5**, 1265 (1962).
264. R. Kuhn, *Schweiz. Med. Wochenschr.*, **87**, 1135 (1957).
265. F. Häfliger and W. Schindler, U.S. Pat. 2,554,736 (1951).
266. R. D. Hoffsommer, D. Taub, and N. L. Wendler, *J. Org. Chem.*, **27**, 4134 (1962); **28**, 1751 (1963); *J. Med. Chem.*, **8**, 555 (1965).
267. E. L. Engelhardt, M. E. Christy, H. C. Zell, C. M. Cylion, M. B. Freedman, and J. M. Sprague, Belgian Pat. 584,061 (1960).
268. G. Merck, *Ann. Chem.*, **95**, 200 (1855).
269. P. D. Baker, *South. Med. J.*, **15**, 579 (1859).
270. A. J. V. Ewins, *Biochem. J.*, **8**, 44 (1914).
271. H. H. Dale, *J. Physiol.* (London), **34**, 163 (1906).
272. W. Murrel, *Lancet*, **1**, 80, 113, 151, 225 (1879).
273. E. Braun-Menéndez, J. C. Fasciolo, L. F. Leloir, and J. M. Munoz, *J. Physiol.* (*London*), **98**, 283 (1940).
274. R. Tiegerstedt and P. G. Bergman, *Skand. Arch. Physiol.*, **8**, 223 (1898).
275. I. H. Page and O. M. Helmer, *J. Exp. Med.*, **71**, 485 (1940).
276. R. K. Kirtikar and B. D. Basu, *Indian Medicinal Plants*, Part II, Sudhindra Nath Basu, Allahabad, 1918, p. 777.
277. S. Siddiqui and R. H. Siddiqui, *J. Indian Chem. Soc.*, **8**, 667 (1931); **9**, 539 (1932); **12**, 37 (1935).
278. R. J. Vakil, *Brit. Heart J.*, **11**, 350 (1949).
279. R. W. Wilkins, *Ann. Intern. Med.*, **37**, 1144 (1952).
280. J. M. Müller, E. Schlittler, and H. J. Bein, *Experientia*, **8**, 338 (1952).

REFERENCES

281. M. M. Rapport, A. A. Green, and I. H. Page, *Science*, **108**, 329 (1948).
282. M. M. Rapport, *J. Biol. Chem.*, **180**, 961 (1949).
283. K. E. Hamlin and F. E. Fischer, *J. Am. Chem. Soc.*, **73**, 5007 (1951).
284. V. Erspamer and B. Asero, *Nature (London)*, **169**, 800 (1952).
285. CIBA A. G., German Pat. 848,818 (1949).
286. J. Druey and B. H. Ringier, *Helv. Chim. Acta*, **34**, 195 (1951).
287. F. Gross, J. Druey, and R. Meier, *Experientia*, **6**, 19 (1950).
288. T. L. Sourkes, *Arch. Biochem. Biophys.*, **51**, 444 (1954).
289. G. A. Stein, H. A. Bronner, and K. Pfister III, *J. Am. Chem. Soc.*, **77**, 700 (1955).
290. K. Pfister and G. A. Stein, U.S. Pat. 2,868,818 (1959).
291. J. A. Oates, L. Gillespie, Jr., S. Udenfriend, and A. Sjoerdsma, *Science*, **131**, 1890 (1960).
292. F. C. Novello and J. M. Sprague, *J. Am. Chem. Soc.*, **79**, 2028 (1957).
293. R. W. Wilkins, *New Engl. J. Med.*, **257**, 1026 (1957).
294. K. Sturm, W. Seidel, R. Weyer, and H. Ruschig, *Chem. Ber.*, **99**, 328 (1966).
295. J. A. Cella and R. C. Tweit, *J. Org. Chem.*, **24**, 1109 (1959).
296. W. Hollander and R. W. Wilkins, *Prog. Cardiovasc. Dis.*, **8**, 291 (1966).
297. R. P. Mull and R. A. Maxwell in *Antihypertensive Agents*, E. Schlittler, Ed., Academic Press, New York, 1967, pp. 115–149.
298. R. A. Maxwell, R. P. Mull, and A. J. Plummer, *Experientia*, **15**, 267 (1959).
299. Abbott Laboratories, British Pat. 906,245 (1962).
300. L. R. Swett, W. B. Martin, J. D. Taylor, G. M. Everett, A. A. Wykes, and Y. C. Gladdish, *Ann. N.Y. Acad. Sci.*, **107**, 891 (1963).
301. J. D. Taylor, A. A. Wykes, Y. C. Gladdish, W. B. Martin, and G. M. Everett, *Fed. Proc.*, **19**, 278 (1960).
302. C. E. Powell and I. H. Slater, *J. Pharmacol. Exp. Ther.*, **122**, 480 (1958).
303. J. W. Black, A. F. Crowther, R. G. Shanks, L. H. Smith, and A. C. Dornhorst, *Lancet* **1**, 1080 (1964).
304. A. F. Crowther and L. H. Smith, Belgian Pat. 640,312 (1964).
305. P. M. Lish, J. H. Weikel, and K. W. Dungan, *J. Pharmacol. Exp. Ther.*, **149**, 161 (1965).
306. B. K. Wasson, W. K. Gibson, R. S. Stuart, H. W. R. Williams, and C. H. Yates, *J. Med. Chem.*, **15**, 651 (1972); German Pats. 1,925,954–1,925,956 (1970).
307. S. H. Ferreira, D. C. Bartelt, and L. J. Greene, *Biochemistry*, **9**, 2538 (1970).
308. M. A. Ondetti, B. Rubin, and D. W. Cushman, *Science*, **196**, 441 (1977).
309. M. A. Antonaccio and D. W. Cushman, *Fed. Proc.*, **40**, 2275 (1981).
310. M. A. Ondetti and D. W. Cushman, *J. Med. Chem.*, **24**, 355 (1981).
311. T. J. MacLagan, *Practitioner*, **19**, 321 (1877); *Lancet*, **2**, 179 (1879).
312. L. Knorr, *Chem. Ber.*, **16**, 2597 (1883).
313. L. Knorr, *Chem. Ber.*, **17**, 2037 (1884).
314. F. Stolz, German Pat. 97,011 (1897).
315. W. Filehne, *Z. Klin. Med.*, **32**, 572 (1897).
316. W. V. Ruyle, L. H. Sarett, and A. Matzuk, South African Pat. 6,701,021 (1968); *C.A.* **70**, 106208j (1969).
317. J. Hannah, W. V. Ruyle, H. Jones, A. R. Matzuk, K. W. Kelly, B. E. Witzel, W. J. Holtz, R. A. Houser, T. Y. Shen, and L. H. Sarett, *Brit. J. Clin. Pharm.*, **4**, Suppl. 1, 7 (1977).
318. Geigy A. G., Swiss Pats. 266,236, 267,222, 269,980 (1950).

319. I. Häfliger, U.S. Pat. 2,745,783 (1956).
320. J. J. Burns, R. K. Rose, T. Cherkin, A. Goldman, A. Schulert, and B. B. Brodie, *J. Pharmacol. Exp. Ther.*, **109**, 346 (1953); **113**, 481 (1955).
321. R. A. Scherrer, in *Antiinflammatory Agents*, Vol. 1, R. A. Scherrer and M. W. Whitehouse, Eds., Academic Press, New York, 1974.
322. J. H. Wilkinson and I. L. Finar, *J. Chem. Soc.*, **1948**, 32.
323. T. Y. Shen and 15 coauthors, *J. Am. Chem. Soc.*, **85**, 488 (1963).
324. T. Y. Shen, in *Clinoril in the Treatment of Rheumatic Disorders*, E. C. Huskisson and P. Franchimont, Eds., Raven Press, New York, 1976, p. 1.
325. T. Y. Shen et al., German Pat. 2,039,426 (1971).
326. T. Y. Shen, *J. Med. Chem.*, **24**, 1 (1981).
327. J. S. Nicholson and S. S. Adams, British Pat. 971,700 (1964).
328. J. R. Vane and S. H. Ferreira, Eds., *Antiinflammatory Drugs*, Springer-Verlag, New York, 1978.
329. T. Y. Shen, in ref. 7c, Chapter 62.
330. H. Brown, *J. Am. Chem. Soc.*, **49**, 958 (1927).
331. J. Forestier, *Bull. Mem. Soc. Méd. Hop. Paris*, **53**, 323 (1929).
332. B. M. Sutton, E. McGusty, D. T. Waltz, and M. S. DiMartino, *J. Med. Chem.*, **15**, 1095 (1972).
333. E. Munthe, *Penicillamine Research in Rheumatic Disease*, Fabritius and Sonner, Oslo, 1976.
334. G. P. Rodnam, C. McEwen, and S. L. Wallace, *Primer on the Rheumatic Diseases*, 7th ed., The Arthritis Foundation, Atlanta, Ga., 1973.
335. R. A. Scherrer and M. W. Whitehouse, Eds., *Antiinflammatory Agents*, Vols. 1 and 2, Academic Press, New York, 1974.
336. P. Bresloff, in *Advances in Drug Research*, Vol. 2, N. J. Harper and A. B. Simmons, Eds., Academic Press, New York, 1977.
337. *Rheumatoid Arthritis: Pathogenic Mechanisms and Consequences in Therapeutics*, W. Muller, H. G. Hanwerth, and K. Fehr, Eds., Academic Press, New York, 1971.
338. *Inflammation: Mechanisms and Control*, I. H. Lepow and P. A. Ward, Eds., Academic Press, New York, 1972.
339. G. Renoux, *Pharm. Ther.*, A, **2**, 397 (1978).
340. J. L. Everett, J. J. Roberts, and W. C. J. Ross, *J. Chem. Soc.*, **1953**, 2390.
341. H. Arnold, F. Bourseaux, and N. Brock, *Naturwissenschaften*, **45**, 64 (1957); *Angew. Chem.*, **70**, 539 (1958).
342. A. C. Allison, *Proc. Roy. Soc. Med.*, **63**, 1077 (1970).
343. J. Pelletier and E. Caventou, *Ann. Chim. Phys.*, **14**, 69 (1820).
344. S. L. Wallace, *Am. J. Med.*, **30**, 439 (1961).
345. J. J. Pfiffner, O. Wintersteiner, and H. M. Vars, *J. Biol. Chem.*, **111**, 585 (1935).
346. S. P. Hench, E. C. Kendall, C. H. Slocumb, and H. F. Polley, *Mayo Clin. Proc.*, **24**, 181 (1949).
347. L. W. O'Neal, *Med. Clin. North Am.*, **52**, 313 (1968).
348. J. Fried and E. F. Sabo, *J. Am. Chem. Soc.*, **75**, 2273 (1953).
349. J. Fried and A. Borman, *Vitam. Horm.*, **16**, 303 (1958).
350. I. Ringler, S. Mauer, and E. Heyder, *Proc. Soc. Exp. Biol. Med.*, **107**, 451 (1961).
351. M. E. Wolff, in reference 7c, Chapter 63.
352. F. P. Nijkamp, R. J. Flower, S. Moncada and J. R. Vane, *Nature* (*London*), **263**, 479 (1975).

REFERENCES

353. M. E. Wolff, J. Bater, P. A. Kollman, D. L. Lee, I. D. Kuntz, E. Bloom, D. Matulich, and J. Morris, *Biochemistry*, **17**, 3201 (1978).
354. O. R. Rodig, in *Medicinal Chemistry*, 3rd ed., Part II, A. Burger, Ed., Wiley-Interscience, New York, 1970, pp. 894–895.
355. G. B. Elion, E. Burgi, and G. H. Hitchings, *J. Am. Chem. Soc.*, **74**, 411 (1952).
356. G. E. Underwood, *Proc. Soc. Exp. Biol. Med.*, **111**, 660 (1962).
357. G. B. Elion, P. A. Furman, J. A. Fife, P. de Miranda, L. Beauchamp, and H. J. Schaeffer, *Proc. Nat. Acad. Sci. U.S.*, **74**, 5716 (1977).
358. T. P. Johnston, G. McCaleb, and J. A. Montgomery, *J. Med. Chem.*, **6**, 669 (1963).
359. J. J. Vavra, C. de Boer, A. Dietz, L. J. Hanka and W. T. Sokolski, *Antibiot. Ann.*, 1959–1960, 230 (1960).
360. D. A. Clarke, R. K. Barclay, C. C. Stock, and C. S. Rondestvedt, Jr., *Proc. Soc. Exp. Biol. Med.*, **90**, 484 (1955).
361. R. Preussman, H. Druckrey, S. Ivankovik, and O. von Hodenberg, *Ann. N.Y. Acad. Sci.*, **163**, 697 (1969).
362. J. A. Montgomery, T. P. Johnston, and Y. F. Shealy, in ref. 7b, pp. 631–633.
363. Y. F. Shealy, C. A. Krauth, and J. A. Montgomery, *J. Org. Chem.*, **27**, 2150 (1962).
364. P. Zeller, H. Gutmann, B. Hegedüs, A. Kaiser, A. Langemann, and M. Müller, *Experientia*, **19**, 129 (1963).
365. W. Bollag and E. Grunberg, *Experientia*, **19**, 130 (1963).
366. S. M. Buckley, C. C. Stock, M. L. Crossley, and C. P. Rhoads, *Cancer Res.*, **10**, 207 (1950).
367. R. H. Blum, R. B. Livingston, and S. K. Carter, *Europ. J. Cancer*, **9**, 195 (1973).
368. B. Rosenberg, E. Reushaw, L. Van Camp, T. Hartwick, and J. Drobnick, *J. Bacteriol.*, **93**, 716 (1967).
369. L. L. Munchausen and R. O. Rahn, *Cancer Chemother. Rep.*, Part I, **59**, 643 (1975).
370. M. Paulshock and J. C. Watts, U.S. Pat. 3,310,469 (1967).
371. G. W. Smith, U.S. Pat. 3,328,251 (1967).
372. W. W. Prichard, U.S. Pat. 3,317,387 (1967).
373. C. E. Hoffman, "Amantadine HCl and Related Compounds", in *Selective Inhibitors of Vital Functions*, W. A. Carter, Ed., CRC Press, Cleveland, 1973, p. 199.
374. D. J. Bauer and P. W. Sadler, *Brit. J. Pharm. Chemother.*, **15**, 101 (1960).
375. R. L. Thompson, J. Davis, P. B. Russell, and G. H. Hitchings, *Proc. Soc. Exp. Biol. Med.*, **84**, 496 (1953).
376. R. L. Thompson, M. L. Price, and S. A. Minton, *Proc. Soc. Exp. Biol. Med.*, **78**, 11 (1951).
377. J. S. Oxford and D. D. Perrin, *J. Gen. Virol.*, **23**, 59 (1974).
378. D. J. Bauer, in *Chemotherapy of Virus Diseases*, Vol. 1, *International Encyclopedia of Pharmacology*, D. J. Bauer, Ed., Pergamon Press, Oxford, 1972, p. 35.
379. A. Isaacs and J. Lindenmann (1957), as quoted in *Interferon: The New Hope for Cancer*, by M. Edelhart and J. Lindenmann, Addison-Wesley, Reading, Mass., 1981. See L. M. Kershner, *C & E News*, Sept. 28, 1981, p. 47.
380. G. Barger and H. H. Dale, *J. Physiol.*, **43**, 499 (1911).
381. J. J. Abel and S. Kubota, *J. Pharmacol. Exp. Ther.*, **13**, 243 (1919).
382. L. Popielski, *Arch. Ges. Physiol.*, **178**, 214 (1920).
383. H. H. Dale, *Lancet*, **2**, 1233, 1285 (1929).
384. F. Leonard and C. P. Huttrer, *Histamine Antagonists*, Chemical Biological Coordination Center, National Research Council, Washington, D.C., 1950.

385. E. Fourneau and D. Bovet, *Arch. Int. Pharmacodyn.*, **46**, 178 (1933).
386. G. Ungar, J. I. Parrot, and D. Bovet, *Compt. Rend. Soc. Biol.*, **124**, 445 (1937).
387. A. M. Staub, *Ann. Inst. Pasteur*, **63**, 400, 420, 485 (1939).
388. B. N. Halpern, *Arch. Int. Pharmacodyn.*, **68**, 339 (1942).
389. M. M. Mosnier, French Pat. 913,161 (1943).
390. G. R. Rieveschl, Jr., U.S. Pat. 2,421,714 (1947).
391. G. R. Rieveschl, Jr., and W. F. Huber, Abstract, 109th Meeting of the American Chemical Society, 1946, p. 50K.
392. CIBA A. G., Swiss Pat. 190,541 (1934).
393. K. Miescher and K. Hoffmann, *Helv. Chim. Acta*, **24**, 458 (1941).
394. L. N. Gay and P. E. Carliner, *Science*, **109**, 359 (1949).
395. D. T. Witiak and R. C. Cavestri, in ref. 7c, Chapter 49.
396. E. R. Loew, *Physiol. Rev.*, **27**, 542 (1947).
397. A. S. F. Ash and H. O. Schild, *Brit. J. Pharmacol.*, **27**, 427 (1966).
398. J. W. Black, W. A. M. Duncan, G. J. Durant, C. R. Ganellin, and M. E. Parsons, *Nature*, **236**, 385 (1972).
399. C. R. Ganellin, *J. Med. Chem.*, **16**, 620 (1973).
400. R. T. Brittain, D. Jack, and B. J. Price, *TIPS*, **2**, 310 (1981).
400a. J. M. Hoffman, A. M. Pietruszkiewicz, C. N. Habecker, B. T. Phillips, W. A. Bolhofer, E. J. Cragoe, Jr., M. L. Torchiana, W. C. Lumma, Jr., and J. J. Baldwin, *J. Med. Chem.*, **26**, 140 (1983).
401. C. R. Ganellin and G. J. Durant, in ref. 7c, pp. 543–545.
402. A. A. Algieri, Abstracts, 183rd National Meeting of the American Chemical Society, Las Vegas, Nev., March 28–April 2, 1982, MEDI No. 8.
402a. C. A. Lipinski, *J. Med. Chem.*, **26**, 1 (1983).
402b. B. M. Conti-Tronconi, C. M. Gotti, M. W. Hunkapiller, and M. A. Raftery, *Science*, **218**, 1227 (1982).
403. E. G. Vernier, in *Evaluation of Drug Activities: Pharmacometrics*. D. R. Laurence and A. L. Bacharach, Eds., Academic Press, New York, 1964, Vol. 1, pp. 301–311. D. D. Bonnycastle, *ibid.*, Vol. 2, pp. 507–520.
404. Mein, *Ann. Chem.*, **6**, 67 (1833).
405. P. L. Geiger and O. Hesse, *Ann. Chem.*, **5**, 43 (1833); **6**, 44 (1833); **7**, 269 (1833).
406. A. Ladenburg, *Ann. Chem.*, **217**, 75 (1883).
407. R. Willstätter, *Chem. Ber.*, **34**, 3163 (1901).
408. J. Gadamer and F. Hammer, *Arch. Pharm.*, **259**, 122 (1921).
409. K. Hess and O. Wahl, *Chem. Ber.*, **55**, 1979 (1922).
410. R. Willstätter and E. Berner, *Chem. Ber.*, **56**, 1079 (1923).
411. Steffens, *Arch. Pharm.*, **262**, 211 (1924).
412. G. Fodór and G. Csepreghy, *Tetrahedron Lett.*, 7, 16 (1959).
413. B. V. Rama Sastry, in ref. 7c, Chapter 44.
414. B. H. Takman and H. J. Adams, *ibid.*, Chapter 51.
415. W. H. McGregor, L. Stein, and J. D. Belluzzi, *Life Sci.*, **23**, 1371 (1978).
416. B. W. Erickson and R. B. Merrifield, in *The Proteins*, 3rd ed., Vol. 2, H. Neurath and R. L. Hill, Eds., Academic Press, New York, 1976, pp. 255–527.
417. J. Meienhofer, in ref. 7b, Chapter 27, and quoted in refs. 37, 43, 44, 52, 53.
418. D. Römer, H. H. Büscher, R. C. Hill, J. Pless, W. Bauer, F. Cardinaux, D. Closse, D. Hauser, and R. Huguenin, *Nature (London)*, **268**, 547 (1977).

REFERENCES

419. B. von Graffenried, E. del Pozo, J. Roubicek, E. Krebs, W. Pöldinger, P. Burmeister, and L. Kerp, *Nature (London)*, **272**, 729 (1978).
420. C. R. A. Wright, *J. Chem. Soc.*, **27**, 1031 (1874).
421. H. Dreser, *Dtsch. Med. Wochenschr.*, **24**, 185 (1898).
422. L. F. Small and R. E. Lutz, *Chemistry of the Opium Alkaloids*, Suppl. No. 103, Public Health Reports, Washington, D.C., 1932.
423. U. Weiss, *J. Am. Chem. Soc.*, **77**, 5891 (1955).
424. M. Freund, E. Speyer, and E. Gutman, *Chem. Ber.*, **53**, 2250 (1920).
425. L. F. Small, H. M. Fitch, and W. E. Smith, *J. Am. Chem. Soc.*, **58**, 1457 (1936).
426. See L. F. Small, N. B. Eddy, E. Mosettig, and C. K. Himmelsbach, *Studies on Drug Addiction*, Suppl. No. 138, Public Health Reports, U.S. Government Printing Office, Washington, D.C., 1938.
427. O. Eisleb, U.S. Pat. 2,167,351 (1930).
428. O. Schaumann, *Arch. Pharmakol. Exp. Pathol.*, **196**, 109 (1940); *Pharmazie*, **4**, 364 (1949).
429. O. Eisleb, quoted by E. C. Kleiderer, J. B. Rice, V. Conquest, and J. H. Williams, *U.S. Dept. Commerce Off. Publ. Board*, Rep. PP-981 (1945).
430. A. Pohland and H. R. Sullivan, *J. Am. Chem. Soc.*, **75**, 4458 (1953).
431. P. A. J. Janssen, *J. Am. Chem. Soc.*, **78**, 3862 (1956).
432. R. Grewe, *Naturwissenschaften*, **33**, 333 (1946).
433. O. Schnider and A. Grüssner, *Helv. Chim. Acta*, **32**, 821 (1949).
434. E. L. May and J. G. Murphy, *J. Org. Chem.*, **20**, 257 (1955).
435. N. B. Eddy and E. L. May, in *Synthetic Analgesics*, Part II, J. Rolfe, Ed., Pergamon Press, London, 1965.
436. K. W. Bentley, D. G. Hardy, and B. Meek, *J. Am. Chem. Soc.*, **89**, 3267, 3273 (1967).
437. L. Lasagna and H. K. Beecher, *J. Pharmacol. Exp. Ther.*, **112**, 356 (1954).
438. S. Shiotani, T. Kometani, Y. Iitaka, and A. Itai, *J. Med. Chem.*, **21**, 153 (1978).
438a. R. J. Kobylecki, A. C. Lane, C. F. Smith, L. P. G. Wakelin, W. B. T. Cruse, E. Egert, and O. Kennard, *J. Med. Chem.*, **25**, 1278 (1982).
439. A. H. Beckett and A. F. Casy, *J. Pharm. Pharmacol.*, **6**, 986 (1954).
440. P. S. Portoghese, *J. Pharm. Sci.*, **52**, 865 (1966).
441. H. H. Loh, T. M. Cho, Y. C. Wu, and E. L. Way, *Life Sci.*, **14**, 2231 (1974).
442. A. M. Ernst, *Arch. Int. Pharmacodyn. Ther.*, **58**, 363 (1938); **61**, 73 (1939).
443. H. Silvestrini and G. Maffii, *Farmaco, Sci. Ed.*, **14**, 440 (1959).
444. K. Takagi, H. Fukada, M. Watanabe, and M. Sato, *Zasshi*, **80**, 1506 (1960).
445. D. Miller, in reference 7c, Chapter 53.
446. R. B. Barber and H. Rapoport, *J. Med. Chem.*, **19**, 1175 (1976).
447. P. Chabrier, R. Guidicelli, and J. Thuillier, *Ann. Pharm. Franç.*, **8**, 261 (1950).
448. B. Weiss, *Am. J. Pharm.*, **131**, 286 (1959).
449. S. Hyman and S. H. Rosenblum, *Ill. Med. J.*, **104**, 257 (1953).
450. J. Knoll, S. Makleit, T. Friedman, L. G. Hársing, Jr., and P. Hadházy, *Arch. Int. Pharmacodyn. Ther.*, **210**, 241 (1974).
451. J. Knoll, S. Fürst, and S. Makleit, *J. Pharm. Pharmacol.*, **27**, 99 (1975).
452. J. W. Lewis, P. A. Major, and D. I. Haddlesey, *J. Med. Chem.*, **16**, 12 (1973).
453. M. Murakami, S. Kawahara, N. Inukai, N. Nagano, H. Iwamoto, and H. Ida, *Chem. Pharm. Bull. (Tokyo)*, **20**, 1706 (1972).

454. I. Monković, H. Wong, A. W. Pircio, Y. G. Perron, I. J. Pachter, and B. Belleau, *Can. J. Chem.*, **53**, 3094 (1975).
455. F. G. Banting and C. H. Best, *J. Lab. Clin. Med.*, **7**, 251 (1922).
456. P. G. Katsoyannis, A. Tometsko, and K. Fukuda, *J. Am. Chem. Soc.*, **85**, 2863 (1963).
457. P. G. Katsoyannis, K. Fukuda, A. Tometsko, K. Suzuki, and M. Tilak, *J. Am. Chem. Soc.*, **86**, 930 (1964).
458. J. Meienhofer, E. Schnabel, H. Bremer, O. Brinkhoff, R. Zabel, W. Sroka, H. Klostermeyer, D. Brandenburg, T. Okuda, and H. Zahn, *Z. Naturforsch.*, **18b**, 1120 (1963).
459. P. G. Katsoyannis, *Am. J. Med.*, **40**, 652 (1966).
460. G. H. Dixon, *Excerpta Med. Found., Int. Congr. Ser.*, No. 58, 1207 (1964).
461. The Shanghai Insulin Research Group, *Sci. Sinica*, **16**, 61 (1973).
462. G. Weitzel, F.-U. Bauer, and K. Eisele, *Z. Physiol. Chem.*, **357**, 187 (1976).
463. R. Obermeier and R. Geiger, *Z. Physiol. Chem.*, **357**, 759 (1976).
464. R. Geiger, *Chem. Ztg. Chem. App.*, **100**, 111 (1976).
465. J. Meienhofer, in reference 7b, pp. 813–831.
466. M. Brown, J. Rivier, and W. Vale, *Fed. Proc.*, **35**, 782 (1976).
467. B. Blank, in reference 7b, pp. 1050–1057.
468. M. Janbon, J. Chaptal, A. Vedal, and J. Schaap, *Montpellier Med.*, **21–22**, 441 (1942).
469. A. Loubatières, *Ann. N.Y. Acad. Sci.*, **71**, 4 (1957).
470. H. Franke and J. Fuchs, *Dtsch. Med. Wochenschr.*, **80**, 1449 (1955).
471. C. K. Wanatabe, *J. Biol. Chem.*, **33**, 253 (1918).
472. E. Frank, M. Nothmann, and A. Wagner, *Klin. Wochenschr.*, **7**, 1996 (1928).
473. G. Ungar, L. Friedman, and S. L. Shapiro, *Proc. Soc. Exp. Biol. Med.*, **95**, 190 (1957), and subsequent papers.
474. K. H. Slotta and R. Tschesche, *Chem. Ber.*, **62B**, 1398 (1929).
475. S. M. Free and J. W. Wilson, *J. Med. Chem.*, **7**, 395 (1964).
476. K. Boček, J. Kopecký, M. Krivuková, and D. Vlachová, *Experientia*, **20**, 667 (1964).
477. J. Kopecký, K. Boček, and D. Vlachová, *Nature (London)*, **207**, 981 (1965).
478. L. F. Hodes, G. F. Hazard, R. I. Geran, and S. Richman, *J. Med. Chem.*, **20**, 469 (1977).
479. W. A. Creasey, *Drug Disposition in Humans: The Basis of Clinical Pharmacology.* Oxford University Press, 1979.
480. D. E. Hathaway, Ed., *Foreign Compound Metabolism in Mammals*, Vol. 5, The Chemical Society, London, 1979.
481. J. W. Bridges and L. F. Chasseaud, Eds., *Progress in Drug Metabolism*, Vol. 4, Wiley, New York, 1980.
482. E. R. Garret and J. L. Hirtz, Eds., *Drug Fate and Metabolism*, Vol. 3, Marcel Dekker, New York, 1980.
483. D. Rosi, G. Peruzzotti, E. W. Dennis, D. A. Berberian, H. Freele, and S. Archer, *Nature (London)*, **208**, 1005 (1965).
484. R. Wolfenden, *Ann. Rev. Biophys.*, **5**, 271 (1976); *Methods Enzymol.*, **46**, 15 (1977).
485. P. M. Schwartz, C. Shipman, Jr., and J. C. Drach, *Antimicrob. Agents Chemother.*, **10**, 64 (1976).
486. P. E. Borondy, T. Chang, E. Maschewske, and A. J. Glazko, *Ann. N.Y. Acad. Sci.*, **284**, 9 (1977).
487. D. V. Santi and G. L. Kenyon, in reference 7a, pp. 376–378.
488. R. H. Abeles and A. L. Maycock, *Acc. Chem. Res.*, **9**, 313 (1976).
489. R. R. Rando, *Science*, **185**, 320 (1974).

REFERENCES

490. C. Walsh, *Horizons Biochem. Biophys.*, **3**, 36 (1977).
491. T. I. Kalman, Ed., *Drug Action and Design: Mechanism-Based Enzyme Inhibitors*, Elsevier/North Holland, Amsterdam, 1979.
492. B. W. Metcalf, *Ann. Rep. Med. Chem.*, **16**, 289 (1981).
493. J. S. Wiseman and R. H. Abeles, *Biochemistry*, **18**, 427 (1979).
494. R. B. Silverman and S. J. Hoffman, *J. Am. Chem. Soc.*, **102**, 884 (1980).
495. C. Paech, J. I. Salach, and T. P. Singer, *J. Biol. Chem.*, **255**, 2700 (1980).
496. A. Burger, C. S. Davis, H. Green, D. H. Tedeschi, and C. L. Zirkle, *J. Med. Pharm. Chem.*, **4**, 571 (1961).
497. A. Burger and S. Nara, *J. Med. Chem.*, **8**, 859 (1965).
498. Y. Osawa, C. Yarborough, and Y. Osawa, *Science*, **215**, 1251 (1982).
499. B. R. Baker, Ed., *Design of Active-Site-Directed Irreversible Enzyme Inhibitors*. Wiley, New York, 1967.
500. H. H. Fox and J. T. Gibas, *J. Org. Chem.*, **18**, 994 (1953).
501. P. Zeller, A. Pletscher, K. F. Gey, H. Gutmann, B. Hegedüs, and O. Straub, *Ann. N.Y. Acad. Sci.*, **80**, 555 (1959).
502. L. Pauling, *Chem. Eng. News*, **24**, 1375 (1946).
503. H. Southworth, *Proc. Soc. Exp. Biol. Med.*, **36**, 58 (1937).
504. T. Mann and D. Keilin, *Nature (London)*, **146**, 164 (1940).
505. H. W. Davenport and A. E. Wilhelmi, *Proc. Soc. Exp. Biol. Med.*, **48**, 53 (1941).
506. R. O. Roblin, Jr. and J. W. Clapp, *J. Am. Chem. Soc.*, **72**, 4890 (1950).
507. W. H. Miller, A. M. Dessert, and R. O. Roblin, Jr., *J. Am. Chem. Soc.*, **72**, 4893 (1950).
508. J. M. Sprague, *Ann. N.Y. Acad. Sci.*, **71**, (4), 328 (1958).
509. K. H. Beyer and J. E. Baer, *Pharmacol. Rev.*, **13**, 517 (1961).
510. C. H. Hansch, *Acc. Chem. Res.*, **2**, 232 (1969).
511. O. Exner, *Collect. Czech. Chem. Commun.*, **32**, 1, 24 (1967).
512. L. Michaelis and M. L. Menten, *Biochem. Z.*, **49**, 333 (1913).
513. J. Ferguson, *Proc. Roy. Soc., Ser. B.*, **127**, 387 (1939).
514. M. Shaffer and F. W. Tilley, *J. Bact.*, **14**, 259 (1927).
515. G. M. Badger, *Nature (London)*, **158**, 585 (1946).
516. Lewis and Randall, *Thermodynamics*, McGraw Hill, New York, 1923, Chapter 22.
517. A. Leo, C. Hansch, and D. Elkins, *Chem. Rev.*, **71**, 525 (1971).
518. W. P. Purcell, G. E. Bass, and J. M. Clayton, *Strategy of Drug Design: A Guide to Biological Activity*. Wiley, New York, 1973.
519. W. Hückel, *Z. Elektrochem.*, **27**, 305 (1921).
520. V. Schomaker and L. Pauling, *J. Am. Chem. Soc.*, **61**, 1769 (1939).
521. H. C. Longuet-Higgins, *Trans. Faraday Soc.*, **45**, 173 (1949).
522. J. Metzger and F. Ruffler, *J. Chim. Phys.*, **51**, 52 (1954).
523. H. G. Grimm, *Z. Elektrochem.*, **31**, 474 (1925); **34**, 430 (1928); **47**, 53, 594 (1934).
524. A. Korolkovas, *Essentials of Molecular Pharmacology: Background for Drug Design*, Wiley, New York, 1970.
525. C. Hansch, *J. Am. Chem.*, **19**, 1 (1976).
526. L. F. Fieser, M. G. Ettlinger, and G. Fawaz, *J. Am. Chem. Soc.*, **70**, 3228 (1948).
527. B. B. Brodie and C. A. Hogben, *J. Pharm. Pharmacol.*, **9**, 345 (1957).
528. P. H. Bell and R. O. Roblin, Jr., *J. Am. Chem. Soc.*, **64**, 2905 (1942).
529. E. B. Leffer, H. M. Spencer, and A. Burger, *J. Am. Chem. Soc.*, **73**, 2611 (1951).

530. C. Hansch, *Intra-Science Chem. Rep.*, **8**, 17 (1974).
531. K. C. Chu, in ref. 7a, Chapter 10.
532. Y. C. Martin, *J. Med. Chem.*, **24**, 229 (1981).
533. T. M. Bustard and R. S. Egan, *Tetrahedron*, **27**, 4457 (1971).
534. R. Bergin and D. Carlström, *Acta Crystallogr., Sect. B.*, **24**, 1506 (1968).
535. J. P. Tollenaere, H. Moereels, and L. A. Raymaekers, *Atlas of the Three-Dimensional Structure of Drugs*, Elsevier/North Holland, Amsterdam, 1979.
536. J. P. Tollenaere, *TIPS*, **2**, 273 (1981).
537. J. H. Oppenheimer, D. Koerner, H. J. Schwartz, and M. I. Surks, *J. Clin. Endocrinol. Metab.*, **35**, 330 (1972).
538. E. C. Jorgensen, in ref. 7c, Chapter 39.
539. Y. Kanda, D. S. Goodman, R. E. Canfield, and F. J. Morgan, *J. Biol. Chem.*, **249**, 6796 (1974).
540. C. C. F. Blake, I. D. A. Swan, C. Rérat, J. Berthou, A. Laurent, and B. Rérat, *J. Mol. Biol.*, **61**, 217 (1971).
541. C. C. F. Blake and S. J. Oatley, *Nature (London)*, **268**, 115 (1977).
542. V. Cody, J. Hazel, D. A. Langs, and W. L. Duax, *J. Med. Chem.*, **20**, 1628 (1977).
543. N. Zenker and E. C. Jorgensen, *J. Am. Chem. Soc.*, **81**, 4643 (1959).
544. A. Windaus, H. Lettré, and F. Schenck, *Ann. Chem.*, **520**, 98 (1935).
545. J. L. Napoli, M. A. Fivizzani, H. K. Schnoes, and H. F. Deluca, Proceedings of the 6th Parathyroid Conference, University of British Columbia, Vancouver, B.C., June 12–17, 1977.
546. H. Kawashima, K. Hoshina, Y. Hashimoto, T. Takeshita, S. Ishimoto, T. Noguchi, N. Ikekawa, M. Morisaki, and H. Orimo, *Fed. Eur. Biochem. Soc. Lett.*, **76**, 177 (1977).
547. R. B. Merrifield, *J. Am. Chem. Soc.*, **85**, 2149 (1963).
548. R. Archer, in *Handbook of Physiology*, Section 7, *Endocrinology*, Part I, American Physiological Society, Washington, D.C., 1974, pp. 119–130.
549. I. Vavra, A. Machova, V. Holecek, J. H. Cort, M. Zaoral, and F. Sorm, *Lancet*, **1**, 948 (1968).
550. M. Manning, L. Balaspiri, M. Acosta, and W. H. Sawyer, *J. Med. Chem.*, **16**, 975 (1973).
551. J. Meienhofer, in ref. 7b, Chapter 27.
552. W. Rittel, R. Maier, M. Bruegger, B. Kamber, B. Riniker, and P. Siebe, *Experientia*, **32**, 246 (1976).
553. T. Morikawa, E. Munekata, S. Sakakibara, T. Noda, and M. Otani, *Experientia*, **32**, 1104 (1976).
554. H. Immer, V. R. Nelson, C. Revesz, K. Sestanj, and M. Götz, *J. Med. Chem.*, **17**, 1060 (1974).
555. R. Hirschmann and D. Veber, in *The Chemistry of Polypeptides*, P. G. Katsoyannis, Ed., Plenum, New York, 1973, pp. 125–142.
556. F. Labrie, M. Savary, D. H. Coy, E. J. Coy, and A. V. Schally, *Endocrinology*, **98**, 289 (1976).
557. J. E. Rivier and W. Vale, *Life Sci.*, **23**, 869 (1978).
558. J. Humphries, Y.-P. Wan, K. Folkers, and C. Y. Bowers, *J. Med. Chem.*, **20**, 967, 1674 (1977).
559. H.-G. Gattner, E. W. Smith, and V. K. Naithani, *J. Physiol. Chem.*, **356**, 1465 (1975).
560. R. Geiger, *Chem.-Ztg., Chem. App.*, **100**, 111 (1976).
561. H. J. Friesen, D. Brandenburg, C. Diaconescu, H.-G. Gattner, V. K. Naithani, J.

REFERENCES

Nowak, H. Zahn, S. Dockerill, S. P. Wood, and T. L. Blundell, in *Peptides 1977: Proceedings of the 5th American Peptide Symposium*, M. Goodman and J. Meienhofer, Eds., Wiley, New York, 1978, pp. 136–140.

562. T. L. Blundell, G. D. Dodson, D. C. Hodgkin, and D. A. Mercola, *Adv. Protein Chem.*, **26**, 279 (1972).
563. D. Brandenburg, W.-D. Busse, H.-G. Gattner, H. Zahn, A. Wollmer, J. Gliemann, and W. Puls, in *Peptides 1972*, H. Hanson and H. D. Jakubke, Eds., North Holland/Elsevier, New York, 1973, pp. 270–283.
564. J. MacLean, *Am. J. Physiol.*, **41**, 250 (1916).
565. W. H. Howell, *Harvey Lectures*, Series XII, 1916–1917.
566. W. H. Howell and E. Holt, *Am. J. Physiol.*, **47**, 328 (1918).
567. K. N. von Kaulla, in ref. 7b, pp. 1085–1096.
568. E. Husemann, *J. Prakt. Chem.*, **155**, 13 (1940).
569. E. Husemann, K. N. von Kaulla, and R. Kappesser, *Z. Naturforsch.*, **1**, 584 (1946).
570. F. W. Schofield, *Can. Vet. Rec.*, **3**, 74 (1932).
571. K. P. Link, R. S. Overman, W. R. Sullivan, C. F. Huebner, and L. D. Scheel, *J. Biol. Chem.*, **138**, 21, 513, 529 (1941); **142**, 941 (1942).
572. R. Anschütz, *Chem. Ber.*, **36**, 465 (1903).
573. J. B. Bingham, O. O. Meyer, and F. J. Pohle, *Am. J. Med. Sci.*, **202**, 563 (1941).
574. K. P. Link, *Circulation*, **19**, 97 (1949).
575. M. A. Stahmann and K. P. Link, U.S. Pat. 2,427,578 (1947).
576. H. Veldstra, P. W. Wiardi, and J. Alberda, *Rec. Trav. Chim.*, **72**, 358 (1953).
577. B. van Zanten and C. N. Nauta, *Arzneimittel-Forsch.*, **14**, 29 (1964).
578. I. Chmielewska and J. Cieslak, *Tetrahedron*, **4**, 135 (1958).
579. E. Lederer, *Biochem. J.*, **93**, 449 (1964).
580. D. Nachmansohn and M. Berman, *J. Biol. Chem.*, **165**, 551 (1946).
581. A. F. Casy, *Prog. Med. Chem.*, **11**, 1 (1975).
582. F. Hucho and J. P. Changeux, *FEBS Lett.*, **38**, 11 (1973).
583. J. B. Stenlake, in ref. 7c, Chapter 46.
584. A. Karlin and E. Bartels, *Biochem. Biophys. Acta*, **126**, 525 (1966).
585. A. K. Cho, W. L. Haslett, and D. J. Jenden, *J. Pharmacol. Exp. Ther.*, **138**, 249 (1962).
586. G. B. Koelle, in *The Pharmacological Basis of Therapeutics*, 5th ed., L. S. Goodman and A. Gilman, Eds., Macmillan, New York, 1975, p. 472.
587. J. G. Cannon, in reference 7c, Chapter 43.
588. P. D. Armstrong, J. G. Cannon, and J. P. Long, *Nature (London)*, **220**, 65 (1968).
589. C. Y. Chiou, J. P. Long, J. G. Cannon, and P. D. Armstrong, *J. Pharmacol. Exp. Ther.*, **166**, 243 (1969).
590. C. Chothia and P. Pauling, *Nature (London)*, **226**, 541 (1970).
591. C. Bernard, *Compt. Rend. Soc. Biol.*, **2**, 195 (1851).
592. H. King, *J. Chem. Soc.*, 1381 (1935); 936 (1947); 265 (1948); *Nature (London)*, **158**, 515 (1946).
593. A. J. Everett, L. A. Low, and S. Wilkinson, *Chem. Commun.*, **1970**, 1020.
594. R. B. Barlow and H. R. Ing, *Nature (London)*, **161**, 718 (1948); *Brit. J. Pharmacol. Chemother.*, **3**, 298 (1948).
595. W. D. M. Paton and E. J. Zaimis, *Nature (London)*, **161**, 718 (1948).
596. D. Bovet, F. Bovet-Nitti, S. Guarino, V. G. Longo, and M. Marotta, *R. C. Ist. Sup. Sanità*, **12**, 106 (1949).

597. B. V. Rama Sastry, in reference 7c, Chapter 34.
598. C. P. Huttrer, C. Djerassi, W. L. Beears, R. L. Mayer, and C. R. Scholz, *J. Am. Chem. Soc.*, **68**, 1999 (1946).
599. D. Bovet, R. Horclois, and F. Walthert, *Compt. Rend. Soc. Biol.*, **138**, 99 (1944).
600. B. N. Halpern and F. Walthert, *Compt. Rend. Soc. Biol.*, **139**, 402 (1945).
601. R. F. Rekker, H. Timmerman, and A. F. Harms, *Arzneim.-Forsch.*, **21**, 688 (1971).
602. W. Th. Nauta, T. Bultsma, R. F. Rekker, and H. Timmerman, *Spec. Contrib. 3rd. Int. Symp. Med. Chem.* (1972), p. 125.
603. A. F. Casy and A. P. Parulkar, *Can. J. Chem.*, **47**, 423 (1969).
604. R. R. Ison and A. F. Casy, *J. Pharm. Pharmacol.*, **23**, 848 (1971).
605. D. T. Witiak and R. C. Cavestri, in reference 7c, Chapter 49.
606. B. N. Halpern, *J. Am. Med. Assoc.*, **129**, 1219 (1945).
607. B. N. Halpern and R. Ducrot, *Compt. Rend. Soc. Biol.*, **140**, 361 (1946).
608. C. P. Huttrer, *Enzymologia*, **12**, 277 (1948).
609. J. D. Dunitz, H. Eser, and P. Strickler, *Helv. Chim. Acta*, **47**, 1897 (1964).
610. C. A. Stone, H. C. Wenger, C. T. Ludden, J. M. Stavorskii, and C. A. Ross, *J. Pharmacol. Exp. Ther.*, **131**, 73 (1961).
611. E. I. Engelhardt, H. C. Zell, W. S. Saari, M. E. Christy, C. D. Colton, C. A. Stone, J. M. Stavorskii, H. C. Wenger, and C. T. Ludden, *J. Med. Chem.*, **8**, 829 (1965).
612. B. H. Takman and H. J. Adams, in reference 7c, Chapter 51, pp. 654–656.
613. H. Braun, *Dtsch. Med. Wochenschr.*, **31**, 1667 (1905).
614. A. Einhorn and E. Uhlfelder, *Ann. Chem.*, **371**, 131 (1909).
615. M. E. Krahl, A. K. Keltch, and G. H. A. Clowes, *J. Pharmacol. Exp. Ther.*, **68**, 330 (1940).
616. J. C. Skou, *Acta Pharmacol. Toxicol.*, **10**, 281 (1954).
617. E. J. Ariëns and A. M. Simonis, *Arch. Int. Pharmocodyn.*, **141**, 309 (1963).
618. N. Löfgren, *Ark. Kemi. Mineral. Geol.*, **22A**, 18 (1946).
619. A. Sekera, J. Sova, and Č. Vrba, *Experientia*, **11**, 275 (1955).
620. K. Miescher, *Helv. Chim. Acta*, **15**, 163 (1932).
621. M. E. Freed, W. F. Bruce, R. S. Hanslick, and A. Mascitti, *J. Org. Chem.*, **26**, 2378 (1961).
623. H. B. Wright and M. B. Moore, *J. Am. Chem. Soc.*, **76**, 4396 (1954).
623. E. Profft, *Chem. Tech.* (Berlin), **3**, 210 (1951); **4**, 241 (1952).
624. A. Hofmann, in *Drugs Affecting the Central Nervous System*, A. Burger, Ed., Marcel Dekker, New York, 1968, pp. 169–235.
625. L. G. Abood, *ibid.*, pp. 127–167.
626. L. Lewin, *Phantastica, Narcotic and Stimulating Drugs: Their Use and Abuse*, Routledge and Kegan Paul, London, 1964.
627. H. Osmond, *Ann. N.Y. Acad. Sci.*, **66**, 418 (1957).
628. T. Leary, R. Metzner, and R. Alpert, *The Psychedelic Experience,* University Books, New Hyde Park, N.Y., 1964.
629. G. E. Rumpf, in *Herbari Amboinensis Actuarium,* Mynard, Uytwerf, et al., Eds., Amsterdam, 1755.
630. G. Sen and K. C. Bose, *Indian Med. World*, **2**, 194 (1931).
631. L. Dorfman, A. Furlenmeier, C. S. Huebner, R. Lucas, H. B. MacPhillamy, J. M. Mueller, E. Schlittler, R. Schwyzer, and A. F. St. André, *Helv. Chim. Acta*, **37**, 59 (1954).

REFERENCES

632. R. B. Woodward, F. E. Bader, H. Bickle, A. J. Frey, and R. W. Kierstad, *J. Am. Chem. Soc.*, **78**, 2023, 2651 (1956).
633. H. J. Bein, *Experientia*, **9**, 107 (1953).
634. N. S. Kline, *Ann. N.Y. Acad. Sci.*, **59**, 107 (1954).
635. W. B. Reid, Jr., J. B. Wright, H. G. Kolloff, and J. H. Hunter, *J. Am. Chem. Soc.*, **70**, 3100 (1948).
636. M. J. VanderBrook, K. H. Olson, M. T. Richmond, and M. H. Kuizenga, *J. Pharmacol. Exp. Ther.*, **94**, 197 (1948).
637. S. Courvoisier, J. Fournel, R. Ducrot, M. Kolsky, and P. Koetchet, *Arch. Int. Pharmacodyn.*, **92**, 305 (1952).
638. W. Y. Cheung, *Science*, **207**, 19 (1980).
639. M. J. Welch, J. C. Aster, M. Ireland, J. Alcala, and H. Maisel, *Science*, **216**, 642 (1982).
640. J.-P. Bourquin, G. Schwab, G. Gamboni, R. Fischer, L. Reusch, S. Guldimann, V. Theus, E. Schenker, and J. Renz, *Helv. Chim. Acta*, **41**, 1072 (1958).
641. C. Kaiser and P. E. Setler, reference 7c, p. 859.
642. E. F. Domino, R. D. Hudson, and G. Zografi, in *Drugs Affecting the Central Nervous System*, Vol. 2, A. Burger, Ed., Marcel Dekker, New York, 1968.
643. M. Gordon, Ed., *Psychopharmacological Agents*, Vols. 1 and 2, Academic Press, New York–London, 1964, 1967.
644. B. Blackwell, in *The Future of Pharmacotherapy: New Drug Delivery Systems*. F. J. Ayd, Ed., International Drug Therapy Newsletter, Baltimore, Md., 1973.
645. G. Stille and H. Hippius, *Pharmakopsychiatr. Neuropsychopharm.*, **4**, 182 (1971).
646. P. A. Janssen, in *Discoveries in Biological Psychiatry*, F. J. Ayd and B. Blackwell, Eds., Lippincott, Philadelphia, 1970, p. 165.
647. B. K. Krueger, J. Forn, and P. Greengard, in *Pre- and Postsynaptic Receptors*, E. Usdin and W. E. Bunney, Eds., Marcel Dekker, New York, 1975, p. 123. See also pp. 207 ff.
648. P. A. J. Janssen, in *Medicinal Chemistry, A Series of Monographs*. Vol. 4, Part II: *Psychopharmacological Agents*, M. Gordon, Ed., Academic Press, New York–London, 1967, p. 199.
649. P. A. J. Janssen, C. van de Westeringh, A. H. M. Jageneau, P. J. A. Demoen, B. K. F. Hermans, G. H. P. van Daele, K. H. L. Schellekens, C. A. M. van der Eycken, and C. J. E. Niemegeers, *J. Med. Pharm. Chem.*, **1**, 281 (1959).
650. K. Nádor and J. Pórszász, *Arzneim.-Forsch.*, **8**, 313 (1958).
651. K. Schoen, I. J. Pachter, and A. Rubin, *Abstracts, 153rd Meeting of the American Chemical Society*, Miami, Fla., April, 1967, p. M-46.
652. R. Kellner, R. T. Rada, A. Engelman, and B. Macaluso, *Curr. Ther. Res. Clin. Exp.*, **20**, 686 (1976).
653. F. G. Henderson, B. L. Martz, and I. H. Slater, *J. Pharmacol. Exp. Ther.*, **122**, 30A (1958).
654. E. C. Greisner, J. Barsky, C. A. Dragstedt, J. A. Wells, and E. A. Zeller, *Proc. Soc. Exp. Biol. Med.*, **84**, 699 (1953).
655. J. Rebhuhn, S. M. Feinberg, and E. A. Zeller, *Proc. Soc. Exp. Biol. Med.*, **87**, 218 (1954).
656. B. B. Brodie and P. A. Shore, *Ann. N.Y. Acad. Sci.*, **66**, 631 (1957).
657. H. P. Loomer, J. C. Saunders, and N. S. Kline, *Am. Psychiatr. Assoc. Rep.*, **8**, 129 (1957).
658. F. Häfliger and V. Burckhardt, in *Psychopharmacological Agents*, M. Gordon, Ed., Vol. 1, Academic Press, New York–London, 1964, Chapter 3.

659. J. F. J. Cade, *Med. J. Australia,* **36,** 349 (1949).
660. H. Thoenen and J. P. Trancer, *Ann. Rev. Pharmacol.,* **13,** 169 (1973).
661. R. M. Kostrzewa and D. M. Jacobowitz, *Pharmacol. Rev.,* **26,** 199 (1974).
662. D. J. Triggle, in reference 7c, p. 245.
663. P. J. Fowler, C. L. Zirkle, E. Macko, P. E. Setler, H. M. Sarau, A. Misher, and D. H. Tedeschi, *Arzneim.-Forsch.,* **27,** 1589 (1977).
664. A. Burger and S. Nara, *J. Med. Chem.,* **8,** 859 (1965).
665. W. J. Kinnard, H. Barry III, N. Watzman, and J. F. Buckley, in *Antidepressant Drugs,* S. Garratini and M. N. G. Dukes, Eds., Excerpta Medica, New York–London, 1967, p. 89.
666. F. Sulser, M. W. Owens, S. J. Strada, and J. V. Dingel, *J. Pharmacol. Exp. Ther.,* **168,** 272 (1969).
667. J. D. Taylor, A. A. Wykes, Y. C. Gladish, and W. B. Martin, *Nature* (London), **187,** 941 (1960).
668. A. L. Maycock, R. H. Abeles, J. L. Salach, and T. P. Singer, *Biochemistry,* **15,** 114 (1976).
669. A. Burger and W. L. Yost, *J. Am. Chem. Soc.,* **70,** 2198 (1948).
670. T. P. Singer, R. W. Von Korff, and D. L. Murphy, Eds., *Monoamine Oxidase: Structure, Function and Altered Functions.* Academic Press, New York–London, 1979.
671. C. L. Zirkle, C. Kaiser, D. H. Tedeschi, R. E. Tedeschi, and A. Burger, *J. Med. Pharm. Chem.,* **5,** 1265 (1962).
672. C. Kaiser and P. E. Setler, in ref. 7c, pp. 997–1067.
673. S. J. Childress, in ref. 7c, pp. 981–996.
674. W. A. Lott and E. J. Pribyl, U.S. Pat. 2,609,386 (1952).
675. H. L. Yale, E. J. Pribyl, W. Braker, J. Bernstein, and W. A. Lott, *J. Am. Chem. Soc.,* **72,** 3716 (1950).
676. F. M. Berger, *J. Pharmacol. Exp. Ther.,* **93,** 470 (1948); **104,** 229, 468 (1952).
677. F. M. Berger and B. J. Ludwig, *J. Pharmacol. Exp. Ther.,* **100,** 27 (1950); U.S. Pat. 2,724,720 (1955).
678. L. O. Randall, C. L. Scheckel, and R. F. Banziger, *Curr. Ther. Res.,* **7,** 590 (1965).
679. L. H. Sternbach, L. O. Randall, R. Banziger, and H. Lehr, in *Drugs Affecting the Central Nervous System,* A. Burger, ed., Marcel Dekker, New York, 1968, p. 237.
680. L. O. Randall, *Dis. Nerv. Syst.,* **21,** Suppl. 2, 7 (1960); **22,** 7 (1961).
681. L. H. Sternbach, G. A. Archer, J. V. Earley, R. I. Fryer, E. Reeder, N. Wasyliw, L. O. Randall, and R. Banziger, *J. Med. Chem.,* **8,** 815 (1965).
682. G. A. Archer et al., Belgian Pat. 629,005 (1963).
683. Hauptmann et al., South African Pat. 6,800,803 (1968); *C.A.,* **70,** 106,579f (1969).
684. C. H. Boehringer Sohn, German Pat. 2,533,924 (1978).
685. M. W. Johns, *Drugs,* **9,** 448 (1975).
686. M. Jouvet, in *Psychopharmacology: A Review of Progress, 1957–1967,* D. H. Efron, Ed., U.S. Government Printing Office, Washington, D.C., 1968, pp. 523–540.
687. M. Jouvet, *Ergeb. Physiol.,* **64,** 166 (1972).
688. J. R. Mouret, P. Bobillier, and M. Jouvet, *Europ. J. Pharmacol.,* **5,** 17 (1968).
689. R. J. Wyatt, *Biol. Psychiatry,* **5,** 33 (1972).
690. J. Matsumoto and M. Jouvet, *Compt. Rend. Soc. Biol.,* **158,** 2137 (1964).
691. D. Dusan-Peyrethon, J. Peyrethon, and M. Jouvet, *Compt. Rend. Soc. Biol.,* **162,** 116 (1968).

REFERENCES

692. O. Liebreich, *Klin. Wochenschr.*, **6**, 325 (1969).
693. S. C. Harvey, in *The Pharmacological Basis of Therapeutics*, 5th ed., L. S. Goodman and A. Gilman, Eds., Macmillan, New York, 1975, Chapter 10.
694. E. Baumann and A. Kast, *Z. Physiol. Chem.*, **14**, 52 (1888).
695. E. Fischer and J. v. Mering, *Ther. Gegenwart.*, **44**, 97 (1903).
696. Farbenfabriken vorm. F. Bayer, German Pat. 247,952 (1911).
697. U. Hoerlein, U.S. Pat. 1,025,872 (1912).
698. O. Juliusburger, *Klin. Wochenschr.*, **49**, 940 (1912).
699. E. Impens, *Deut. Med. Wochenschr.*, **38**, 945 (1912).
700. A. Hauptmann, *Münch. Med. Wochenschr.*, **59**, 1907 (1912).
701. H. A. Shonle and A. Moment, *J. Am. Chem. Soc.*, **45**, 243 (1923).
702. F. Sandberg, *Acta Physiol. Scand.*, **24**, 7 (1951).
703. V. Vercello, *Arch. Exp. Pathol. Pharmakol.*, **16**, 265 (1883).
704. J. A. Vida, in reference 7c, pp. 803–808.
705. Taub and Kropp, German Pat. 537,366 (1929).
706. L. C. Mark, *Clin. Pharmacol. Ther.*, **4**, 504 (1963).
707. C. M. Samour, J. F. Reinhard, and J. A. Vida, *J. Med. Chem.*, **14**, 187 (1971).
708. J. A. Vida, W. R. Wilber, and J. F. Reinhard, *J. Med. Chem.*, **14**, 191 (1971).
709. E. I. Isaacson and J. N. Delgado, in ref. 7c, pp. 838–839.
710. Heyden Co., German Pats. 309,508, 310,426, 310,427, 335,994 (1914).
711. W. T. Read, *J. Am. Chem. Soc.*, **44**, 1746 (1922).
712. T. J. Putnam and H. H. Merritt, *Science*, **85**, 525 (1937).
713. H. H. Merritt and T. J. Putnam, *Arch. Neurol. Psychiatr.*, **39**, 1003 (1938).
714. W. J. Close and M. A. Spielman, in *Medicinal Chemistry*, Vol. 5, W. H. Hartung, Ed., Wiley, New York, 1961, p. 1.
715. H. Biltz, *Chem. Ber.*, **41**, 1379 (p. 1391) (1908).
716. M. A. Spielman, A. O. Gieszler, and W. J. Close, *J. Am. Chem. Soc.*, **70**, 4189 (1948).
717. J. W. Clark-Lewis, *Chem. Rev.*, **58**, 63 (1958).
718. M. A. Spielman, *J. Am. Chem. Soc.*, **66**, 1244 (1944).
719. H. H. Frey and B. H. Kretschmer, *Arch. Int. Pharmacodyn. Ther.*, **193**, 181 (1971).
720. J. Y. Bogue and H. C. Carrington, *Brit. J. Pharmacol.*, **8**, 230 (1953).
721. P. Weichert, *Zentralbl. Pharm. Pharmakother. Laboratoriumsdiagn.*, **111**, 899 (1972); *C.A.*, **78**, 24095w (1973).
722. H. H. Frey and W. Loescher, *Arzneim.-Forsch.*, **26**, 299 (1976).
723. B. Ferrandes, C. Cohen-Added, J. L. Benoit-Guyod, and P. Eymard, *Biochem. Pharmacol.*, **23**, 3363 (1974).
724. R. U. Ostrovskaya and J. Schmidt, *J. Farmakol. Toksikol. (Moscow)*, **36**, 179 (1973).
725. F. H. Johnson, H. Eyring, and M. J. Polissar, *The Kinetic Basis of Molecular Biology*, Wiley, New York, 1954.
726. M. J. Croucher, J. F. Collins, and B. S. Meldrun, *Science*, **216**, 899 (1982).
727. L. Terenius, *Acta Pharmacol. Toxicol.*, **32**, 317 (1973).
728. C. B. Pert and S. H. Snyder, *Science*, **179**, 1011 (1973).
729. E. J. Simon, J. M. Hiller, and I. Edelman, *Proc. Nat. Acad. Sci., U.S.*, **70**, 1947 (1973).
730. H. H. Loh, T. M. Cho, Y. C. Wu, R. A. Harris, and E. L. Way, *Life Sci.*, **16**, 1811 (1975).

731. S. K. Sharma, M. Nirenberg, and W. A. Klee, *Proc. Nat. Acad. Sci., U.S.*, **72**, 590 (1974).
732. J. Traber, K. Fischer, S. Latziri, and B. Hamprecht, *FEBS Lett.*, **49**, 260 (1974).
733. J. Hughes, *Brain Res.*, **88**, 295 (1975).
734. L. Terenius and W. Wahlström, *Acta Physiol. Scand.*, **94**, 74 (1975).
735. J. Hughes, T. W. Smith, H. W. Kosterlitz, L. A. Fothergill, B. A. Morgan, and H. R. Morris, *Nature*, **258**, 577 (1975).
736. G. W. Pasternak, R. Goodman, and S. H. Snyder, *Life Sci.*, **16**, 1765 (1975).
737. C. H. Li, *Nature,* **201**, 924 (1964).
738. J. M. Polak, S. N. Sullivan, S. R. Bloom, P. Facer, and A. G. E. Pearse, *Lancet*, **1**, 972 (1977).
739. F. Bloom, E. Battenberg, J. Rossier, N. Ling, J. Leppalvoto, T. M. Vargo, and R. Guillemin, *Life Sci.*, **20**, 43 (1977).
740. M. Ross, R. Dingledine, B. M. Cox, and A. Goldstein, *Brain Res.*, **124**, 513 (1977).
741. P. W. Schiller, C. F. Yam, and M. Lis, *Biochemistry*, **16**, 1831 (1977).
742. M. R. Johnson and G. M. Milne, reference 7c, Chapter 52.
743. R. Grewe, *Naturwissenschaften*, **33**, 333 (1946).
744. R. Grewe and A. Mondon, *Chem. Ber.*, **81**, 279 (1948).
745. L. F. Small, H. M. Fitch, and W. E. Smith, *J. Am. Chem. Soc.*, **58**, 1457 (1936).
746. H. J. C. Yeh, R. S. Wilson, W. A. Klee, and A. E. Jacobson, *J. Pharm. Sci.*, **65**, 903 (1976).
747. M. T. Freund and E. Speyer, *J. Prakt. Chem.*, **94**, 135 (1916).
748. U. Weiss, *J. Am. Chem. Soc.*, **77**, 5091 (1955).
749. R. J. J. Ch. Lousberg and U. Weiss, *Experientia*, **30**, 1440 (1974).
750. K. W. Bentley, D. G. Hardy, and P. A. Major, *J. Chem. Soc.*, 2385 (1969).
751. W. M. Benson, P. L. Stefko, and L. O. Randall, *J. Pharmacol. Exp. Ther.*, **109**, 189 (1953).
752. N. B. Eddy, H. Besendorf, and B. Pellmont, *Bull. Narcotics*, **10**, 23 (1958).
753. J. Hellerbach, O. Schnider, H. Besendorf, and B. Pellmont, *Synthetic Analgesics, Part IIA: Morphinans*, Pergamon Press, Oxford, 1966.
754. D. C. Palmer and M. J. Strauss, *Chem. Res.*, **77**, 1 (1977).
755. L. S. Harris and A. K. Pierson, *J. Pharmacol. Exp. Ther.*, **143**, 141 (1964).
756. A. Ziering and J. Lee, *J. Org. Chem.*, **12**, 911 (1947).
757. P. A. J. Janssen and N. B. Eddy, *J. Med. Pharm. Chem.*, **2**, 31 (1960).
758. J. Cass and W. S. Frederik, *Curr. Ther. Res.*, **3**, 97 (1961).
759. W. F. M. van Bever, C. J. E. Niemegeers, and P. A. Janssen, *J. Med. Chem.*, **17**, 1047 (1974).
760. A. H. Beckett, *J. Pharm. Pharmacol.*, **8**, 848 (1956).
761. A. Gero, *Science*, **119**, 112 (1954).
762. J. G. Henkel, K. H. Bell, and P. S. Portoghese, *J. Med. Chem.*, **17**, 124 (1974).
763. A. Pohland and H. R. Sullivan, *J. Am. Chem. Soc.*, **75**, 4458 (1953).
764. N. B. Eddy, *Chem. Ind.*, 1462 (1959).
765. A. Hunger, J. Kebrle, A. Rossi, and K. Hoffman, *Experientia*, **13**, 400 (1957).
766. G. deStevens, Ed., *Analgetics,* Academic Press, New York, 1965, Chapter 8.
767. A. Brossi, H. Bessendorf, L. A. Pirk, and A. Rheiner, in reference 766, Chapter 6.
768. H. C. Li, *J. Econ. Bot.*, **28**, 437 (1974).

REFERENCES

769. F. V. Rossi, *Am. J. Pharm.*, **142,** 161 (1970).
770. R. S. Wilson and E. L. May, *J. Med. Chem.*, **17,** 475 (1974).
771. H. G. Pars, F. E. Granchelli, J. K. Keller, and R. Razdan, *J. Am. Chem. Soc.*, **88,** 3664 (1966), and subsequent papers in *J. Med. Chem.*
772. H. L. Herzog, A. Nobile, S. Tolksdorf, W. Charney, E. G. Hershberg, and P. L. Perlman, *Science,* **121,** 176 (1955).
773. H. L. Herzog, C. C. Payne, M. A. Jevnik, D. Gould, E. L. Shapiro, E. P. Oliveto, and E. B. Hershberg, *J. Am. Chem. Soc.*, **77,** 4781 (1955).
774. For a review, see *New Drugs,* American Medical Association, Chicago, 1967.
755. H. J. Lee and M. R. I. Soliman, *Science,* **215,** 991 (1982).
776. T. Y. Shen, in reference 7c, Chapter 62.
777. *Ibid.,* pp. 1225–1226.
778. J. R. Vane, *Nature New Biol.*, **231,** 232 (1971).
779. S. H. Ferreira, S. Moncada, and J. R. Vane, *Nature New Biol.*, **231,** 237 (1971).
780. J. B. Smith and A. L. Willis, *Nature New Biol.*, **231,** 235 (1971).
781. W. V. Ruyle, L. H. Sarett and A. Matzuk, South African Pat. 6,701,021 (1968); *C.A.* **70,** 106208j (1969).
782. J. Hannah, W. V. Ruyle, H. Jones, A. R. Matzuk, K. W. Kelly, B. E. Witzel, W. J. Holtz, R. A. Houser, T. Y. Shen, and L. H. Sarett, *Brit. J. Clin. Pharm.*, **4,** Suppl. 1, 7 (1977).
783. Geigy A. G., Swiss Pats. 266,236, 267,222, 269,980 (1950).
784. I. Häfliger, U.S. Pat. 2,745,783 (1956).
785. J. J. Burns, R. K. Rose, T. Cherkin, A. Goldman, A. Schulert, and B. B. Brodie, *J. Pharmacol. Exp. Ther.*, **109,** 346 (1953); **113,** 481 (1955).
786. R. A. Scherrer, in *Antiinflammatory Agents,* Vol. 1, R. A. Scherrer and M. W. Whitehouse, Eds., Academic Press, New York, 1974.
787. J. H. Wilkinson and I. L. Finar, *J. Chem. Soc.*, **1948,** 32.
788. T. Y. Shen and 15 coauthors, *J. Am. Chem. Soc.*, **85,** 488 (1963).
789. T. Y. Shen, in *Clinoril in the Treatment of Rheumatic Disorders,* E. C. Huskisson and P. Franchimont, Eds., Raven Press, New York, 1976, p. 1.
790. T. Y. Shen et al., German Pat. 2,039,426 (1971).
791. T. Y. Shen, *J. Med. Chem.*, **24,** 1 (1981).
792. J. S. Nicholson and S. S. Adams, British Pat. 971,700 (1964).
793. J. R. Vane and S. H. Ferreira, Eds., *Antiinflammatory Drugs,* Springer-Verlag, New York, 1978.
794. T. Y. Shen, in ref. 7c, Chapter 62.
795. R. A. Scherrer and M. W. Whitehouse, Eds., *Antiflammatory Agents,* Vols. 1 and 2, Academic Press, New York, 1974.
796. G. J. Roth, N. Stanford, and P. W. Majerus, *Proc. Natl. Acad. Sci. U.S.*, **72,** 3073 (1975).
797. J. W. Burch, N. Stanford, and P. W. Majerus, *J. Clin. Invest.*, **61,** 314 (1978).
798. J. R. Vane, *Agents Actions,* **8,** 430 (1978).
799. J. G. Lombardino, E. H. Wiseman, and W. M. McLamore, *J. Med. Chem.*, **14,** 1171 (1971).
800. J. G. Lombardino and E. H. Wiseman, *TIPS,* **2,** 132 (1981).
801. R. A. Scherrer, in *Antiinflammatory Agents,* R. A. Scherrer and M. W. Whitehouse, Eds., Vol. 1, p. 45, Academic Press, New York–London, 1974.

802. E. L. Huskisson, *Semin. Arthritis Rheum.*, **7**, 1 (1977).
803. T. Y. Shen, in *Antiinflammatory Drugs*, J. R. Vane and S. H. Ferreira, Eds., Springer-Verlag, New York, 1978, p. 323.
804. F. Stolz, German Pat. 97,011 (1897).
805. W. Filehne, *Z. Klin. Med.*, **32**, 572 (1897).
806. L. Knorr, *Chem. Ber.*, **16**, 2597 (1883).
807. J. J. Burns, T. F. Yu, A. Ritterbrand, J. M. Perel, A. B. Gutman, and B. B. Brodie, *J. Pharmacol. Exp. Ther.*, **119**, 418 (1957).
808. E. C. Huskisson, in *Antiinflammatory Drugs*, J. R. Vane and S. H. Ferreira, Eds., Springer-Verlag, New York, **1978**, p. 399.
809. G. Renoux, *Pharm. Ther., A*, **2**, 397 (1978).
810. L. A. Runge, R. S. Pinals, S. H. Lourie, and R. H. Tomar, *Arthritis Rheum.*, **20**, 1445 (1977).
811. J. Scott, P. A. Dieppe, and E. C. Huskisson, *Ann. Rheum. Dis.*, **37**, 259 (1978).
812. I. H. Lepow and P. A. Ward, Eds., *Inflammation: Mechanisms and Control,* Academic Press, New York, 1972.
813. I. A. Jaffe, *Arthritis Rheum.*, **13**, 435 (1970).
814. E. C. Huskisson and E. D. Hart, *Ann. Rheum. Dis.*, **31**, 402 (1972).
815. B. M. Sutton, E. McGusty, D. T. Waltz, and M. S. DiMartino, *J. Med. Chem.*, **15**, 1095 (1972).
816. D. T. Walz, M. S. DiMartino, and L. W. Chakrin, *J. Pharmacol. Exp. Ther.*, **197**, 142 (1976).
817. K. Brune, *Agents Actions*, **7**, Suppl. 3, 149 (1977).
818. A. C. Allison, *Proc. Roy. Soc. Med.*, **63**, 1077 (1970).
819. Y. H. Chang, *Arthritis Rheum.*, **20**, 1135 (1977).
820. I. Spilberg, *Arthritis Rheum.*, **18**, 129 (1975).
821. H. Umezawa, in *Methods in Enzymology,* Vol. 45, L. Lorand, Ed., Academic Press, New York–London, 1976, p. 678.
822. A. Adam, M. Devyo, V. Souvannavong, P. Lefrancier, J. Chody, and E. Lederer, *Biochem. Biophys. Res. Commun.*, **72**, 339 (1976).
823. A. Reugger, M. Kuhn, H. Lichti, H. R. Loosli, R. Huguenin, C. Quiquerez, and A. Wartburg, *Helv. Chim. Acta*, **59**, 1075 (1976).
824. M. Gabor, in *Antiinflammatory Drugs*, J. R. Vane and S. H. Ferreira, Eds., Springer-Verlag, New York, 1978, p. 698.
825. G. Vogel, M. K. Marek, and R. Oertner, *Arzneim.-Forsch.*, **20**, 699 (1970).
826. M. W. Whitehouse, *Prog. Drug. Res.*, **8**, 321 (1965).
827. R. T. Buckler and D. L. Garling, in reference 7b, pp. 1155–1159.
828. K. C. Nicolaou and J. B. Smith, *Ann. Rep. Med. Chem.*, **14**, 178 (1979).
829. K. C. Nicolaou, W. E. Barnette, G. P. Gasic, and R. L. Magolda, *J. Am. Chem. Soc.*, **99**, 7736 (1977).
830. C. L. Bundy and J. M. Baldwin, *Tetrahedron Lett.*, 1371 (1978).
831. R. E. Counsell and R. Brueggemeier, in reference 7b, Chapter 28.
832. T. F. Gallagher and F. C. Koch, *J. Pharmacol. Exp. Ther.*, **55**, 97 (1935).
833. R. I. Dorfman, in *Methods in Hormone Research,* Vol. 2, A. Dorfman, Ed., Academic Press, New York, 1962. p. 275.
834. E. Eisenberg and G. S. Gordan, *J. Pharmacol. Exp. Ther.*, **99**, 38 (1950).
835. L. G. Hershberger, E. G. Shipley, and R. K. Meyer, *Proc. Soc. Exp. Biol. Med.*, **83**, 175 (1953).

REFERENCES

836. S. C. Lyster, G. H. Lund, and R. O. Stafford, *Endocrinology,* **58,** 781 (1956).
837. M. E. Wolff, G. Zanati, G. Shanmagasundarum, S. Gupte, and G. Aadahl, *J. Med. Chem.,* **13,** 531 (1970).
838. J. A. Edwards and A. Bowers, *Chem. Ind. (London),* 1962 (1961).
839. K. V. Yorka, W. L. Truett, and W. S. Johnson, *J. Org. Chem.,* **27,** 4580 (1962).
840. K. Miescher, E. Tschapp, and A. Wettstein, *Biochem. J.,* **30,** 1977 (1936).
841. K. C. James, *Experientia,* **28,** 479 (1972).
842. A. J. Solo, N. Bejba, P. Hebborn, and M. May, *J. Med. Chem.,* **18,** 165 (1975).
843. R. Deghenghi and M. L. Givner, ref. 7b, Chapter 29.
844. E. A. Doisy, C. D. Veler, and S. A. Thayer, *Am. J. Physiol.,* **90,** 329 (1929).
845. A. Butenandt, *Naturwissenschaften,* **17,** 879 (1929).
846. E. A. Doisy, S. A. Thayer, L. Levin, and J. M. Curtis, *Proc. Soc. Exp. Biol. Med.,* **28,** 288 (1930).
847. G. F. Marrian, *J. Soc. Chem. Ind. (London),* **49,** 515 (1930); *Biochem. J.,* **24,** 435, 1021 (1930).
848. M. N. Huffman, D. W. MacCorquodale, S. A. Thayer, E. A. Doisy, G. V. Smith, and O. W. Smith, *J. Biol. Chem.,* **134,** 591 (1940).
849. H. H. Inhoffen, W. Longemann, W. Hohlweg, and A. Serini, *Chem. Ber.,* **71,** 1024 (1938).
850. E. Allen and E. A. Doisy, *J. Am. Med. Assoc.,* **81,** 819 (1923).
851. C. Clauberg and Z. Ustum, *Zentralbl. Gynäkol.,* **62,** 1745 (1938).
852. J. A. Hogg and J. Korman, in *Medicinal Chemistry,* Vol. 2, F. F. Blicke and C. M. Suter, Eds., Wiley, New York, 1956, p. 34.
853. E. C. Dodds, L. Goldberg, W. Lawson, and R. Robinson, *Nature (London),* **141,** 247 (1938).
854. F. von Wessely, *Angew. Chem.,* **53,** 197 (1940).
855. J. Masson, *Rev. Can. Biol.,* **3,** 491 (1944).
856. U. V. Solmssen, *Chem. Rev.,* **37,** 481 (1945).
857. E. Allen et al., U.S. Pat. 2,914,563 (1959).
858. A. Ercoli and R. Gardi, *Chem. Ind. (London),* 1037 (1961).
859. R. Wiechert and F. Neumann, German Pat. 1,189,991 (1965); *C.A.,* **63,** 1842e (1965).
860. S. Fang, K. M. Anderson, and S. Liao, *J. Biol. Chem.,* **244,** 6584 (1969).
861. J. P. Bennett, *Chemical Contraception,* Columbia University Press, New York, 1974.
862. C. R. Kay, *Oral Contraception Study,* Pitman Medical, New York, 1974.
863. George Washington University Medical Center, *Population Reports: Oral Contraceptives,* Series A, No. 2, Washington, D.C., 1975.
864. R. I. Dorfman, reference 7b, pp. 999–1043.
865. See G. K. Bass and L. S. Stuart, in *Disinfection, Sterilization and Preservation,* C. A. Lawrence and S. S. Block, Eds., Lea and Febiger, Philadelphia, Pa., 1968, Chapter 9.
866. R. Koch, *Klin. Wochenschr.,* **19,** 221 (1882).
867. W. I. B. Beveridge, *The Art of Scientific Investigation,* Random House, New York, 1950.
868. S. Rideal and J. T. A. Walker, *J. Roy. Sanit. Inst.,* **24,** 424 (1903).
869. D. F. Spooner and G. Sykes, in *Methods in Microbiology,* Vol. 7B, J. R. Norris and D. W. Ribbons, Eds., Academic Press, New York, 1972, Chapter 4.
870. See D. C. Schroder, *Chem. Rev.,* **55,** 186 (1952).
871. F. D. Chattaway, *J. Chem. Soc.,* **87,** 145 (1905).

872. H. D. Dakin, J. B. Cohen, and J. Kenyon, *Brit. Med. J.*, **1**, 160 (1916).
873. W. E. Knox, P. K. Stumpf, and D. E. Green, *J. Bacteriol.*, **55**, 451 (1948).
874. M. Kosugi, J. J. Kaminski, S. H. Selk, I. H. Pitman, N. Bodor, and T. Higuchi, *J. Pharm. Sci.*, **65**, 1743 (1976).
875. H. A. Shelanski and M. V. Shelanski, *J. Int. Coll. Surg.*, **25**, 727 (1956).
876. H. C. King and P. B. Price, *Surg. Gynecol. Obstet.*, **116**, 361 (1963).
877. B. J. Stern, *Selected Papers*, Citadel Press, New York, 1959, p. 368.
878. C. E. Coulthard, J. Marshall, and F. L. Pyman, *J. Chem. Soc.*, 280 (1930).
879. H. Bechhold and P. Ehrlich, *Z. Physiol. Chem.*, **47**, 173 (1906).
880. E. G. Klarmann, L. W. Gates, V. A. Shternov, and P. H. Cox, *J. Am. Chem. Soc.*, **54**, 3315 (1932).
881. W. C. Harden and J. H. Brewer, *J. Am. Chem. Soc.*, **59**, 2379 (1937).
882. W. S. Gump, U.S. Pat. 2,250,408 (1941).
883. R. M. Shuman, R. W. Leech and E. C. Alvord, Jr., *Morbid. Mortal.*, **22**, 93 (1973).
884. I. R. Shimi, S. Shoukry and S. Zaki, *Antimicrob. Agents Chemother.*, **9**, 580 (1976).
885. J. B. Adams and M. Hobbs, *J. Pharm. Pharmacol.*, **10**, 507 (1958).
886. W. F. von Oettingen, *Therapeutic Agents of the Quinoline Group*, Chemical Catalog Co., New York, 1933, p. 33.
887. G. R. Pettit and A. K. Das Gupta, *Chem. Ind.* (*London*), 1016 (1962).
888. K. Sigg, *Schweiz. Med. Wochenschr.*, **77**, 123 (1947).
889. A. Claus, *Arch. Pharm.*, **231**, 704 (1893).
890. P. Muehlens and W. Menck, *Muench. Med. Wochenschr.*, **68**, 802 (1921).
891. M. G. Schultz, *J. Am. Med. Assoc.*, **220**, 273 (1972).
892. Anon., *Lancet* (*1*), **1968**, 679.
893. R. G. Fargher, L. O. Galloway, and M. E. Robert, *J. Text. Ind.*, **21**, 245 (1935).
894. J. Bindler, U.S. Pat. 2,703,332 (1955).
895. H. C. Stecker, U.S. Pat. 3,041,236 (1962).
896. J. Buechi, J. B. Hansen, and S. A. Tammilehto, *Pharm. Acta Helv.*, **1971**, 602.
897. T. B. Johnson and F. W. Lane, *J. Am. Chem. Soc.*, **43**, 348 (1921).
898. E. G. Klarman, L. W. Gates, and V. A. Shternov, *J. Am. Chem. Soc.*, **53**, 3397 (1931).
899. C. E. Coulthard and G. Sykes, *Pharm. J.*, **137**, 79 (1936).
900. C. R. Phillips and S. Kaye, *Am. J. Hyg.*, **50**, 270 (1949).
901. R. R. Ernst, *Dev. Biol. Stand.*, **23**, 40 (1974).
902. R. K. Hoffman and B. Warshawsky, *Appl. Microbiol.*, **6**, 358 (1958).
903. A. Einhorn and M. Goettler, *Ann. Chem.*, **343**, 207 (1908).
904. G. Domagk, *Dtsch. Med. Wochenschr.*, **61**, 829 (1935); German Pats. 680,599, 682,441 (1935).
905. R. Shelton, M. G. VanCampen, C. H. Tilford, H. C. Lang, L. Nisonger, F. J. Bandelin, and H. L. Rubenkoenig, *J. Am. Chem. Soc.*, **68**, 753 (1946).
906. G. Hartman and W. Bosshard, U.S. Pat. 2,581,336 (1952).
907. O. Dressel, *J. Chem. Ed.*, **38**, 620 (1961); **39**, 320 (1962).
908. D. J. Beaver, D. P. Roman, and P. J. Stoffel, *J. Am. Chem. Soc.*, **79**, 1236 (1957).
909. W. E. Frick and W. Stammbach, U.S. Pat. 3,214,468 (1965).
910. J. N. Ashley, J. J. Barker, A. J. Ewins, G. Newberry, and A. D. Self, *J. Chem. Soc.*, 103 (1942).
911. K. Schern, *Zent. Bakteriol. Parasitenkd., Abt. I, Orig.*, **96**, 356, 440 (1925).

912. N. Von Jancsó and H. Von Jancsó, *Z. Immunitätsforsch.*, **81**, 1 (1935).
913. E. Frank, M. Northmann, and A. Wagner, *Klin. Wochenschr.*, **5**, 2100 (1926).
914. E. M. Lourie and W. Yorke, *Ann. Trop. Med. Parasitol.*, **31**, 435 (1937).
915. K. Schern and R. Artagaveytia-Allende, *Arch. Soc. Biol. Montevideo*, **6**, 244 (1935); *Z. Immunitätsforsch.*, **89**, 21 (1936).
916. J. N. Ashley, J. J. Barker, A. J. Ewins, G. Newberry, and A. D. Self, *J. Chem. Soc.*, 103 (1942).
917. H. Jensch, U.S. Pat. 2,673,197 (1964); *Arzneim.-Forsch.*, **5**, 634 (1955).
918. W. R. Thrower and F. C. Valentine, *Lancet*, **1**, 133 (1943).
919. F. H. S. Curd and F. L. Rose, *J. Chem Soc.*, 729 (1946).
920. R. B. Barlow and H. R. Ing, *Brit. J. Pharmacol. Chemother.*, **3**, 298 (1948).
921. W. D. M. Paton and E. J. Zaimis, *Brit. J. Pharmacol. Chemother.*, **4**, 381 (1949); **6**, 155 (1951).
922. R. L. Rose and G. Swain, *J. Chem. Soc.*, 4422 (1956).
923. G. E. Davies, J. Francis, A. R. Martin, F. L. Rose, and G. Swain, *Brit. J. Pharmacol.*, **9**, 192 (1954).
924. R. Hall, *Prog. Biochem.*, **2**, 24 (1967).
925. A. Davies, *Rep. Prog. Appl. Chem.*, **58**, 494 (1973).
926. S. Nakamura and H. Umezawa, *J. Antibiot.* (Tokyo) Ser. A, **9**, 66 (1955).
927. R. Despois, S. Pinnert-Sindico, L. Ninet, and J. Preud'Homme, *G. Microbiol.*, **2**, 76 (1956).
928. D. R. McCalla and D. Voutsinos, *Mutat. Res.*, **26**, 3 (1974).
929. M. C. Dodd, W. B. Stillman, M. Roys, and C. Crosby, *J. Pharmacol. Exp. Ther.*, **82**, 11 (1944).
930. K. Hayes, U.S. Pat. 2,610,181 (1952).
931. R. M. Jacob et al., to Rhone Poulenc, U.S. Pat. 2,944,061 (1960).
932. M. Bock, *Arzneim.-Forsch.*, **11**, 587 (1961).
933. D. Edwards, M. Dye, and H. Carne, *J. Gen. Microbiol.*, **76**, 135 (1973)
934. G. Y. Lesher, E. J. Froelich, M. D. Gruett, J. H. Bailey, and R. P. Brundage, *J. Med. Pharm. Chem.*, **5**, 1063 (1962).
935. A. Bauernfeind, *Antibiot. Chemother.* (Basel), **17**, 122 (1971).
936. G. J. Bourguignon, M. Levitt, and R. Sternglass, *Antimicrob. Agents Chemother.*, **4**, 479 (1973).
937. D. Kaminsky and R. I. Meltzer, *J. Med. Chem.*, **11**, 160 (1968).
938. J. W. Churchman, *J. Exp. Med.*, **16**, 221 (1912); *J. Am. Med. Assoc.*, **85**, 1849 (1925).
939. *Color Index*, 2nd ed., **3**, 3357 (1957).
940. Fischer and Fischer, *Ann. Chem.*, **194**, 276, 290 (1878).
941. A. D. Wolfe, R. G. Allison, and F. E. Hahn, *Biochemistry*, **11**, 1569 (1972).
942. N. E. Sharpless, C. L. Greenblatt, and W. H. Jennings, *Trans. N.Y. Acad. Sci.*, **35**, 187 (1973).
943. E. W. Stearn and A. E. Stearn, *J. Bacteriol.*, **11**, 345 (1926).
944. Kalle A. G., German Pat. 253,884 (1912).
945. A. E. Chichibabin and O. A. Zeide, *J. Russ. Phys. Chem. Soc.*, **46**, 1216 (1914).
946. C. H. Browning and W. Gilmour, *J. Pathol. Bacteriol.*, **18**, 144 (1913).
947. A. Albert and B. Ritchie, *Org. Syn.*, **22**, 5 (1942).
948. T. J. Franklin and G. A. Snow, *Biochemistry and Antimicrobial Action*, Wiley, New York, 1975, p. 84.

949. Ehrlich Centennial Symposium, *Ann. N.Y. Acad. Sci.*, **59**, 141 (1964).
950. P. Ehrlich, *Chem. Ber.*, **42**, 17 (1909); *Z. Angew. Chem.*, **23**, 2 (1910); *Lancet*, **2**, 445 (1913).
951. P. Ehrlich, *Arch. Microskop. Anat.*, **13**, 263 (1877).
952. H. Kolbe, *Ann. Chem.*, **113**, 125 (1860).
953. H. Hörlein, O. Dressel, and R. Kothe, quoted by F. Mietzsch, *Chem. Ber.*, **71A**, 15 (1938).
954. H. Hörlein, *Proc. Roy. Soc. Med.*, **29**, 313 (1935).
955. M. Heidelberger and W. A. Jacobs, *J. Am. Chem. Soc.*, **41**, 2131 (1919).
956. F. Mietzsch and J. Klarer, German Pat. 607,537 (1935).
957. M. Heidelberger, Letter to R. Dripps, June 1, 1972, quoted from *Retrospectroscope*, by J. H. Cromroe, Jr., Von Gehr Press, Menlo Park, Calif., 1978, p. 66.
958. G. Domagk, *Dtsch. Med. Wochenschr.*, **61**, 250 (1935).
959. J. Tréfouël, Mme. J. Tréfouël, F. Nitti, and D. Bovet, *Compt. Rend. Soc. Biol.*, **120B**, 756 (1935).
960. L. Colebrook and M. Kenny, *Lancet*, **1**, 1379 (1936).
961. G. A. H. Butle, W. H. Grey, and D. Stephenson, *Lancet*, **1**, 1286 (1936).
962. A. T. Fuller, *Lancet*, **1**, 194 (1937).
963. P. Gelmo, *J. Prakt. Chem.*, **77**, 369 (1908).
964. E. H. Northey, *The Sulfonamides and Allied Compounds*, American Chemical Society Monograph Series, Reinhold, New York, 1948.
965. A. J. Ewins and Phillips, to May and Baker Laboratories, British Pats. 512,145 (1939), 530,187 (1940).
966. L. E. H. Witby, *Lancet*, **2**, 1210 (1938).
967. H. Erlenmeyer and W. Würgler, *Helv. Chim. Acta*, **25**, 249 (1942).
968. A. H. Tracy and R. C. Elderfield, *J. Org. Chem.*, **6**, 54 (1940).
969. H. Erlenmeyer and H.v. Meyenburg, *Helv. Chim. Acta*, **20**, 204 (1937).
970. R. J. Fosbinder and L. A. Walter, *J. Am. Chem. Soc.*, **61**, 2032 (1939).
971. W. G. Christiansen, U.S. Pat., 2,242,237 (1941).
972. Skrimshire, U.S. Pat. 2,230,962 (1941).
973. C. F. Hübner, U.S. Pat. 2,447,702 (1948).
974. J. Vonkennel, J. Kimming, and B. Korth, *Z. Klin. Med.*, **138**, 695 (1940); *Klin. Wochenschr.*, **20**, 2 (1941).
975. J. P. Bourque and J. Joyal, *Can. Med. Assoc. J.*, **68**, 337 (1953).
976. G. W. Anderson, H. E. Faith, H. W. Marson, P. S. Winnek, and R. O. Roblin, Jr., *J. Am. Chem. Soc.*, **64**, 2902 (1942).
977. H. M. Wuest and M. Hoffer, U.S. Pat. 2,430,094 (1947).
978. E. M. Yow, *Am. Pract. Digest Treat.*, **4**, 521 (1953).
979. R. O. Roblin, Jr., J. H. Williams, P. S. Winnek, and J. P. English, *J. Am. Chem. Soc.*, **62**, 2002 (1940).
980. W. T. Caldwell and H. B. Kime, *J. Am. Chem. Soc.*, **62**, 2365 (1940).
981. W. T. Caldwell, E. C. Kornfeld, and C. K. Donnell, *J. Am. Chem. Soc.*, **63**, 2188 (1941).
982. J. M. Sprague, L. W. Kissinger, and R. M. Lincoln, *J. Am. Chem. Soc.*, **63**, 3028 (1941).
983. R. O. Roblin, P. S. Winnek, and J. P. English, *J. Am. Chem. Soc.*, **64**, 567 (1942).
984. H. Horstmann, T. Knott, W. Scholtan, E. Schraufstatter, A. Walter, and U. Wörffel, *Arzneim.-Forsch.*, **11**, 682 (1961).
985. W. Klötzer and H. Bretschneider, *Monatsh. Chem.*, **87**, 136 (1956).

REFERENCES

986. H. Bretschneider, W. Klötzer, and G. Spiteller, *Monatsh. Chem.*, **92**, 128 (1961).
987. B. Fust and E. Böhni, in *Antibiotic Medicine in Clinical Therapy*, Vol. 6, Suppl. 1, p. 3 (1959).
988. G. Hitzenberger and K. H. Spitzky, *Med. Klin. (Munich)*, **57**, 310 (362).
989. M. Reber, G. Rutishauser, and H. Tholen, in *Clearance-Untersuchungen am Menschen mit Sulfamethoxazol und Sulforthodimethoxin*, 3rd International Congress of Chemotherapy, Stuttgart, 1963, Vol. 1, H. P. Kuemmerle and R. Preziosi, Eds. Thieme, Stuttgart, 1964, p. 648.
990. L. Weinstein, M. A. Madoff, and C. M. Samet, *New Engl. J. Med.*, **263**, 793, 842, 900 (1960).
991. B. Camerino and G. Palamidessi, *Gazz. Chim. Ital.*, **90**, 1815 (1960).
992. A. Eichhorn, *Zbl. Pharm. Pharmakother. Laboratoriumsdiagn.*, **109**, 145 (1970); through *Ann. Rep. Med. Chem.*, 109 (1970).
993. R. G. Shepherd, in *Medicinal Chemistry*, 3rd ed., A. Burger, Ed., Wiley-Interscience, New York, 1970, p. 255.
994. C. L. Fox, Jr. and H. M. Rose, *Proc. Soc. Exp. Biol. Med.*, **50**, 142 (1942).
995. F. C. Schmelkes, O. Wyss, H. C. Marks, B. J. Ludwig, and F. B. Stranskov, *Proc. Soc. Exp. Biol. Med.*, **50**, 145 (1942).
996. J. K. Seydel, E. Krüger-Thiemer, and E. Wempe, *Z. Naturforsch.*, **15b**, 620 (1960).
997. A. Rastelli, P. G. DeBenedetti, G. C. Battistuzzi, and A. Albasini, *J. Med. Chem.*, **18**, 963 (1975).
998. J. K. Seydel, *Mol. Pharmacol.*, **2**, 259 (1966).
999. J. K. Seydel, in *Physico-Chemical Aspects of Drug Action*, E. J. Ariens, Ed., Pergamon Press, New York, 1968, p. 169.
1000. J. K. Seydel, *J. Med. Chem.*, **14**, 724 (1971).
1001. E. R. Garrett, J. B. Mielck, J. K. Seydel, and H. J. Kessler, *J. Med. Chem.*, **12**, 740 (1969).
1002. M. Yamazaki, N. Kakeya, T. Morishita, A. Kamada, and A. Aoki, *Chem. Pharm. Bull. (Tokyo)*, **18**, 702 (1970).
1003. T. Fujita and C. Hansch, *J. Med. Chem.*, **10**, 991 (1967).
1004. J. Rieder, *Arzneim.-Forsch.*, **13**, 81, 89, 95 (1963).
1005. W. Scholtan, *Arzneim.-Forsch.*, **14**, 348 (1964); **18**, 505 (1968).
1006. E. Krüger-Thiemer, W. Diller, and P. Bünger, *Antimicrob. Agents Chemother.*, 183 (1965).
1007. K. Irmscher, D. Gabe, K. Jahnke, and W. Scholtan, *Arzneim.-Forsch.*, **16**, 1019 (1966).
1008. R. T. Williams and D. V. Parke, *Ann. Rev. Pharmacol.*, **4**, 85 (1964).
1009. H. Nogami, A. Hasegawa, M. Hanano, and K. Imaoka, *Yakugaku Zassi*, **88**, 893 (1968).
1010. R. H. Adamson, J. W. Bridges, M. R. Kibby, S. Walker, and R. T. Williams, *Biochem. J.*, **118**, 41 (1970).
1011. G. Zbinden, in *Molecular Modification in Drug Design*, R. F. Gould, Ed., American Chemical Society, Washington, D.C., 1964, p. 25.
1012. M. L. Moore and C. S. Miller, *J. Am. Chem. Soc.*, **64**, 1572 (1942).
1013. U. P. Basu, *J. Indian Chem. Soc.*, **26**, 130 (1949).
1014. J. S. Lockwood, *J. Am. Med. Assoc.*, **111**, 2259 (1938).
1015. T. C. Stamp, *Lancet*, **2**, 10 (1939).
1016. D. D. Woods, *Brit. J. Exp. Pathol.*, **21**, 74 (1940).
1017. F. R. Selbie, *Brit. J. Exp. Pathol.*, **21**, 90 (1940).
1018. G. M. Findlay, *Brit. J. Exp. Pathol.*, **21**, 356 (1940).

1019. S. D. Rubbo and J. M. Gillespie, *Nature,* **146,** 838 (1940).
1020. R. Kuhn and K. Schwartz, *Chem. Ber.,* **74B,** 1617 (1941).
1021. K. C. Blanchard, *J. Biol. Chem.,* **140,** 919 (1941).
1022. H. McIllwain, *Brit. J. Exp. Pathol.,* **23,** 265 (1942).
1023. S. D. Rubbo, M. Maxwell, R. A. Fairbridge, and J. M. Gillespie, *Austral. J. Exp. Biol. Med. Sci.,* **19,** 185 (1941).
1024. P. Fildes, *Lancet,* **1,** 955 (1940).
1025. R. Tschesche, *Z. Naturforsch.,* **26b,** 10 (1947).
1026. G. M. Brown, *J. Biol. Chem.,* **237,** 536 (1962).
1027. R. Weisman and G. M. Brown, *J. Biol. Chem.,* **239,** 326 (1964).
1028. N. Anand in reference 7b, pp. 15–20.
1029. J. Greenberg, *J. Pharmacol. Exp. Ther.,* **97,** 484 (1949).
1030. J. Greenberg and E. M. Richeson, *J. Pharmacol. Exp. Ther.,* **99,** 444 (1950).
1031. G. H. Hitchings and J. J. Burchall, *Adv. Enzymol.,* **27,** 417 (1965).
1032. N. Rist, *Nature,* **146,** 838 (1940); *Compt. Rend. Soc. Biol.,* **130,** 972 (1939).
1033. W. H. Feldman, H. C. Hinshaw, and H. E. Moses, *Am. Rev. Tuberc.,* **45,** 303 (1942); *Proc. Staff Meet. Mayo Clin.,* **15,** 695 (1940).
1034. B. C. Jain, B. H. Iyer, and P. C. Guha, *Science and Culture,* **11,** 568 (1946); *C.A.,* **40,** 4687 (1946).
1035. E. V. Cowdry and C. Ruangsiri, *Arch. Pathol.,* **32,** 632 (1941).
1036. S. G. Browne, *Advan. Pharmacol. Chemother.,* **7,** 211 (1969).
1037. E. F. Elslager, *Prog. Drug. Res.,* **18,** 99 (1974).
1038. M. I. Smith, E. L. Jackson, and H. Bauer, *Ann. N.Y. Acad. Sci.,* **52,** 704 (1949–1950).
1039. C. Levaditi, *Compt. Rend. Soc. Biol.,* **135,** 1109 (1941).
1040. R. Donovick, A. Bayan, and D. Hamre, *Am. Rev. Tuberc.,* **66,** 219 (1952).
1041. G. Brownlee, A. F. Green, and M. Woodbine, *Brit. J. Pharmacol.,* **3,** 15 (1948).
1042. P. J. White and D. D. Woods, *J. Gen. Microbiol.,* **40,** 243 (1965).
1043. M. Landy, N. W. Larkum, E. J. Oswald, and F. Streightoff, *Science,* **97,** 265 (1943).
1044. E. M. Wise, Jr., and M. M. Abou-Donia, *Proc. Natl. Acad. Sci. U.S.,* **72,** 2621 (1975).
1045. T. Mann and D. Keilin, *Nature,* **146,** 164 (1940).
1046. M. Tishler, in *Molecular Modification in Drug Design,* Advances in Chemistry Series, Vol. 45, American Chemical Society, Washington, D.C., 1964, p. 1.
1047. K. H. Beyer and J. E. Baer, *Pharmacol. Rev.,* **13,** 517 (1961).
1048. P. W. Feit, O. B. T. Nielson, and H. Bruun, *J. Med. Chem.,* **15,** 437 (1972).
1049. M. Janbon, J. Chaptal, A. Vedel, and J. Schaap, *Montpellier Med.,* **21–22,** 441 (1942).
1050. H. Franke and J. Fuchs, *Dtsch. Med. Wochenschr.,* **80,** 1449 (1955).
1051. K. H. Beyer, H. F. Russo, E. K. Tilson, A. K. Miller, W. F. Verwey, and S. R. Gass, *Am. J. Physiol.,* **166,** 625 (1951).
1052. L. H. Schmidt, *Am. Rev. Tuberc.,* **74,** 138 (1956).
1053. C. C. Shephard, *J. Exp. Med.,* **112,** 445 (1960).
1054. R. J. W. Rees, *J. Exp. Pathol.,* 45 (1960).
1055. W. F. Kirchheimer and E. E. Storrs, *Int. J. Lepr.,* **39,** 693 (1960).
1056. R. L. Harned, P. H. Hidy, and E. K. LaBaw, *Antibiot. Chemother.,* **5,** 204 (1955).
1057. A. C. Finlay, G. L. Hobby, F. Hochstein, T. M. Lees, T. F. Lenert, J. A. Menas, S. Y. P'An, P. P. Regna, Y. B. Routein, B. A. Sabin, K. B. Tat, and Y. H. Kane, *Am. Rev. Respir. Dis.,* **63,** 1 (1951).

REFERENCES

1058. H. Umezawa, M. Ueda, K. Maeda, K. Yagishita, S. Korido, Y. Okami, R. Utahara, Y. Osato, K. Nitta, and T. Tacheuchi, *J. Antibiot. (Tokyo),* Ser. A, **10,** 181 (1957).
1059. E. P. Herr, Jr. and M. O. Redstone, *Ann. N.Y. Acad. Sci.,* **135,** 940 (1966).
1060. P. Sensi, A. M. Greco, and R. Ballotta, *Antibiot. Ann.,* 262 (1959–1960).
1061. W. Oppolzer, V. Prelog, and P. Sensi, *Experientia,* **20,** 336 (1964).
1062. J. Leitich, W. Oppolzer, and V. Prelog, *Experientia,* **20,** 343 (1964).
1063. N. Bergamini and G. Fowst, *Arzneim.-Forsch.,* **796,** 953 (1965).
1064. P. Sensi, *Pure Appl. Chem.,* **35,** 383 (1973).
1065. P. Sensi, N. Maggi, S. Furesz, and G. Maffii, *Antimicrob. Agents Chemother.,* **1966,** 699.
1066. N. Maggi, C. R. Pasqualucci, R. Ballotta, and P. Sensi, *Chemotherapia,* **11,** 285 (1966).
1067. R. Pallanza, V. Arioli, S. Furez, and G. Bolzoni, *Arzneim.-Forsch.,* **17,** 529 (1967).
1068. C. C. Shepard, quoted in *Science,* **215,** 1085 (1982).
1069. G. Hartmann, K. O. Honikel, F. Knussel, and J. Nuesch, *Biochem. Biophys. Acta,* **145,** 843 (1967).
1070. H. Umezawa, S. Mizuno, H. Yamazaki, and K. Nitta, *J. Antibiot. (Tokyo),* **21,** 234 (1968).
1071. F. Bernheim, *J. Bacteriol.,* **41,** 385 (1941).
1072. A. K. Saz and F. Bernheim, *J. Pharmacol. Exp. Ther.,* **73,** 78 (1941).
1073. A. K. Saz, F. R. Johnston, A. Burger, and F. Bernheim, *Am. Rev. Tuberc.,* **18,** 40 (1943).
1074. A. Burger, E. L. Wilson, C. O. Brindley, and F. Bernheim, *J. Am. Chem. Soc.,* **67,** 1416 (1945).
1075. German Pat. 50835 (1889); U.S. Pat. 427,564 (1890).
1076. C. Ratledge and B. J. Marshall, *Biochem. Biophys. Acta,* **279,** 58 (1972).
1077. C. Ratledge and K. A. Brown, *Am. Rev. Resp. Dis.,* **106,** 774 (1972).
1078. R. Behnisch, F. Mietzsch, and H. Schmidt, *Am. Rev. Tuberc.,* **61,** 1 (1950).
1079. N. P. Buu-Hoi, *Int. J. Lepr.,* **22,** 16 (1954).
1080. T. F. Davy, *Lepr. Rev.,* **29,** 25 (1958).
1081. R. L. Mayer, *Rev. Méd. France,* Nov.–Dec. 1941; *C.A.,* **36,** 5199 (1942).
1082. E. Huant, *Gazette des Hôpitaux,* Aug. 15, 1945.
1083. S. Kushner, R. T. Dalalian, R. T. Cassell, J. L. Sanjurjo, D. L. McKenzie, and J. Subbarow, *J. Org. Chem.,* **13,** 834 (1948).
1084. D. L. McKenzie, S. Malone, S. Kushner, J. J. Oleson, and Y. Subbarow, *J. Lab. Clin. Med.,* **33,** 1249 (1948).
1085. H. H. Fox, *J. Org. Chem.,* **17,** 542, 547 (1952).
1086. F. Grumbach, N. Rist, D. Libermann, M. Moyeau, S. Cals, and S. Clavel, *Compt. Rend.,* **242,** 2187 (1956).
1087. H. H. Fox, *Science,* **116,** 129 (1952).
1088. G. Winder, P. B. Collins, and D. Whelan, *J. Gen. Microbiol.,* **66,** 379 (1971).
1089. J. P. Thomas, C. O. Baughn, R. G. Wilkinson, and R. G. Shepherd, *Am. Rev. Resp. Dis.,* **83,** 891 (1961).
1090. R. G. Wilkinson, M. B. Cantrall, and R. G. Shepherd, *J. Med. Chem.,* **5,** 835 (1962).
1091. V. C. Barry, J. G. Belton, J. B. O'Sullivan, and T. Twomey, *J. Chem. Soc.,* 896 (1956).
1092. V. C. Barry, J. G. Belton, M. L. Conalty, and T. Twomey, *Nature (London),* **162,** 622 (1948).
1093. L. Pasteur and J. Joubert, *Compt. Rend. Hebd. Seances Acad. Sci. Paris,* **85,** 101 (1877).

1094. F. Wrede and E. Strack, *J. Physiol. Chem.*, **181**, 58 (1929).
1095. W. I. B. Beveridge, *Seeds of Discovery*, W. W. Norton, New York, 1981.
1096. E. Chain, H. W. Florey, A. D. Gardner, N. G. Heatley, M. A. Jennings, J. Orr-Ewing, and A. G. Sanders, *Lancet*, **2**, 226 (1940).
1097. E. P. Abraham, E. Chain, C. M. Fletcher, A. D. Gardner, M. A. Jennings, and H. W. Florey, *Lancet*, **2**, 177 (1941).
1098. H. W. Florey, E. Chain, N. G. Heatley, M. A. Jennings, A. G. Sanders, E. P. Abraham, and M. E. Florey, *Antibiotics*, Vol. 2, Oxford University Press, London, 1949.
1099. H. T. Clarke, J. R. Johnson, and R. Robinson, Eds., *The Chemistry of Penicillin*, Princeton University Press, Princeton, N.J., 1949.
1100. A. J. Moyer and R. D. Coghill, *J. Bacteriol.*, **51**, 79 (1946).
1101. R. D. Coghill, *Chem. Eng. News*, **22**, 588 (1944).
1102. R. D. Coghill and R. S. Koch, *Chem. Eng. News*, **23**, 2310 (1945).
1103. J. C. Sheehan and K. R. Henery-Logan, *J. Am. Chem. Soc.*, **81**, 2912, 5838 (1959).
1104. Q. F. Soper, C. W. Whitehead, O. K. Behrens, J. J. Corse, and R. G. Jones, *J. Am. Chem. Soc.*, **70**, 2849 (1948), and previous papers cited therein.
1105. K. Sakaguchi and S. Murao, *J. Agric. Chem. Soc. Jpn.*, **23**, 411 (1950).
1106. F. R. Batchelor, F. P. Doyle, J. H. C. Nayler, and C. N. Rolinson, *Nature (London)*, **183**, 257 (1959).
1107. C. A. Claridge, A. Gourevitch, and J. Lein, *Nature (London)*, **187**, 237 (1960).
1108. F. R. Batchelor and J. Cameron Wood, *Nature (London)*, **195**, 1000 (1962).
1109. J. R. E. Hoover and G. L. Dunn, in reference 7b, Chapter 15.
1110. G. N. Rolinson and S. Stevens, *Brit. Med J.*, **2**, 191 (1961); D. M. Brown and P. Acred, *ibid.*, **2**, 197 (1961); E. T. Knudsen, G. N. Rolinson, and S. Stevens, *ibid.*, **2**, 198 (1961).
1111. F. P. Doyle, G. R. Fosker, J. H. C. Nayler, and H. Smith, *J. Chem. Soc.*, 1440 (1962).
1112. E. F. Gale, E. Cundliffe, P. E. Reynolds, M. H. Richmond, and M. J. Waring, in *The Molecular Basis of Antibiotic Action*, Wiley, New York, 1972, p. 49.
1113. J. T. Park and M. J. Johnson, *J. Biol. Chem.*, **179**, 585 (1949); J. T. Park, *ibid.*, **194**, 897 (1952).
1114. J. T. Park and J. L. Strominger, *Science*, **125**, 99 (1957).
1115. D. J. Tipper and J. L. Strominger, *Proc. Nat. Acad. Sci. U.S.*, **54**, 1133 (1965); *J. Biol. Chem.*, **243**, 3169 (1968).
1116. G. Brotzu, *Lav. Ist. Igiene Cagliari* (1948).
1117. G. G. F. Newton and E. P. Abraham, *Nature (London)*, **175**, 548 (1955).
1118. G. G. F. Newton and E. P. Abraham, *Biochem. J.*, **58**, 103 (1954).
1119. R. B. Morin, B. G. Jackson, E. H. Flynn, and R. W. Roeske, *J. Am. Chem. Soc.*, **84**, 3400 (1962).
1120. H. W. O. Weissenburger and M. G. Van der Hoeven, U.S. Pats. 3,499,909 (1970), 3,575,970 (1971); Belgian Pat. 719,712 (1969) (to Glaxo Laboratories Ltd.).
1121. B. Fechtig, H. Peter, H. Bickel, and E. Vischer, *Helv. Chim. Acta*, **51**, 1108 (1968).
1122. S. Karaday, S. H. Pines, L. M. Weinstock, F. E. Roberts, G. S. Brenner, A. M. Hoinowski, T. Y. Cheng, and M. Sletzinger, *J. Am. Chem. Soc.*, **94**, 1410 (1972).
1123. B. Shimizu, M. Kaneko, M. Kimura, and S. Sugawara, *Chem. Pharm. Bull. (Tokyo)*, **24**, 2629 (1976).
1124. C. W. Hale, G. G. F. Newton, and E. P. Abraham, *Biochem. J.*, **79**, 403 (1961).
1125. J. L. Spencer, E. H. Flynn, R. W. Roeske, F. Y. Siu, and R. R. Chauvette, *J. Med. Chem.*, **9**, 746 (1966).
1126. W. E. Wick, *Appl. Microbiol.*, **15**, 765 (1967).

REFERENCES

1127. A. W. Bauer, W. M. Kirby, J. C. Sherris, and M. Turck, *Am. J. Clin. Pathol.*, **45**, 493 (1966).
1128. J. E. Dolfini, H. E. Appelgate, G. Bach, H. Basch, J. Bernstein, J. Schwartz, and F. L. Weisenborn, *J. Med. Chem.*, **14**, 117 (1971).
1129. R. R. Chauvette, E. H. Flynn, B. G. Jackson, E. R. Lavagnino, R. B. Morin, R. A. Mueller, R. P. Pioch, R. W. Roeske, C. W. Ryan, J. L. Spencer, and E. van Heyningen, *J. Am. Chem. Soc.*, **84**, 3401 (1962).
1130. W. S. Boniece, W. E. Wick, D. H. Holmes, and C. E. Redman, *J. Bacteriol.*, **84**, 1292 (1962).
1131. J. S. Kahan, F. M. Kahan, R. Goegelman, S. A. Currie, M. Jackson, E. O. Stapley, T. W. Miller, A. K. Miller, D. Hendlin, S. Mochales, S. Hernandez, and H. B. Woodruff, Abstracts, 16th Interscience Conference on Antimicrobial Agents and Chemotherapy, Chicago, 1976, No. 227.
1132. T. T. Howarth, A. G. Brown, and T. J. King, *J. Chem. Soc., Chem. Commun.*, 266 (1976).
1133. T. Kamiya, in *Recent Advances in the Chemistry of β-Lactam Antibiotics*, F. Plavac, R. W. Ratcliffe, K. Hoogsteen, and B. G. Christensen, Eds. The Chemical Society (London), Cambridge, England, 1976.
1134. R. B. Sykes and C. M. Cimarusti, quoted in *Science*, **213**, 1238 (1981).
1135. J. Davies, D. S. Jones, and H. G. Khorana, *J. Mol. Biol.*, **18**, 48 (1966).
1136. S. A. Waksman and H. Lechevalier, *Science*, **109**, 305 (1949).
1137. K. L. Rinehart, Jr., M. Hichens, J. L. Foght, and W. S. Chilton, *Antimicrob. Agents Chemother.*, **163**, 193 (1963).
1138. K. L. Rinehart, Jr., M. Hichens, A. D. Argoudelis, W. S. Chilton, H. E. Carter, M. P. Georgiadis, C. P. Schaffner, and R. T. Schillings, *J. Am. Chem. Soc.*, **84**, 3218 (1962).
1139. M. J. Cron, D. L. Evans, F. M. Palermiti, D. F. Whitehead, I. R. Hopper, P. Chu, and R. V. Lemieux, *J. Am. Chem. Soc.*, **80**, 4741 (1958).
1140. S. Umezawa, Y. Ito, and S. Fukatsu, *J. Antibiot., Ser. A*, **11**, 162 (1958).
1141. M. Maeda, M. Murase, H. Mawatari, and H. Umezawa, *J. Antibiot., Ser. A*, **11**, 163 (1958).
1142. H. Maehr and C. P. Schaffner, *J. Am. Chem. Soc.*, **89**, 6787 (1967).
1143. D. J. Cooper and M. D. Yudin, *Chem. Commun.*, 821 (1967).
1144. N. E. Allen, quoted in *C & E News*, Aug. 31, 1981, p. 9.
1145. R. J. Dubos, *J. Exp. Med.*, **70**, 1 (1939).
1146. R. J. Dubos and R. D. Hotchkiss, *J. Exp. Med.*, **73**, 629 (1941).
1147. J. D. Gregory and L. C. Craig, *J. Biol. Chem.*, **172**, 839 (1948).
1148. G. F. Gause and M. G. Brazhnikova, *Nature (London)*, **154**, 703 (1944).
1149. A. H. Gordon, A. J. P. Martin, and R. L. M. Synge, *Biochem. J.*, **41**, 596 (1947).
1150. R. Schwyzer and P. Sieber, *Helv. Chim. Acta*, **40**, 624 (1957).
1151. B. A. Johnson, H. S. Anker, and F. Meleney, *Science*, **102**, 376 (1942).
1152. P. G. Stansly and M. E. Schlosser, *J. Bacteriol.*, **54**, 549 (1947).
1153. H. Brockmann and G. Schmidt-Kastner, *Chem. Ber.*, **88**, 57 (1955).
1154. M. M. Shemyakin, E. I. Vinogradova, M. Y. Feigina, N. A. Aldanova, Y. B. Shvetsov, and L. A. Fonina, *Zh. Obsch. Khim.*, **36**, 1391 (1966).
1155. F. E. Hunter, Jr. and L. S. Schwartz, in *Antibiotics, Vol. 1: Mechanism of Action*, D. Gottlieb and P. D. Shaw, Eds. Springer-Verlag, New York, 1967, p. 631.
1156. H. Umezawa, T. Aoyagi, H. Morishima, M. Matsuzaki, M. Hamada, and T. Takeuchi, *J. Antibiot.*, **23**, 259 (1970).

1157. M. J. Antonaccio and D. W. Cushman, *Fed. Proc.*, **40**, 2275 (1981).
1158. B. M. Duggar, *Ann. N.Y. Acad. Sci.*, **51**, 177 (1948); U.S. Pat. 2,482,055 (1949).
1159. A. C. Finlay, G. L. Hobby, S. Y. P'an, P. P. Regna, J. P. Routien, D. B. Seely, G. M. Shull, B. A. Sobin, I. A. Solomons, J. W. Vinson, and J. H. Kane, *Science*, **111**, 85 (1950).
1160. P. P. Minieri, M. C. Firman, A. G. Mistretta, A. Abbey, C. E. Bricker, N. E. Rigler, and H. Sokol, *Antibiot. Ann. 1953–1954*, 81 (1954).
1161. K. Gerzon, E. H. Flynn, M. V. Sigal, P. F. Wiley, R. Monahan, and U. C. Quarck, *J. Am. Chem. Soc.*, **78**, 6396 (1956).
1162. D. R. Harris, S. G. McGearchin, and H. H. Mills, *Tetrahedron Lett.*, 679 (1965).
1163. P. F. Wiley, M. V. Sigal, O. Weaver, R. Monahan, and K. Gerzon, *J. Am. Chem. Soc.*, **79**, 6070 (1957).
1164. P. F. Wiley, R. Gale, C. W. Pettinga, and K. Gerzon, *J. Am. Chem. Soc.*, **79**, 6074 (1957).
1165. F. A. Hochstein, H. Els, W. D. Celmer, B. L. Shapiro, and R. B. Woodward, *J. Am. Chem. Soc.*, **82**, 3225 (1960).
1166. W. D. Celmer and D. C. Hobbs, *Carbohydr. Res.*, **1**, 137 (1965).
1167. W. D. Celmer, *J. Am. Chem. Soc.*, **87**, 1797 (1965).
1168. R. L. Wagner, F. A. Hochstein, K. Murai, N. Messina, and P. P. Regna, *J. Am. Chem. Soc.*, **75**, 4684 (1953).
1169. R. B. Woodward, *Angew. Chem.*, **69**, 50 (1957).
1170. R. B. Woodward, L. S. Weiler, and P. C. Dutta, *J. Am. Chem. Soc.*, **87**, 4662 (1965).
1171. J. D. Dutcher, G. Boyak, and S. Fox, *Antibiot. Ann. 1953–1954*, 191 (1954).
1172. J. Vandeputte, J. L. Wachtel, and E. T. Stiller, *Antibiot. Ann., 1955–1956*, 587 (1956).
1173. R. L. Harned, P. H. Hiddy, and E. K. LaBaw, *Antibiot. Chemother.*, **5**, 204 (1955).
1174. F. A. Kuehl, Jr., F. J. Wolf, N. R. Trenner, R. L. Peck, R. P. Buhs, I. Putter, R. Ormond, J. E. Lyons, L. Chaiet, E. Howe, B. D. Hunnewell, G. Downing, E. Newstead, and K. Folkers, *J. Am. Chem. Soc.*, **77**, 2344 (1955).
1175. P. W. Brian, P. J. Curtis, and H. G. Hemming, *Trans. Brit. Mycol. Soc.*, **38**, 305 (1955); **39**, 173 (1956).
1176. J. C. Gentles, *Nature (London)*, **182**, 476 (1958).
1177. V. Prelog, *Pure Appl. Chem.*, **7**, 551 (1963).
1178. W. Oppolzer, V. Prelog, and P. Sensi, *Experientia*, **20**, 336 (1964).
1179. H. Hoeksema, B. Bannister, R. D. Birkenmeyer, F. Kagan, B. J. Magerlein, F. A. MacKellar, W. Schroeder, G. Slomp, and R. R. Herr, *J. Am. Chem. Soc.*, **86**, 4223 (1964).
1180. P. W. K. Woo, H. W. Dion, and Q. R. Bartz, *Tetrahedron Lett.*, 2617, 2625 (1971).
1181. T. Oda, T. Mori, and Y. Kyotani, *J. Antibiot. (Tokyo)*, **24**, 503 (1971).
1182. T. Mori, Y. Kyotani, I. Watanabe, and T. Oda, *J. Antibiot. (Tokyo)*, **25**, 149 (1972).
1183. U. Hollstein, in ref. 7b, Chapter 16.
1184. R. B. Woodward and 47 coauthors, *J. Am. Chem. Soc.*, **103**, 3210, 3213, 3215 (1981).
1185. S. A. Waksman and H. B. Woodruff, *Proc. Soc. Exp. Biol. Med.*, **45**, 609 (1940).
1186. C. E. Dagliesh and A. R. Todd, *Nature (London)*, **164**, 830 (1949).
1187. C. E. Dagliesh, A. W. Jonson, A. R. Todd, and L. C. Vining, *J. Chem. Soc.*, 2946 (1950).
1188. H. Brockmann and N. Grubhofer, *Naturwissenschaften*, **36**, 376 (1949).
1189. A. Cerami, E. Reich, D. C. Ward, and I. H. Goldberg, *Proc. Natl. Acad. Sci. U.S.*, **57**, 1030 (1967).

1190. E. Reich and I. H. Goldberg, *Prog. Nucleic Acid Res.*, **3**, 183 (1964).
1191. G. Cassinelli and P. Orezzi, *Giorn. Microbiol.*, **11**, 167 (1963); *C.A.*, **62**, 9482b (1965).
1192. A. DiMarco, G. Canevazzi, A. Grein, P. Orezzi, and M. Gaetani, Belgian Pat. 639,897 (1964); *C.A.*, **62**, 16922a (1965).
1193. E. M. Acton, A. Fujiwara, and D. W. Henry, *J. Med. Chem.*, **17**, 659 (1974).
1194. F. Arcamone, G. Cassinelli, G. Fantini, A. Grein, P. Orezzi, C. Poli, and C. Spalla, *Biotechnol. Bioeng.*, **11**, 1101 (1969).
1195. F. Arcamone, G. Franceschi, and S. Penco, German Pat. 1,917,874 (1969); *C.A.*, **73**, 45799r (1970).
1196. G. Bonadonna, S. Monfardini, M. DeLena, F. Fossati-Bellani, and C. Beretta, *Cancer Res.*, **30**, 2572 (1970).
1197. G. R. Pettit, J. J. Einck, C. L. Herald, R. H. Ode, R. B. Von Drelle, P. Brown, M. G. Brazhnikova, and G. F. Gause, *J. Am. Chem. Soc.*, **97**, 7387 (1975).
1198. T. Hata, J. Koga, Y. Sano, K. Kanamori, A. Matsumae, R. Sugawara, T. Hoshi, and T. Shima, *J. Antibiot. (Tokyo), Ser. A*, **7**, 107 (1954).
1199. T. Takita, Y. Muraoka, T. Nakatani, A. Fujii, Y. Umezawa, H. Naganawa, and H. Umezawa, *J. Antibiot. (Tokyo), Ser. A*, **31**, 801 (1978).
1200. H. Umezawa, in *Cancer Medicine*, J. F. Holland and E. Frei III, Eds., Lea and Febiger, Philadelphia, Pa., 1973, p. 817.
1200a. Y. Aoyagi, K. Katano, H. Suguna, J. Primeau, L.-H. Chang, and S. M. Hecht, *J. Am. Chem. Soc.*, **104**, 5537 (1982).
1201. S. M. Kupchan, Y. Komoda, W. A. Court, G. J. Thomas, R. M. Smith, A. Karim, C. J. Gilmore, R. C. Haltiwanger, and R. F. Bryan, *J. Am. Chem. Soc.*, **94**, 1354 (1972).
1202. W. J. Ross, reference 7b, Chapter 20.
1203. F. Hawking, in *Experimental Chemotherapy*, Vol. 1, R. J. Schnitzer and F. Hawking, Eds., Academic Press, New York–London, 1963, pp. 129–256.
1204. Zeidler, *Chem. Ber.*, **7**, 1180 (1874).
1205. P. Müller, Ed., *DDT,* 3 vols., Birkhäuser Verlag, Basel–Stuttgart, 1955.
1206. B. Soloway, U.S. Pat. 2,676,131 (1954).
1207. J. Hyman, British Pat. 618,432 (1949).
1208. F. A. Gunther, U.S. Pat. 2,218,148 (1940); *Chem. Ind.*, **1946**, 399.
1209. R. B. Krauss, *J. Am. Chem. Soc.*, **36**, 961 (1914).
1210. W. W. Lewers and A. Lowy, *Ind. Eng. Chem.*, **17**, 1289 (1925).
1211. F. K. Kleine and W. Fischer, *Dtsch. Med. Wochenschr.*, **48**, 1693 (1923).
1212. E. Fourneau, J. Tréfouël, Mme. J. Tréfouël, and J. Vallée, *Compt. Rend.*, **178**, 675 (1924).
1213. A. Adams, J. N. Ashley, and H. Bader, *J. Chem. Soc.*, 3739 (1956).
1214. H. G. Plimmer and N. R. Bateson, *Proc. Roy. Soc., Ser. B*, **80**, 477 (1908).
1215. H. W. Thomas, *Brit. Med. J.*, **1**, 1140 (1905).
1216. P. Ehrlich and A. Bertheim, *Chem. Ber.*, **40**, 3292 (1907).
1217. W. A. Jacobs and M. Heidelberger, *J. Am. Chem. Soc.*, **41**, 1810 (1919).
1218. E. Fourneau, A. M. Navarro-Martin, and J. Tréfouël, *Ann. Inst. Pasteur*, **37**, 551 (1923).
1219. W. Yorke and F. Murgatroyd, *Ann. Trop. Med.*, **24**, 449 (1930).
1220. E. A. H. Friedheim, *Ann. Inst. Pasteur*, **65**, 108 (1940); *J. Am. Chem. Soc.*, **66**, 1775 (1944).
1221. E. A. H. Friedheim, U.S. Pats. 2,659,723 (1953), 2,772,303 (1956).
1222. L. Banda, *Chem. Ber.*, **45**, 1787 (1912).

1223. H. Jensch, *Angew. Chem.*, **50**, 891 (1957).
1224. F. H. S. Curd and D. G. Davey, *Nature (London)*, **163**, 89 (1949); *Brit. J. Pharmacol.*, **5**, 25 (1950).
1225. H. Jensch, *Arzneim.-Forsch.*, **5**, 634 (1955); *Med. Chem.*, **6**, 134 (1958).
1226. J. N. Ashley, S. S. Berg, and R. D. MacDonald, *J. Chem. Soc.*, 4525 (1960).
1227. World Health Organization, *African Trypanosomiasis*, Technical Report Series No. 434, Geneva, 1969.
1228. A. B. Clarkson and F. H. Brohn, *Science*, **194**, 204 (1976).
1229. E. A. Steck, *The Chemotherapy of Protozoan Diseases*, Vol. 2, Walter Reed Army Institute of Research, Washington, D.C., 1972.
1230. C. H. Browning, G. T. Morgan, J. V. M. Robb, and L. P. Walls, *J. Pathol. Bacteriol.*, **46**, 203 (1938).
1231. G. Woolfe, *Ann. Trop. Med. Parasitol.*, **46**, 285 (1952).
1232. F. Evens, K. N. Niemeegers, and A. Packchanian, *Am. J. Trop. Med. Hyg.*, **6**, 665 (1957).
1233. M. Bock, A. Haberkorn, H. Herlinger, K. H. Mayer, and S. Petersen, *Arzneim.-Forsch.*, **22**, 1564 (1972).
1234. P. P. Levine, *Cornell Vet.*, **29**, 309 (1939); *J. Parasitol.*, **26**, 233 (1940).
1235. M. Mitrovic and J. C. Bauernfriend, *Poult. Sci.*, **46**, 402 (1967).
1236. P. Yvore, *Rec. Med. Vet.*, **144**, 1059 (1968).
1237. J. F. Ryley and M. J. Betts, in *Advances in Pharmacology and Chemotherapy*, Vol. 2, S. Garratini, A. Goldin, F. Hawking, and I. J. Kopin, Eds. Academic Press, New York–London, 1973, pp. 221–293.
1238. A. C. Cuckler, M. Garcillo, C. Malanger, and E. C. McManus, *Poult. Sci.*, **39**, 1241 (1960).
1239. J. Enzeby and C. Garan, *Bull. Soc. Sci. Vet. Lyon*, **71**, 235 (1969).
1240. S. Kantor, R. L. Kennett, Jr., E. Waletzky, and A. S. Tomkufcik, *Science*, **168**, 373 (1970).
1241. R. F. Shumard and M. E. Callender, *Antimicrob. Agents Chemother.*, 369 (1967).
1242. S. R. M. Bushby and F. C. Copp, *J. Pharm. Pharmacol.*, **7**, 112 (1955).
1243. A. C. Cuckler, A. B. Kupferberg, and N. Millman, *Antibiot. Chemother.*, **5**, 540 (1955).
1244. S. Nakamura and H. Umezawa, *J. Antibiot. (Tokyo)*, Ser. A, **9**, 66 (1955).
1245. R. Despois, S. Pinnert-Sindico, L. Ninet, and J. Preud'Homme, *G. Microbiol.*, **2**, 76 (1956).
1246. C. Cosar and F. Julon, *Ann. Inst. Pasteur (Paris)*, **96**, 238 (1959).
1247. H. L. Howes, J. E. Lynch, and J. L. Kolvin, *Antimicrob. Agents Chemother.*, 261 (1970).
1248. P. N. Giraldi, V. Mariothi, G. Nannini, G. P. Tosolini, E. Drachi, W. Logemann, I. de Carneri, and G. Monti, *Arzneim.-Forsch.*, **20**, 52 (1970).
1249. E. A. Steck, *Prog. Drug Res.*, **18**, 289 (1974).
1250. H. G. Smyly and C. W. Young, *Proc. Soc. Exp. Biol. Med.*, **21**, 354 (1924).
1251. G. Vianna, *Arch. Brasil. Med.*, **2**, 426 (1912).
1252. M. Ardehali, *Int. J. Dermatol.*, **13**, 26 (1974).
1253. E. F. Cappucino and R. A. Stauber, *Proc. Soc. Exp. Biol. Med.*, **101**, 742 (1959).
1254. R. K. Farmer, *J. Comp. Pathol. Ther.*, **60**, 294 (1950).
1255. A. C. Cuckler and C. M. Malanga, *Proc. Soc. Exp. Biol. Med.*, **92**, 483 (1956).
1256. J. M. S. Lucas, *Vet. Res.*, **73**, 465 (1961).

REFERENCES

1257. V. K. Bhagwat and F. L. Pyman, *J. Chem. Soc.*, **127**, 1832 (1925).
1258. R. S. Desowitz and H. J. C. Watson, *Ann. Trop. Med. Parasitol.*, **47**, 324 (1953).
1259. K. Butler, H. L. Howes, J. E. Lynch, and D. K. Pirie, *J. Med. Chem.*, **10**, 891 (1967).
1260. J. Kollonitsch, U.S. Pat. 3,450,710 (1967).
1261. E. H. Peterson, *Poult. Sci.*, **47**, 1245 (1968).
1262. F. Lösch, *Arch. Path. Anatom.*, **65**, 196 (1875).
1263. G. Woolfe, in *Experimental Chemotherapy*, Vol. 1, R. J. Schnitzer and F. Hawking, Eds., Academic Press, New York–London, 1963, pp. 355–443.
1264. R. R. Burtner and J. M. Brown, *J. Am. Chem. Soc.*, **73**, 897 (1951); **75**, 2334 (1953).
1265. S. Purchas, *Hakluytus Posthumus*, W. Stansby for H. Fetherstone, London, 1625.
1266. E. S. Docker, *Lancet*, **2**, 113, 169 (1858).
1267. S. K. Simon, *J. Am. Med. Assoc.*, **53**, 1526 (1909).
1268. G. Dock, *N.Y. Med. J.*, **90**, 49 (1909).
1269. L. Rogers, *Brit. Med. J.*, **2**, 405 (1912).
1270. M.-M. Janot, R. H. F. Manske, and H. L. Holmes, *The Alkaloids*, Vol. 3, Academic Press, New York, London, 1953, pp. 363–394.
1271. R. P. Evstigneeva, R. S. Livshits, M. S. Bainova, L. I. Zakharkin, and N. A. Preobrazhenskii, *J. Gen. Chem. USSR*, **22**, 1467 (1952).
1272. E. E. van Tamelen, P. E. Aldrich, and J. B. Hester, Jr., *J. Am. Chem. Soc.*, **79**, 4817 (1957).
1273. A. R. Battersby and S. Cox, *Chem. Ind.*, **1957**, 983.
1274. H. T. Openshaw and N. Whittaker, *J. Chem. Soc.*, 1461 (1963).
1275. A. Brossi, M. Baumann, and O. Schnider, *Helv. Chim. Acta*, **42**, 1515 (1959), and subsequent papers.
1276. J. Herrero, A. Brossi, M. Faust, and J. R. Frey, *Ann. Biochem. Exp. Med.*, **20** (Suppl.), 475 (1960).
1277. F. Blanc, Y. Nosny, M. Armengaud, M. Sankale, M. Martin, and G. Charmot, *Presse Méd.*, **69**, 1548 (1961).
1278. A. P. Grollman, *Proc. Nat. Acad. Sci. U.S.*, **56**, 1867 (1966).
1279. R. T. Shillings and C. P. Schaffner, *Antimicrob. Agents Chemother.*, **1961**, 274.
1280. K. O. Courtney, P. E. Thompson, R. Hodgkinson, and J. R. Fitzsimmons, *Ann. Biochem. Exp. Med.*, **20**, 449 (1960).
1281. G. W. Raiziss and B. C. Fisher, *J. Am. Chem. Soc.*, **48**, 1323 (1926).
1282. L. F. Hewitt and H. King, *J. Chem. Soc.*, 817 (1926).
1283. C. Levaditi, *Lancet*, **209**, 593 (1925).
1284. P. W. Brown, *J. Am. Med. Assoc.*, **105**, 1319 (1935).
1285. Farbwerke Meister, Lucius, and Bruening, German Pat. 213,155.
1286. H. H. Anderson and A. C. Reed, *Calif. West. Med.*, **35**, 439 (1931).
1287. CIBA A. G., German Pat. 117,767 (1899).
1288. V. Papesch and R. R. Burtner, *J. Am. Chem. Soc.*, **58**, 1314 (1936).
1289. A. Küster, *Klin. Wochenschr.*, **41**, 1125 (1904).
1290. P. Mühlens and W. Week, *Münch. Med. Wochenschr.*, **68**, 802 (1921).
1291. H. H. Anderson, N. A. David, and D. A. Koch, *Proc. Soc. Exp. Biol. Med.*, **28**, 484 (1931).
1292. H. H. Anderson and D. A. Koch, *Proc. Soc. Exp. Biol. Med.*, **28**, 838 (1931).
1293. A. C. Tenney, *Ill. Med. J.*, **70**, 145 (1936).

1294. J. H. Burckhalter, U.S. Pat. 2,746,963 (1956).
1295. N. G. Latour and R. E. Reeves, *Exp. Parasitol.*, **17**, 203 (1965).
1296. J. H. Burckhalter, R. I. Leib, Y. S. Chough, and R. F. Tietz, *J. Med. Chem.*, **6**, 89 (1963).
1297. J. W. Reinertson and P. E. Thompson, *Proc. Soc. Exp. Biol. Med.*, **76**, 518 (1951).
1298. World Health Organization, *Amoebiasis*, Technical Report Series No. 421, Geneva, 1969.
1299. J. H. Burckhalter, F. H. Tendick, E. M. Jones, W. F. Holcomb, and A. L. Rawlins, *J. Am. Chem. Soc.*, **68**, 1894 (1946).
1300. E. W. Dennis and D. A. Berberian, *Antibiot. Chemother.*, **4**, 554 (1954).
1301. A. R. Surrey and R. A. Cutler, *J. Am. Chem. Soc.*, **76**, 578 (1954).
1302. A. R. Surrey, *J. Am. Chem. Soc.*, **76**, 2214 (1976).
1303. E. W. Dennis and D. A. Berberian *Antibiot. Chemother.*, **4**, 554 (1954).
1304. W. Logeman, L. Almirante, and I. de Carneri, *Farmaco (Pavia)*, **11**, 926 (1956).
1305. A. R. Surrey and J. R. Mayer, *J. Med. Pharm. Chem.*, **3**, 409, 419 (1961).
1306. N. W. Bristow, P. Oxley, G. A. H. Williams, and G. Woolfe, *Trans. Roy. Soc. Trop. Med. Hyg.*, **50**, 182 (1956).
1307. A. C. Cuckler, A. B. Kupferberg, and N. Millman, *Antibiot. Chemother.*, **5**, 540 (1955).
1308. Hubbard and Steahly, U.S. Pats. 2,573,641, 2,573,656, 2,573,657 (1951).
1309. CIBA A. G., Belgian Pat. 632,989 (1963); *C.A.*, **61**, 1873d (1964).
1310. F. Kradolfer and R. Jarumilinta, *Ann. Trop. Med. Parasitol.*, **59**, 210 (1965).
1311. C. Cosar, P. Ganter, and L. Julon, *Presse Méd.*, **69**, 1069 (1961).
1312. S. J. Powell, A. J. Wilmot, and R. Elsdon-Dew, *Ann. Trop. Med. Parasitol.*, **61**, 511 (1967).
1313. M. W. Miller, H. L. Howes, Jr., R. V. Kasubick, and A. R. English, *J. Med. Chem.*, **13**, 849 (1970).
1314. W. J. Ross, in reference 7b, Chapter 19.
1315. H. M. Selzer, *Med. Today*, **3**, 110 (1969).
1316. J. Schneider, *Bull. Soc. Pathol. Exot. Fil.*, **54**, 84 (1961).
1317. S. Bassily, Z. Farid, J. W. Mikhail, D. C. Kent, and J. S. Lehman, *J. Trop. Med. Hyg.*, **73**, 15 (1970).
1318. V. P. Pai, F. F. Wadia, M. Darge, C. R. Sule, and S. S. Kale, *J. Assoc. Phys. Ind.*, **22**, 531 (1974).
1319. R. B. Burrows and W. G. Hahmes, *Am. J. Trop. Med. Hyg.*, **1**, 626 (1952).
1320. G. H. F. Nuttall and S. Hadwen, *Parasitology*, **2**, 156, 229, 236 (1909).
1321. W. Kikuth, *Zentralbl. Bakteriol. Parasitenkd.*, Abt. 1 (orig.), **135**, 135 (1935).
1322. E. Enigk and U. Reusse, *Z. Tropenmed. Parasitol.*, **6**, 141 (1957).
1323. J. N. Ashley, S. S. Berg, and J. M. S. Lucas, *Nature (London)*, **185**, 461 (1960).
1324. C. G. L. Beveridge, J. W. Thwaite, and G. Shepherd, *Vet. Rec.*, **72**, 383 (1960).
1325. R. A. Barrett, E. Beveridge, P. L. Bradley, C. G. D. Brown, S. R. M. Bushby, M. L. Clark, R. A. Neal, R. Smith, and J. K. H. Wilde, *Nature (London)*, **206**, 1350 (1965).
1326. W. A. Summers, *Proc. Soc. Exp. Biol. Med.*, **66**, 509 (1947); *Am. J. Trop. Med. Hyg.*, **2**, 1037 (1953).
1327. W. Roehl, *Arch. Schiffs Tropen-Hyg.*, **30**, 311 (1926); *Naturwissenschaften*, **14**, 1156 (1926).
1328. P. Kopanaris, *Arch. Schiffs Tropen-Hyg.*, **15**, 586 (1911).
1329. E. Brumpt, *Compt. Rend.*, **200**, 783 (1935).

REFERENCES

1330. I. H. Vincke and M. Lips, *Ann. Soc. Belg. Méd. Trop.*, **30**, 79 (1948).
1331. T. S. Osdene, P. B. Russell, and L. Rane, *J. Med. Chem.*, **10**, 431 (1967).
1332. F. Y. Wiselogle, *A Survey of Antimalarial Drugs, 1941–1945*. Edwards, Ann Arbor, Mich., 1946.
1333. T. R. Sweeney and R. E. Strube, in ref. 7b, Chapter 18.
1334. C. S. Jang, F. Y. Fu, C. Y. Wang, K. C. Huang, G. Lu, and T. C. Chou, *Science*, **103**, 59 (1946).
1335. B. R. Baker, R. E. Schaub, F. J. McEvoy, and J. H. Williams, *J. Org. Chem.*, **17**, 132 (1952).
1336. See review by R. B. Turner and R. B. Woodward, in *The Alkaloids*, Vol. 3, R. H. F. Manske and H. L. Holmes, Eds., Academic Press, New York–London, 1953, pp. 1–63.
1337. R. B. Woodward and W. E. Doering, *J. Am. Chem. Soc.*, **66**, 849 (1944); **67**, 860 (1945).
1338. W. Schulemann, *Proc. Roy. Soc. Med.*, **25**, 897 (1932).
1339. W. Kikuth and W. Menk, *Chemotherapie der wichtigsten Tropenkrankheiten: Die Chemotherapie der Malaria*, S. Hirzel, Leipzig, 1943, p. 45.
1340. See P. Mühlens, *Naturwissenschaften*, **14**, 1162 (1926).
1341. W. Schulemann, F. Schönhöfer, and A. Wingler, *Klin. Wochenschr.*, **11**, 381 (1932).
1342. A. S. Alving, T. N. Pullman, B. Craige, Jr., R. Jones, Jr., C. M. Whorton, and L. Eichelberger, *J. Clin. Invest.*, **27**, 34 (1948).
1343. W. C. Cooper, A. V. Myatt, T. Hernandez, G. M. Jeffery, and G. R. Coatney, *Am. J. Trop. Med. Hyg.*, **2**, 949 (1953).
1344. N. L. Drake, J. Van Hook, J. Garman, R. Hayes, R. Johnson, G. Kelley, S. Melamed, and R. Peck, *J. Am. Chem. Soc.*, **68**, 1529 (1946).
1345. R. C. Elderfield, C. B. Kremer, S. M. Kupchan, O. Birstein, and G. Cortes, *J. Am. Chem. Soc.*, **69**, 1258 (1947).
1346. R. C. Elderfield, E. F. Claflin, H. Mertel, O. McCurdy, R. Mitch, C. Ver Nooy, B. Wark, and I. Wempen, *J. Am. Chem. Soc.*, **77**, 4819 (1955).
1347. A. J. Lysenko, A. A. Churnosova, A. Godzova, G. Fastovskaja, and E. Zalznova, *Med. Parasitol. Parasit. Dis.* (USSR), **24**, 132, 137, 147 (1955); *Trop. Dis. Bull.*, **53**, 16 (1956).
1348. H. Mauss and F. Mietzsch, *Klin. Wochenschr.*, **12**, 1276 (1933).
1349. H. Jensch and O. Eisleb, U.S. Pat. 1,782,727 (1930).
1350. W. Schulemann et al., U.S. Pat. 1,889,704 (1932).
1351. F. Schönhöfer, *Z. Physiol. Chem.*, **274**, 1 (1942).
1352. O. Y. Magidson and A. M. Grigorowsky, *Chem. Ber.*, **69**, 396, 537 (1936).
1353. J. Schneider, P. Decourt, and D. Mechali, *Proceedings of the 4th International Congress of Tropical Medicine and Maladies*, Washington, D.C., 1948, Vol. 1, p. 756.
1354. A. R. Surrey and H. F. Hammer, *J. Am. Chem. Soc.*, **68**, 113 (1946).
1355. E. A. Steck, L. L. Hallock, and A. J. Holland, *J. Am. Chem. Soc.*, **68**, 129, 132 (1946).
1356. B. Riegel, G. R. Lappin, C. J. Albisetti, B. H. Adelson, R. M. Dodson, L. G. Ginger, and R. H. Baker, *J. Am. Chem. Soc.*, **68**, 1229 (1946).
1357. G. R. Coatney, *Am. J. Trop. Med. Hyg.*, **12**, 121 (1963).
1358. V. H. Gladkikh, O. I. Kellina, and Y. V. Korogodina, *Med. Parazitol. Parazit. Bolezn.*, **28**, 443 (1959); *C.A.*, **54**, 6956 (1960).
1359. A. D. Ainley and H. King, *Proc. Roy. Soc., Ser. B*, **125**, 60 (1938).
1360. G. A. H. Buttle, T. A. Henry, and J. W. Trevan, *Biochem. J.*, **28**, 426 (1934).
1361. C. C. Cheng, *J. Pharm. Sci.*, **60**, 1596 (1971).

1362. J. Mead and J. B. Koepfli, *J. Biol. Chem.*, **145**, 507 (1944).
1363. F. E. Kelsey, F. K. Oldham, W. Cantrell, and E. M. K. Geiling, *Nature (London)*, **157**, 440 (1946).
1364. R. E. Lutz, P. S. Bailey, M. T. Clark, J. F. Codington, A. J. Deinet, J. A. Freek, G. H. Harnest, N. H. Leake, F. A. Martin, R. J. Rowlett, J. M. Salsbury, N. H. Shearer, J. D. Smith, and J. W. Wilson, *J. Am. Chem. Soc.*, **68**, 1813 (1946).
1365. R. M. Pinder and A. Burger, *J. Med. Chem.*, **11**, 267 (1968).
1366. C. J. Ohnmacht, A. R. Patel, and R. E. Lutz, *J. Med. Chem.*, **14**, 926 (1971).
1367. R. E. Strube, *J. Trop. Med. Hyg.*, **78**, 171 (1975).
1368. G. R. Coatney, W. C. Cooper, M. B. Eddy, and J. Greenberg, Public Health Monograph No. 9, Government Printing Office, Washington, D.C., 1953, Chapter 10.
1369. A. Albert, *The Acridines: Their Preparation, Properties and Uses,* Arnold, London, 1951.
1370. F. H. S. Curd and F. L. Rose, *J. Chem. Soc.*, 343 (1946).
1371. F. H. S. Curd and F. L. Rose, *J. Chem. Soc.*, 362 (1946); 574, 586 (1948).
1372. F. H. S. Curd, D. G. Davey, and F. L. Rose, *Ann. Trop. Med. Parasitol.*, **39**, 208 (1945).
1373. F. H. S. Curd and F. L. Rose, *J. Chem. Soc.*, 729 (1946).
1374. A. F. Crowther and A. A. Levi, *Brit. J. Pharmacol.*, **8**, 93 (1953).
1375. E. A. Falco, G. H. Hitchings, P. B. Russell, and H. Vanderwerf, *Nature (London)*, **164**, 1133 (1949).
1376. G. H. Hitchings, G. B. Elion, H. Vanderwerf, and E. A. Falco, *J. Biol. Chem.*, **174**, 765 (1948).
1377. E. A. Falco, L. G. Goodwin, G. H. Hitchings, I. M. Rollo, and P. B. Russell, *Brit. J. Pharmacol.*, **6**, 185 (1951).
1377a. E. Marshall, *Science*, **219**, 466 (1983).
1378. P. J. Islip, in ref. 7b, Chapter 21.
1379. J. B. Christopherson, *Lancet*, 325 (1918).
1380. M. Ron. Pedrique and N. Ercoli, *Bull. WHO.*, **45**, 411 (1971).
1381. H. Mauss, *Chem. Ber.*, **81**, 19 (1948).
1382. W. Kikuth and R. Goennert, *Ann. Trop. Med. Parasitol.*, **42**, 256 (1948).
1383. D. Rosi, G. Peruzzotti, E. W. Dennis, D. A. Berberian, H. Freele, and S. Archer, *Nature (London)*, **208**, 1005 (1965).
1384. D. Rosi, E. W. Dennis, D. A. Berberian, H. Freele, B. F. Tullar, and S. Archer, *J. Med. Chem.*, **10**, 867 (1967).
1385. S. Archer and A. Yarinsky, *Prog. Drug Res.*, **16**, 11 (1972).
1386. E. F. Elslager, D. F. Worth, and J. D. Howells, German Pat. 1,876,086 (1969).
1387. R. Foster and B. L. Cheetham, *Trans. Roy. Soc. Trop. Med. Hyg.*, **67**, 674 (1973).
1388. R. Foster, B. L. Cheetham, and D. F. King, *ibid.*, **67**, 685 (1973).
1389. H. C. Richards and R. Foster, *Nature (London)*, **222**, 581 (1969).
1390. C. A. R. Baxter and H. C. Richards, *J. Med. Chem.*, **14**, 1033 (1971).
1391. C. R. Lambert, H. Wilhelm, H. Streibel, F. Kradolfer, and P. Schmidt, *Experientia*, **20**, 452 (1964).
1392. C. H. Robinson, E. Bueding, and J. Fisher, *Mol. Pharmacol.*, **6**, 604 (1970).
1393. H. P. Striebel, *Experientia*, **32**, 457 (1976).
1394. E. Bueding, C. L. Liu, and S. H. Rogers, *Brit. J. Pharmacol.*, **46**, 480 (1972).
1395. D. Posthuma and W. J. Vaatstra, *Biochem. Pharmacol.*, **20**, 1133 (1971).
1396. A. L. Bartlet, *Brit. J. Pharmacol.*, **58**, 395 (1976).

REFERENCES

1397. S. D. Ross, M. Markarian, and M. Schwartz, *J. Am. Chem. Soc.*, **75**, 4967 (1953).
1398. H.-Y. Chang and C. T. Yuan Sheng Wu Hua Hsueh Yu Sheng Wu Wu Li Hsueh Pao, **6** (3), 273 (1966); through *C.A.*, **66**, 1240 (1967).
1399. M. Davis, D. E. Wright, and J. M. S. Lucas, in *Veterinary Pesticides*, SCI Monograph No. 33, Society of Chemical Industry, London, 1969, pp. 25, 34.
1400. E. Druckrey and H. Metzger, *J. Med. Chem.*, **16**, 436 (1973).
1401. H. Metzger and D. Duewel, *Int. J. Biochem.*, **4**, 133 (1973).
1402. D. and T. Rowlands, *Pesticide Sci.*, **4**, 883 (1973).
1403. M. Harfenist, *Pesticide Sci.*, **4**, 871 (1973).
1404. G. A. C. Gough and H. King, *J. Chem. Soc.*, 669 (1930).
1405. T. H. Maren, *J. Am. Chem. Soc.*, **68**, 1864 (1946).
1406. E. A. H. Friedheim and R. Cavier, *Bull. Soc. Pathol. Exot.*, **66**, 531 (1973).
1407. E. A. H. Friedheim, *Bull. WHO.*, **50**, 572 (1974).
1408. W. Raether and G. Laemmler, *Ann. Trop. Med. Parasitol.*, **65**, 107 (1971).
1409. R. Hewitt, E. White, S. Kushner, W. S. Wallace, H. W. Stewart, and Y. Subba Row, *Ann. N.Y. Acad. Sci.*, **50**, 128 (1948).
1410. D. W. Gibson, D. H. Connor, H. L. Brown, H. Fuglsang, J. Anderson, B. O. L. Duke, and A. A. Buck, *Am. J. Trop. Med. Hyg.*, **25**, 74 (1976).
1411. S. Kushner, L. M. Brancone, R. I. Hewitt, W. L. McEwen, Y. Subba Row, H. W. Stewart, R. J. Turner, and J. J. Denton, *J. Org. Chem.*, **13**, 144 (1948).
1412. R. Hewitt, W. S. Wallace, E. White, and Y. Subba Row, *J. Parasitol.*, **34**, 237 (1948).
1413. G. Mourequand, E. Roman, and J. Coisnard, *J. Méd. Lyon*, **32**, 189 (1951).
1414. C. Fayard, *Ascaridiose et piperazine*, Thèse de Paris, 1947.
1415. W. Riedl, *Angew. Chem.*, **67**, 184 (1955).
1416. E. Schraufstaetter and R. Goennert, British Pat. 824,345 (1959).
1417. H. Von Ruschig, J. Koenig, D. Duewel, and H. Loewe, Brit. Pat. 1,124,613 (1968); *Arzneim.-Forsch.*, **23**, 1745 (1973).
1418. R. B. Burrows, C. J. Hatton, W. G. Lillis, and G. R. Hunt, *J. Med. Chem.*, **14**, 87 (1971).
1419. E. Lieber and R. Slutkin, *J. Org. Chem.*, **27**, 2214 (1962).
1420. D. K. Hass and J. A. Collins, *J. Vet. Res.*, **35**, 103 (1974).
1421. R. R. Whetstone and D. Harman, U.S. Pat. 2,956,073 (1960).
1422. H. Thomas and P. Andrews, *Pesticide Sci.*, **8**, 556 (1977).
1423. H. Thomas and R. Goennert, *Res. Vet. Sci.*, **24**, 20 (1978).
1424. J. D. Kendall and H. D. Edwards, U.S. Pat. 2,412,815 (1946).
1425. E. F. Elslager and D. F. Worth, U.S. Pat. 2,925,417 (1960).
1426. W. C. Austin, W. Courtney, J. C. Danilewicz, D. H. Morgan, L. H. Conover, H. L. Howes, Jr., J. E. Lynch, J. W. McFarland, R. L. Cornwell, and V. J. Theodorides, *Nature (London)*, **212**, 1273 (1966).
1427. J. W. McFarland and H. L. Howes, Jr., *J. Med. Chem.*, **15**, 365 (1972).
1428. E.-L. Lee, N. Iyngkaran, A. W. Grieve, M. J. Robinson, and A. S. Dissanaike, *Am. J. Trop. Med. Hyg.*, **25**, 563 (1976).
1429. J. E. Lynch and B. Nelson, *J. Parasitol*, **45**, 659 (1959).
1430. H. D. Brown, A. R. Matzuk, I. R. Ilves, L. H. Petersen, S. A. Harris, L. H. Sarett, J. R. Egerton, J. J. Yakstis, W. C. Campbell, and A. C. Cuckler, *J. Am. Chem. Soc.*, **83**, 1764 (1961).
1431. R. K. Prichard, *Int. J. Parasitol.*, **3**, 409 (1973).

1432. D. J. Tocco, R. P. Buhs, H. D. Brown, A. R. Matzuk, H. E. Mertel, R. E. Harman, and N. R. Trenner, *J. Med. Chem.*, **7**, 399 (1964).
1433. R. D. Hoff, M. H. Fisher, R. J. Bochis, A. Lusi, F. Waksumski, J. R. Egerton, J. J. Yakstis, A. C. Cuckler, and W. C. Campbell, *Experientia*, **26**, 550 (1970).
1434. P. Actor, E. L. Anderson, C. J. DiCuollo, R. J. Ferlauto, J. R. E. Hoover, J. F. Pagano, I. R. Ravin, S. F. Scheidy, R. J. Stedman, and V. J. Theodorides, *Nature (London)*, **215**, 321 (1967).
1435. J. P. Brugmans, D. C. Thienpont, I. van Wijngaarden, O. F. J. Vanparijs, V. L. Schuermans, and H. L. Lauwers, *J. Am. Med. Assoc.*, **217**, 313 (1971).
1436. P. A. J. Janssen, *Prog. Drug. Res.*, **20**, 347 (1976).
1437. R. W. Sidwell, in *Chemotherapy of Infectious Disease*, H. H. Gadebush, Ed., CRC Press, Cleveland, 1976, p. 31.
1438. M. Suffness and J. D. Douros, *TIPS*, **2**, 307 (1981).
1439. B. Moss, in *Selective Inhibitors of Viral Functions*, W. A. Carter, Ed., CRC Press, Cleveland, 1973, p. 313.
1440. E. Heller, in *RNA Polymerase and Transcription: Proceedings of the 1st International Lepetit Symposium*, L. Silvestri, Ed., North Holland, Amsterdam, 1969, p. 287.
1441. J. H. Subak-Sharpe, T. H. Pennington, T. F. Szilagyi, M. C. Timbury, and J. F. Williams, *ibid.*, p. 739.
1442. R. W. Sidwell and J. T. Witkowski, in ref. 7b, Chapter 23, pp. 578–579.
1443. S. A. Waksman and H. B. Woodruff, *Proc. Soc. Exp. Biol. Med.*, **45**, 609 (1940).
1444. H. Umezawa, in *Cancer Medicine*, J. F. Holland and E. Frei III., Eds., Lea & Febiger, Philadelphia, 1973, Sect. XIII-6.
1445. A. DiMarco and L. Lenaz, *ibid.*, p. 826.
1446. A. DiMarco in *Antineoplastic and Immunosuppressive Agents*, Part II, A. C. Sartorelli and D. G. Johns, Eds., Springer-Verlag, Heidelberg, 1975, p. 593.
1447. M. Israel, E. J. Modest, and E. Frei III, *Cancer Res.*, **35**, 1365 (1975).
1448. R. B. Livingston and S. K. Carter, *Single Agents in Cancer Chemotherapy*, IFI/Plenum, New York, 1970.
1449. G. P. Bodey and E. J. Freireich, *Abstr. Am. Assoc. Cancer Res.*, **17**, 128 (1976).
1450. M. Edelhart and J. Lindenmann, *Interferon: The New Hope for Cancer*, Addison-Wesley, Reading, Mass., 1981.
1451. W. E. Stewart II and L. S. Lin, *Pharmacol. Ther.*, **6**, 443 (1979).
1452. G. Gregoriadis, *Pharmacol. Ther.*, **10**, 103 (1980).
1453. A. Isaacs and J. Lindenmann, *Proc. Roy. Soc., Ser. B*, **147**, 258 (1957).
1454. A. Isaacs, J. Lindenmann, and R. C. Valentine, *Proc. Roy. Soc., Ser. B*, **147**, 268 (1957).
1455. S. Pestka, S. Maeda, and T. Staehelin, *Ann. Rep. Med. Chem.*, **16**, 229 (1981).
1456. E. R. Andrews, R. W. Fleming, J. M. Grisar, J. C. Kihm, D. L. Wenstrup, and G. D. Mayer, *J. Med. Chem.*, **17**, 882 (1974).
1457. R. F. Krueger and G. D. Mayer, *Science*, **169**, 1213, 1214 (1970).
1458. E. DeClerg, in *Selective Inhibitors of Viral Functions*, W. A. Carter, Ed., CRC Press, Cleveland, 1973, p. 177.
1459. R. R. Crenshaw, G. M. Luke, and P. Simonoff, *J. Med. Chem.*, **19**, 262 (1976).
1460. J. Dausset, *Science*, **213**, 1469 (1981).
1461. J. Klein, *Science*, **203**, 516 (1979).
1462. H. T. Orr, J. A. Lopez de Castro, P. Parham, H. Ploegh, and J. L. Strominger, *Proc. Natl. Acad. Sci., U.S.*, **76**, 4395 (1979).

1463. J. A. Montgomery, T. P. Johnston, and Y. F. Shealy, in ref. 7b, Chapter 24, pp. 600–602.
1464. S. A. Fusari, T. H. Haskell, R. P. Frohardt, and Q. R. Bartz, *J. Am. Chem. Soc.,* **76,** 2881 (1954).
1465. H. W. Dion, S. A. Fusari, Z. L. Jakubowski, J. G. Zora, and Q. R. Bartz, *J. Am. Chem. Soc.,* **78,** 3075 (1956).
1466. D. R. Seeger, D. B. Cosulich, J. M. Smith, Jr., and M. E. Hultquist, *J. Am. Chem. Soc.,* **71,** 1753 (1949).
1467. S. F. Zakrzewski, *J. Biol. Chem.,* **238,** 4002 (1963).
1468. B. R. Baker and B.-T. Ho, *J. Pharm. Sci.,* **54,** 1261 (1965).
1469. J. J. Burchall and G. H. Hitchings, *Mol. Pharmacol.,* **1,** 126 (1965).
1470. G. M. Timmis, D. G. I. Felton, H. O. J. Collier, and P. L. Huskinson, *J. Pharm. Pharmacol.,* **9,** 46 (1957).
1471. G. H. Hitchings, G. B. Elion, E. A. Falco, P. B. Russell, and H. Vander Werff, *Ann. N.Y. Acad. Sci.,* **52,** 1318 (1950).
1472. B. Roth, E. A. Falco, G. H. Hitchings, and S. R. M. Bushby, *J. Med. Pharm. Chem.,* **5,** 1103 (1962).
1473. H. C. Carrington, A. F. Crowther, D. G. Davey, A. A. Elvi, and F. L. Rose, *Nature (London),* **168,** 1080 (1951).
1474. E. J. Modest, G. E. Foley, M. M. Pechet, and S. Farber, *J. Am. Chem. Soc.,* **74,** 855 (1952).
1475. A. Piscala and F. Šorm, *Collect. Czech. Chem. Commun.,* **29,** 2060 (1964).
1476. S. Jasinska, F. Link, D. Blaškovič, and B. Rada, *Acta Virol.,* **6,** 17 (1962).
1477. G. A. Galegov, R. M. Bikbulatov, K. A. Vanag, and R. M. Shen, *Vop. Virusol.,* **13,** 18 (1968).
1478. G. P. Khare, R. W. Sidwell, J. H. Huffman, R. L. Tolman, and R. K. Robins, *Proc. Soc. Exp. Biol. Med.,* **140,** 880 (1972).
1479. W. M. Shannon, G. Arnett, and F. M. Schabel, Jr., *Antimicrob. Agents Chemother.,* **2,** 159 (1972).
1480. W. M. Shannon, *Ann. N.Y. Acad. Sci.,* **284,** 472 (1977).
1481. M. C. Wang and A. Bloch, *Biochem. Pharmacol.,* **21,** 1063 (1972).
1482. O. P. Babbar and B. L. Chowdhury, *J. Sci. Ind. Res.,* **21C,** 312 (1962).
1483. R. W. Sidwell, J. H. Huffman, G. P. Khare, L. B. Allen, J. T. Witkowski, and R. K. Robins, *Science,* **177,** 705 (1972).
1484. C. Heidelberger, in *Cancer Medicine,* J. F. Holland and E. Frei III, Eds., Lea and Febiger, Philadelphia, 1937, p. 768.
1485. P. V. Dannenberg and C. Heidelberger, *Biochemistry,* **15,** 1331 (1976).
1486. C. Heidelberger, D. G. Parsons, and D. C. Remy, *J. Med. Chem.,* **7,** 1 (1964).
1487. W. H. Prusoff and D. C. Ward, *Biochem. Pharmacol.,* **25,** 1233 (1976).
1488. C. Heidelberger, *Ann. N.Y. Acad. Sci.,* **225,** 317 (1975).
1489. J. S. Evans, E. A. Musser, G. D. Mengel, K. R. Fosblad, and J. H. Hunter, *Proc. Soc. Exp. Biol. Med.,* **106,** 350 (1961).
1490. R. W. Talley and V. K. Vaitkevicius, *Blood,* **21,** 352 (1963).
1491. F. M. Schabel, Jr. and J. A. Montgomery, in *Chemotherapy of Virus Diseases, Vol. 1, International Encyclopedia of Pharmacology and Therapeutics,* D. J. Bauer, Ed., Pergamon, Oxford, 1972, p. 231.
1492. G. B. Elion, E. Burgi, and G. H. Hitchings, *J. Am. Chem. Soc.,* **74,** 411 (1952).

1493. G. B. Elion and G. H. Hitchings, *J. Am. Chem. Soc.*, **77,** 1676 (1955).
1494. R. Schwartz and W. Dameshek, *Nature (London)*, **183,** 1682 (1959).
1495. G. H. Hitchings and G. B. Elion, *Account. Chem. Res.*, **2,** 202 (1969).
1496. G. H. Hitchings and G. B. Elion, U.S. Pat. 3,056,785 (1962).
1497. G. B. Elion, *Proc. Roy. Soc. Med.*, **65,** 257 (1972).
1498. V. Meyer, *Chem. Ber.*, **1,** 1725 (1887).
1499. P. Ehrlich, *Collected Papers of Paul Ehrlich,* Vol. 1, F. Himmelweit, Ed., Pergamon Press, London, 1956, p. 596.
1500. L. Hektoen and H. J. Corper, *J. Infect. Dis.*, **28,** 279 (1921).
1501. H. Arnold and F. Bourseaux, *Angew. Chem.*, **70,** 539 (1958).
1502. H. Arnold, F. Bourseaux, and N. Brock, *Naturwissenschaften,* **45,** 64 (1958); *Arzneim.-Forsch.*, **11,** 143 (1961).
1503. G. Gomori, *Proc. Soc. Exp. Biol. Med.*, **69,** 407 (1948).
1504. A. Kint and L. Verlinden, *N. Engl. J. Med.*, **291,** 308 (1974).
1505. W. Ludwig, U.S. Pat. 2,937,211 (1960); 3,053,907 (1962).
1506. W. L. Davies, R. R. Grunert, R. F. Haff, J. W. McGahen, E. M. Neumayer, M. Paulshock, J. C. Watts, T. R. Woods, E. C. Hermann, and C. E. Hoffman, *Science,* **144,** 862 (1964).
1507. J. G. Whitney, W. A. Gregory, J. C. Kaner, J. R. Roland, J. A. Snyder, R. E. Benson, and E. C. Hermann, *J. Med. Chem.*, **13,** 254 (1970).
1508. A. Takatsuki and G. Tamura, *J. Med. Chem.*, **15,** 536 (1972).
1509. W. B. Flagg, F. J. Stanfield, R. F. Haff, R. C. Stewart, R. J. Stedman, J. Gold, and R. J. Ferlauto, *Antimicrob. Agents Chemother.*, **1968,** 194 (1969).
1510. M. Baggliolini, B. Dewald, and H. Aebi, *Biochem. Pharmacol.*, **18,** 2187 (1969).
1511. F. K. Hess and K. R. Freter, in ref. 7b, Chapter 25.
1512. T. H. Althuis, *New Engl. J. Med.*, **303,** 1004 (1980); Abstracts, 183rd Meeting of the American Chemical Society, Las Vegas, Nevada, March 31, 1982, Medi-52.

INDEX

Abstinence drugs, 33–34
Acetanilide, 16
Acetarsol, 204
Acetarsone, 206, 207
Acetazolamide, 80–81, 141, 178
Acetohexamide, 68
Acetominophen, 16
Acetophenetidine, 16
Acetylcholine, 19, 20, 55–58, 102–105, 117
Acetyl coenzyme A, 103
Acetyl salicylic acid, 15, 43, 86, 150
Acidity, and hypnotic activity, 137
Acinitrazole, 202, 203
Acriflavine, 16, 199, 209
ACTH, 24
Actinomadura carminata, 195
Actinomycins, 194, 195, 232
Activation, metabolism and, 71–72
Activity, insulin, enhancement of, 67
Acyclovir, 49
Adamantamine, 51
Additive model, 70
Adiphenine, 53
Adrenaline, *see* Epinephrine
Adrenersic amines, 19–20, 41
Adrenoreceptors, types of, 32
Adriamycin, 195, 232
Afridol violet, 16, 165, 197, 198
Alkaloids, 205–206
   in amebiasis, 205–206
   botanical, 14
   calabash, 105
   cinchona, 26, 211–212
   curare, 105, 106
   ephedrine, 20
   indole, 34
   laudanosine, 147–148
   morphine, 60–61, 143–145
   pyrrolidine, 8
   rauwolfia, 34, 39
   solanaceous, 57–58, 108–109
   veratrum, 38
   vinca, 232
Alkanesulfonates, 49
Alkylating agents, 31, 240–241
   as antiinflammatory agents, 46, 152
   as antiviral and antitumor agents, 48, 49
Alkylating groups, and irreversible inhibition, 75
Allergic reactions, mediators of, 110
Allen-Doisy rat test, 157
N-Allylnorcodeine, 30
N-Allylnormeperidine, 84
N-Allylnormorphine, 30, 64, 84
Amantadine, 242
Amebiasis, 190–191, 205–209
Amicarbalide, 209
Amides, 182
Amidines, antibacterial, 165–167
Amidoxime, 40
Aminacrine, 170
Amines, adrenergic, 19–20
Aminitrozole, 208
p-Aminobenzenesulfonamide ion, 174
p-Aminobenzoic acid, 26, 177, 178
γ-Aminobutyric acid, 117
Aminoglycosides, 190–191
β-Aminoketones, 124
p-Aminophenol, 16
Aminopterin, 32, 49, 235
Aminopyrimidines, 217–219
Aminopyrine, 42, 151
Aminoquinolines, 199–200, 213–214
p-Aminosalicylic acid, 31, 180
Aminothiazoles, 181
Amitriptyline, 29, 36, 130
Ammonium salts, quarternary, 164–165
Amobarbital, 22
Amodiaquine, 214–215
Amonoquinolines, 30

Amoscanate, 222, 226
Amphetamine, 83, 126
Amphotericin B, 194, 202, 204
Ampicillin, 187
Amprolium, 202
Amylocaine, 17, 18, 113–114
Anagesterone, 161
Analgetics, 16–17, 58–68
   abstinence drugs, 33–34
   centrally-acting antitussives, 65–66
   defined, 58
   endogenous peptides, 142–143
   mechanism of action, 141–142
   opiate receptors, 64–65
   opioid, 143–145
   structure modification, 145–148
Analogs, transition-state, 9, 72–73
Analysis, confirmational, 90–91
Anaplasmosis, 210
Ancylostoma duodenale, 227
Andosterone, 25
Androgens, 154–156
Androstanes, 158–159
Anesthetics, 17–19, 58, 113–116
Angiotensin, 38
   -converting enzyme, 41–42
Anilides, 19
Antagonists:
   metabolite, 49
   narcotic, 64
Anthelmintics, 220
   azo dyes, 169
   broad spectrum, 227–230
   for cestode infections, 226–227
   for filariasis, 224–226
   for schistosomiasis, 220–222
   for trematode infections, 222–224
Anthiomaline, 204
Anthracycline antibiotics, 195
Antianxiety agents, 131
   benzodiazepines, 132–134
   1,3-propanediols, 131–132
Antibacterials:
   chemotherapeutic, 170–178
   topical, 159, 163
Antibiotics, 183–196
   aminoglycoside, 190–191
   antitumor, 194–196
   azo dyes as, 16
   chloramphenicol, 192–193
   defined, 183
   b-lactam, 184–189
   penicillins, 184–187
   peptide, 191–192
   tetracyclines, 192

Anticholinergics, 55–58, 107–109
Anticoagulants, 100–102
Anticoccidal agents, 202
Anticonvulsants, 134–135, 139–145
Antidepressants, 125
   biochemical hypotheses of, 127–128
   discovery of, 125–127
   monoamine oxidase inhibitors, 125–126, 128–130
   test methods of, 128
   tricyclic, 126–127, 130–131
Antifertility agents, 159, 160, 161
Antifungal agents:
   Amphotericin B, 194, 202, 204
   azo dyes, 169
   nitroheterocyclic, 168
Antihistaminics, 28, 29, 53–54, 109–113
Antihyperglycemic agents, 67–69
Antihypertensive agents, 38–42
Antiinfectious agents, 159, 162–170
   antibacterials, topical, 159, 163
   antimicrobials, general, 159, 163
   antimicrobials, topical and systemic, 167–171
   dyestuffs, 169–170
   halogens, 163
   nitroheterocycles, 167–168
   phenols, 163–164
   pyridonecarboxylic acids, 168–169
   quarternary ammonium salts, 164–165
   topical antibacterials and general antimicrobials, 159, 163
   ureas, amidines, and biguanides, 165–167
Antiinflammatory agents:
   nonsteroidal, 42–46, 150–153
   steroidal, 46–47, 148–150
Antimalarials, *see* Malaria
Antimetabolites:
   antiinflammatory activity, 152
   antineoplastic and antiviral activity, 234–240
Antimicrobials, 26–28
   dyestuffs, 169–170
   general, 159, 163
   nitroheterocycles, 167–168
   pyridonecarboxylic acids, 168–169
   topical and systemic, 167–171
Antimonials, 204, 220–211
Antiplasmodial agents, *see* Malaria
Antiprotozoal agents, 196–219
   for amebiasis, 205–209
   aminoglycosides as, 190–191
   for coccidiosis, 202
   dyestuffs as, 169, 171
   for leishmaniasis and histomoniasis, 203–204
   for malaria, 210–219
   miscellaneous, 209–210

for trichomoniasis, 203
for trypanosomal infections, 196–202
Antipsychotic agents, see Neuroleptic agents
Antipyrine, 151
Antiseptics, 159, 162–167
Antispasmodics, 56–58, 107–109, 110
Antitrypanosomal agents, 165, 166
Antitubercular agents, 27
Antitussives, centrally acting, 65–66
Antiviral and antineoplastic agents, 31–32, 47–52, 230–231
   alkylating agent as, 240–241
   antibiotics, 194–196
   antimetabolites, 234–240
   immunostimulants, 232–234
   miscellaneous, 241–242
   natural compounds, 231–232
Arachidonic acid, 44, 47, 58, 80
Arecoline, 104
Arsenamide, 224, 225
Arsenicals:
   in amebiasis, 205, 206–207
   antitrypanosomal activity, 198–199
   in filariasis, 224, 225
Arsphenamine, 16
Arylacetic acids, 43
N-Arylanthranilic acids, 43
Arylpropionic acids, 43
Ascariasis, 220, 228
Asparginase, 244
Aspersillus scleroticum, 221
Aspirin, 15, 43, 86, 150
Atabrine, 213–214
Atoxyl, 198
Atropa belladonna, 57–58
Atropine, 10, 12, 56, 57, 108
Auranofin, 45
Aurothioglucose, 45
5-Azacytidine, 236
Azaserine, 235
Azathioprine, 240
6-Azauridine, 236
Azidomorphine, 65–66
Aziridines, 49
Aziridium ions, 75
Azo dyes, 16, 165–167, 169–170, 170–173
Azole nucleosides, as antiviral agents, 237
Azomycin, 168, 202, 203, 209
Az-threonam, 189

Babesiosis, 209–210
Bacillus brevis, 191
Bacitacin, 31
Bacitracin, 191
Bacteriostasis, 169

BAL, 198
Balandidiasis, 209
Barbital, 22
Barbiturates, 22, 136–137, 138, 139
Barium chloride, 52
Batrachotoxin, 59
Beclotiamine, 202
Benzalkonium chloride, 165
Benzene, ring equivalents and, 29
Benzimidazoles, 147
Benzimidazolylbenzimidazole, 224
Benzocaine, 17, 18, 113
Benzodiazepines, 35, 132–134, 138
2-(N-piperidinomethyl)-1,4-Benzodioxane, 52
Benzomorphans, 79
Benzopurpurin, 197
1-Benzylindole-3-acetic acid, 43
Benzyl paraben, 164
Berberine, 206
Bialamicol, 207
Bicyclamine, 242–243
Biguanides, 165–167
Binding affinity, 47
Bioisosteric modification, 53, 234, 235
Bioisosterism, 28–30, 84–87
Biological guidance, in molecular modification, 80–82
Biological properties, substitution and, 82–84
Biosterism, steroids, 155
"Bis"-ins, 167
Bisphenyls, 164
Bithionol, 223
Bitin-S, 223
Bitoscante, 226–227
Bleomycins, 195–196, 232
Bond angles, and bioisosteric replacement, 86
Botanical alkaloids, 14
Bothrops jararaca, 41
Bradykinin, 110
   -potentiating peptides, 41–42
Brilliant green, 169
Bromides, and sedation, 135
Bruceantin, 233
Bunamidine, 226
Buprenorphine, 63–64
Burimide, 54
Butirosin, 194
Butorphanol, 66
Butyrophenones, 34, 117, 123–124

Cadinene, 8
Calabash alkaloids, 105
Calcitonin, 98
Calcium metabolism, D vitamins and, 95–96
Calmodulin, 120

Cambendazole, 229
Cannabis sativa, 148
Capreomycin, 179
Captopril, 41, 42
Caramiphen, 66
Carbamates, 137
Carbanilides, 166
Carbarsone, 206, 207
Carbomycin, 194
Carbonic anhydrase, 80, 178
Carbonium ions, 48, 50, 75
Carbon tetrachloride, 223
Carbonyl compounds, 49–50
Carboxypeptidase A, kininase and, 41
Carbutamide, 68, 178
Carcinomycin, 195
Carmustine, 244
Carzinophilin, 195
Catecholamines, 117
Cephaelis ipecacuanha, 205
Cephalosporins, 187–188
Cerebroside sulfate, 65
Cestode infections, 226–227
Cetrimonium bromide, 165
Cetylpyridinium chloride, 165
Chagas' disease, 196
Chain branching, in molecular modification, 82–84
Chaulmoogric acid, 80
Chemotherapeutic agents, antibacterials, 162, 170–178
Chemotherapy, *see* antiviral and antineoplastic agents
Chinchona alkaloids, 211–212
Chiniofon, 206
Chinosol, 164
Chloral hydrate, 135
Chlorambucil, 45, 46
Chloramine T, 163
Chloramphenicol, 30, 31, 168, 183
Chlorazol fast pink, 165
Chlorbetamide, 208
Chlordane, 196
Chlorguanide, 72, 218–219
Chlorhexidine, 167
Chlorine, 163
Chlorisondamine, 33
Chlormadinone, 160
Chloroform, 22, 223
α-Chloro ketones, 75
Chlorophene, 163
Chlorophenothane, 196
Chlorophenoxamide, 208
p-Chlorophenyalanine, 244
Chloroquines, 45, 152, 207, 214, 217

Chlorpromazine, 34, 35, 118, 119–120
Chlorpropamide, 68
Chlorprothixene, 112
Chlorthiazepoxide, 35
Chlorthiazide, 40, 81
Choline acetylase, 103
Choline ether analogs, 41
Chondodendron, 105
Chromatography, 97
Cichona alkaloids, 26
Cimetidine, 29, 55
Clamoxyquin, 207
Clavaceps purpurea, 8
Clavulanic acid, 189
Clioquinol, 164
Clobazam, 133–134
Clofazimine, 183
Cloflucarban, 166
Clomiphene, 157, 158
Clonazepam, 133
Clonorchis sinensis, 223
Clopazine, 122–123
Cocaine, 10, 17, 18, 78, 113
Coccidiosis, 202
Codeine, 10, 60, 65, 143–144
Coformycin, 72
Colchicine, 46, 152
Conessine, 206
Confirmational analysis, 90–91
Congo red, 197
Contraceptives, 159, 160, 161
Coprine, 73
Coprinus attramentarius, 34
Cordycepin, 202
Corticosteroids, 25, 42, 46–47, 148–149, 240
Coxcomb test, 154
Cresylic acids, 163
Crotalaria, 8
Crystal violet, 169
Cupreines, 172
Curare, 11, 32, 105, 106
Cyanamide, 34
Cyanine dyes, as anthelmintics, 227
Cyanoguanidine, 29, 54–55
Cyclazocine, 63–64
Cyclic ketals, 102
Cyclic purinosyl phosphates, 13
Cyclic ureides, 22
Cycloguanil, 204
Cyclophosphamide, 45, 46, 240–241
Cyclopropylamines, 36
Cycloserine, 179, 194
Cyclosporin A, 152
Cycooctylamine, 242–243
Cyproterone acetate, 158

INDEX

Cytidine derivatives, 237
Cytomegalovirus, phonoacetic acid and, 52

Danitracen, 128
Dapsone, 27, 178, 179
Darbazine, 50
Datura stramonium, 57-58
Daunomycin, 195
Deactivation, metabolism and, 71-72
Deazaguanine, 239
Decamethonium, 32, 106
Delephene, 162
Depression, endogenous, 35
Derivitization, peptide analogs and, 99
Desipramine, 83, 130
Desoxycorticosterone acetate-salt hypertension, 42
Dexamethasone, 149
Dextran sulfate, 101
Dextromethorphan, 66
Dextromoramide, 63
Dextropropoxyphene, 63, 147
Diacetazotol, 170
Dialkylaminoalkyl groups, 58
Diamidines, 200-201, 209
Diaminazene, 200
Diamphenethide, 224
Diaverdine, 202
Diazepam, 35, 132-133, 140
Diazepoxide, 133
Dibenamine, 33
Dibucaine, 115
Dicarboxylic amino acid analogs, 141
Dichlophen, 226
cis-Dichlorodiammineplatinum, 50-51
Dichloroisoproterenol, 40
Dicrocoelium dendriticum, 223
Dicumarol, 101
Dieldrin, 196
Diels-Alder addition reactions, 144
Diethylcarbamazine, 225
Diethylstilbestrol, 157
Diflunisal, 43
Digitalis, 10
Digitoxin, 10
Digoxin, 10
Dihydrocupreine, 172
Dihydrofolate reductase, 177, 202, 217
Dihydrofolate synthetase, 173
Diloxanide furoate, 208
Dimemorphan, 66
Dimenhydrinate, 53, 111
Dimercaprol, 198
Dimethisoquin, 116
Dimethisoquine, 18

N,N-Dimethylethylenediamine, 53
Dimetridazole, 204
Diminazene, 167
Diminazine, 209
Diphenadione, 101, 102
Diphenhydramine, 53, 111, 119
Diphenylhydantoin, 30, 135
Diphenyl oxidase, 182
Dirofilaria immitis, 224
Disamide, 141
Disinfectants, 159, 162-167
Disophenol, 223
Dissociation constant, 86, 88, 136
Distamycin, 232
Distribution coefficients, 83
Disulfiram, 34
Dithiazinine iodide, 227-228
Diuretics, 40, 178
Doclovos, 227
Dolantin, 62
Domiphen bromide, 165
DON, 235
Dopamine, 120, 127
Doxorubicin, 195, 232
Dracunculus medinensis, 226
Dryopteris filix mas, 226
Duanomycin, 232
D vitamins, 95-96
Dyclonine, 115-116
Dyestuffs, 15-16, 165-166, 169-170, 197-198

Electron density, substitution and, 21
Electron distribution, and bioisosteric replacement, 86
Electronic orbital distribution, 88
Emetine, 205-206
Endogenous depression, 35
Endogenous peptides, 142-143
Endorphins, 20, 59, 60, 142-143
Enhancement factor, in substitution, 47
Enkephalins, 59
Entamoeba histolytica, *see* Amebiasis
Enterobius vermicularis, 227
Entramin, 203, 208
Enzyme inhibition, *see* Inhibition
Ephedrine alkaloids, 20
Epilepsy, 140
Epinephrine, 19, 20, 39
Epoxides, 49
Ergot, 38
Erythromycin, 193
Erythroxylon coca, 17
Escherichia coli, 175
Escin, 152
Estradiol, 25

Estrogens, 25, 156–157
Ethambutol, 182–183
Ethionamide, 27, 182
Ethisterone, 74
p-Ethoxyacetanilide, 16
Ethoxzolamide, 141
Ethyl alcohol, 135
Ethylene oxide, 164
Ethyl p-aminobenzoate, 17
Ethynodiol diacetate, 160
Ethynylestradiol, 160
Etonitazene, 147–148
Etorphine, 63–64
Eucaine, 17
Eupatorium, 8

F151, 224, 225
Falicaine, 115–116
Fasciolicides, 223–224
Fatty acids, as lead compounds, 79–80
Fenamates, 43, 151
Fenethazine, 112
Fentamyl, 146
Filariacides, 224–226
Filicin, 226
Flavonoids, 152
Flufenamic acids, 43, 44
9-α-Fluorocortisol, 46, 47
Fluorotracen, 128
5-Fluorouracil, 32, 49, 238
Fluperazine, 120
Flurazepam, 133
Folic acid, 26, 177
    analogs of, 235–236
    antagonists of, 210
    dihydrofolate reductase, 202, 217
    dihydrofolate synthetase, 173
Follicle-stimulating hormone, 24
Formaldehyde, 164
Fowler's solution, 198
Free and Wilson method, 89, 235
Fuchsine, 169
Fungal infections, *see* Antifungal agents
Furapromidium, 222
Furazolidone, 209
Furosemide, 40, 178

Ganglionic blocking agents, 32
Genatropine, 108
Genoscopolamine, 108
Gentamicins, 191
Gentian violet, 16, 169
Giardiasis, 168, 209
Glaucarubin, 206
Glaucoma, 178

Gloxazone, 210
Glucantime, 204
Glucosulfone, 178, 179
D-Glucuronic acid, in heparin, 100
Glycoproteins, as drug receptors, 13
Glyzyrrhetinic acid, 152
Gonadoliberin, 98
Gramidicin, 191
Gramine, 19
Grimm's hybrid displacements, 85
Griseofulvin, 194
Guanethidine, 40

H-33258, 225
Haemonchus contortus, 227
Halogens, 163
Haloperidol, 34, 35, 123–124
Halotestin, 155
Haloxon, 226
Hammet sigma value, 88, 89
Hammet substituent constant, 21
Hansch analysis, 70, 78, 88, 89
Heartworm, 224
Heparin and heparinoid anticoagulants, 100–101
Heroin, 60, 61
Herpes virus, 49, 52
Hetol, 223
Hexachloroethane, 223
Hexachlorophene, 164
Hexamethonium, 32, 33, 106
Hexamethylenetetramine, 164
Hexamethylolamine, 50
n-Hexylresorcinol, 164
Histamine, 19, 20, 52
Histamine receptor antagonists, 52–55
Histomoniasis, 203–204
HN2, *see* Mechlorethamine
Holarrhena antidysenterica, 206
Homoharringtonine, 233
Homoisotwistane, 242–243
Homologation, in molecular modification, 82–84
Homovanillic acid, 120
Hormones, 22–24. *See also* Corticosteroids; Steroids
Hybridization, 86
Hycanthone, 72, 221
Hydantoins, 139
Hydnocarpic acid, 80
Hydralazine, 39
Hydrastis canadensis, 206
Hydrazine, 50
Hydrazinophthalazines, 39
Hydrochlorothiazide, 178

INDEX

Hydrocodone, 61
Hydrogen bonding, and bioisosteric replacement, 86
Hydromorphone, 61
p-Hydroxyacetanilide, 16
5-Hydroxytryptamine, 20, 39, 52, 110, 117, 127, 134–135
Hyoscyamine, 57, 108
Hyoscyamus niger, 57–58
Hypnotics, 22, 134–138

Iatrochemistry, 6
Ibuprofen, 43
L-Iduronic acid, 100
Imidazole ring, 52
Imidocarb, 210
Imipramine, 29, 36, 83, 126
Immunoactive agents, 31, 152, 196
  antiinflammatory, 152
  antiviral and antitumor, 232–234
Indole alkaloids, 34
Indoles, 39
Indomethacin, 43, 44, 150
Inhibition, enzyme:
  irreversible, 74–76
  suicide enzyme, 9, 73–74
  transition-state analogs and, 72–73
  see also specific enzymes and substrates
Inorganic gold sodium thiosulfate, 45
Insulin, 67
Interaction model, 70
Interferon, 51, 233
Iodine, 6, 163
Iodochlorhydroxyquin, 207
Ipecac alkaloids, 205–206
Iprindole, 128
Iproniazid, 36, 76–77, 125, 126
Ipronidazole, 204
Irreversible inhibition, 74–76
Isoelectric distribution, 84
Isoniazid, 27, 76, 125, 180–181, 182
Isopentaquine, 213
Isoproterenol, 20, 41
Isosterism, 53

Kala-azar, 203–204
Kanamycins, 179, 191
Kcat inhibitors, 9, 73–74
Kininase II, 41–42

b-Lactam antibiotics, 184–189
Laudanosine alkaloids, 147–148
Lead compounds:
  bioisosterism and, 28
  role of, 8–9

Leishmaniasis, 203–204
Leprosy, 27, 178, 179
Leuconostoc, 101
Leu-enkephalin, 59, 142–143
Leukemias, 195, 232
Leukotriene C, 47
Levamisole, 45, 153, 230, 241
Levopropoxyphene, 66
Levorphanol, 63
Lidocaine, 18, 19, 115
Lincomycin, 194
Lindane, 196
Linear free energy method, 88
Linear multiple regression model, 70
Lipid solubility, and bioisosteric replacement, 86
Lipid theory, 21
Lipophilicity, 82, 88
β-Lipoprotein, 142
Lipoproteins, as drug receptors, 13
Lithium, 36
Litomosoides carinii, 224
Lividomycin, 194
LSD, 30
Lucathone, 72, 221
Luteinizing hormone, 34, 134
Lymphomas, 232
Lynestrenol, 161
Lysosome hydrolases, 44

McFayden-Stevens reaction, 182
Magnus test, 53, 107
Malachite green, 169
Malaria, 171–172, 210–219
  aminopyrimidines, 217–219
  quinine, 211–212
  synthetic antimalarials, 199–200, 212–217
Mannich base, 124
Mauve, 171
Maytansine, 196, 233
Mebendazole, 229
Mecamylamine, 33
Mechlorethamine, 31
Medicinal chemistry, 5
Medroxyprogesterone, 160
Megestrol acetate, 161
Meglumine, 204
Melarsen, 198
Mepacrine, 226
Meperidine, 30, 57, 225
  analgetic properties, 62, 146–147
  analogs of, 123
Mephenesin, 32
Mephesin, 131–132
Mephobarbital, 139

Mepivicaine, 18, 19
Meprobamate, 32, 132
6-Mercaptopurines, 49, 239
Mestranol, 160
Metabolic analogs, 49
Metabolism, drug, 71–72
Metabolites:
   antagonists of, 49
   molecular modification and, 79
   see also specific drugs
Metapone, 144
Met-enkephalin, 59, 142–143
Metformin, 69
Methadols, 147
Methadone, 33, 62–63, 146–147
Methdilazine, 119
Methenamine, 164
Methisazone, 51
Methodology, 15–24
Methonium compounds, 106
Methotrexate, 32, 49, 152, 235–236
L-α-Methyldopa, 39–40
Methylene blue, 16, 197, 212–213
4-Methylhistamine, 54
Methylhydrazine, 50
9-Methylstreptimidone, 232
Metiamide, 54
Metopone, 61
Metrifonate, 222
Metronidazole, 168, 203, 209, 226
Mianserin, 128
Michaelis-Menten equation, 82–83
Miracils, 221
Mithrimycin, 244
Molar refractivity, 89
Molecular modification, 70–71, 76
   bioisosterism, 29, 84–87
   biological guidance in, 80–82
   biological properties, separation of, 76–78
   homologation and chain branching in, 82–84
   methodology of, 78–84
Molindone, 124
Monensin, 202
Monoamine oxidase, 36
Monoamine oxidase inhibitor antidepressants, 125–126, 128–130
Morantel, 228–229
Morphanins, 63, 66, 145–146
Morphine:
   as analgetic, 60–61, 143–145
   analogs of, 60–61
   as antitussive, 65
Muramyl dipeptide, 152
Muscarine, 55, 56, 103–105
Muscle relaxants, 32

Mustard gas, 31
Mycobacterial infections, 178–183, 190
Mydriatics, 56–58
Myrcene, 8

Nagana, 196
Nagana red, 197
Nalidixic acid, 168–169
Nalorphine, 20, 21
Naloxone, 20, 21, 64, 84, 142
Naltrexone, 64
Nandrolone, 156
Narcotic antagonists, 64
Narcotics, 21. *See also specific drugs*
Natural products:
   amebicides, 206
   with antiviral and antitumor activity, 231–232
   role of, 7–8
   see also Alkaloids
Necator americanus, 227
Neomycin, 190, 191
Neuroleptic agents, 84, 116–117
   butyrophenones, 123–124
   other tricyclic systems, 121–123
   phenothiazines, 118–121
   reserpine, 117–118
Neuromuscular blocking agents, 105–107
Neurotransmitters, 116–117. *See also specific transmitters*
Niacin, 23
Niacinamide, 23, 27, 182
Niclosamide, 226
Nicotine, 56, 103
Nicotinic receptors, 56
Nifurtimox, 202
Nimorazole, 209
Niridazole, 208, 209, 222
Nirvanol, 139
Nitrazepam, 133, 140
Nitrimidazine, 209
Nitrofural derivatives, 204
Nitrofurantoin, 168
Nitrofurazone, 30, 168
Nitrogen mustard, 31, 49, 75, 240
Nitroheterocyclics, 167–168, 201–202, 222
Nitroimidazoles, 204, 209
Nitrosoureas, 49
Nitrothiazole, 204, 222
Nitroxynil, 223
Nocardicins, 179, 189
Nonsteroidal antiinflammatory agents, 42–46, 150–153
Norepinephrine, 19, 20, 39, 127
Norethindrone, 160

Norethisterone, 74
Norgesterone, 161
Norgestrel, 160
Nortestosterone, 156
Nortryptyline, 130
Noscapine, 66
N-substitution, sulfanilamides, 174–176, 177
Nuclear magnetic resonance, 91, 97
Nucleic acids:
   antiviral and antineoplastic agents and, 31–32, 237–238
   as drug receptors, 13
Nucleocidin, 202
Nucleophiles, and irreversible inhibition, 75
Nucleoside analogs, 49, 237, 238
Nystatin, 194

Octanol-water partition coefficient, 88
Octaphonium chloride, 165
Oleandromycin, 193–194
Onchocerciasis, 224, 225
Opiate receptors, 64–65, 142
Opioids, 59–60, 135, 143–145
Opisthorchis, 223
Opium, 135
Optochin, 172
Oral anticoagulants, 101–102
Oripavines, 63–64, 144
Ormetroprim, 202
Orphan drugs, 243–244
Orpiment, 198
Orsanine, 198
Orthocaine, 17
Orthoform, 17
Oxamniquine, 222
Oxantel, 228–229
Oxazepam, 132–133
Oxethazine, 115
Oxicams, 151
Oxine, 164
Oxolinic acid, 169
Oxophenarsine, 16
Oxotremorine, 104
Oxymorphone, 61, 144
Oxyphenbutazone, 43
Oxytetracycline, 167, 209

PAB-folic acid antagonists, 210
Pamaquine, 213
Papaverine, 10
Papaver, 60, 65
Paraldehyde, 22
Paramethadione, 140
Parbendazole, 229
Pargyline, 40, 129

Parkinsonism, chlorpromazine and, 120
Paromomycins, 190–191, 206
Partition coefficients, 78, 89
Penicillamine, 45, 152, 221, 242–243
Penicillins, 26, 183, 184–187
Penicillin V, 186
Penicillium griseofulvum, 194
Pentachloroethane, 223
Pentamidine, 167, 204
Pentamoxane, 124
Pentaquine, 213
Pentazocine, 63, 146
Pentosans, 101
Pepstatin, 152, 192
Peptides, 96–100
   analogs, 98–100
   antibiotic, 191–192
Pethidine, *see* Meperidine
Phanquone, 207
Pharmacodynamic drugs, 30
Pharmacology, and medicinal chemistry, 6–7
Phenacemide, 140
Phenanthrene, 61
Phenacetin, 16
Phenanthridium salts, 201
Phenazocine, 63
Phenazopyridine, 169, 170
Phenbenzamine, 53, 110–111
Phenethanolamines, 41
Phenethylamine, 83, 125–126
Phenformin, 68–69
Phenidione, 101, 102
Phenobarbital, 22, 136, 139
Phenols, 162, 163–164
Phenothiazines, 30, 112, 117, 118–121, 126
Phenoxybenzamine, 33
Phenoxymethylpenicillin, 186
Phenoxypropanolamines, 41
Phensuximide, 140
N-Phenyethylnormorphine, 20
Phenylbutazone, 43, 151–152
2-Phenylcycloproylamine, 130
Phenylindanedione, 101
Phenytoin, 30, 135, 139
Pholcodine, 65
Phonoacetic acid, 52
Phosphatidyl serine, 65
Phospholipase, A2, 47
Physical absorption, 82
Phytoalexins, 8
Pilocarpine, 12, 52, 104
Piperazine, 225
Piperoxan, 110, 124
Piroplasmosis, 209
Piroxicam, 151

Plant products, *see* Alkaloids; Natural products
Plasmochin, 213
Plasmodial infections, *see* Malaria
Podophyllotoxins, 232
Polymyxins, 191
Polysaccharide anticoagulants, 100
Povidone-iodine, 163
Pramoxine, 115–116
Praziquantel, 227
Prednisone, 149–150
Primaquine, 83, 213
Primidone, 140
Probenecid, 178
Procaine, 17, 18, 113–114
Procarbazine, 50, 242–243
Prochlorperazine, 120
Prodine, 146
Pro-drugs, drug development and, 15
Proflavin, 170
Progestational agents, 158–159, 160–161
Progesterone, 25
Prolactin, 24, 134
Promazine, 118, 119, 120
Promethazine, 34, 112, 119
Prontosil, 169, 173, 174
Propamidine, 166, 167
1,3-Propanediols, 131–132
Propanolol, 414
b-Propiolactone, 164
6-n-Propylthiuracil, 24
Prostaglandins, 80, 110
    analgetics and, 17
    analogs of, 152–153
    and pain perception, 58
    salicylates and, 17, 44, 86, 150–151
Prostaglandin synthetase, 44
Protamin, 67
Proteins, 96–100
Proteus, 175, 191, 192
Proton tautomerism, 54, 55, 217
Prototype compounds, 6
Protozoal infections, *see* Antiprotozoal agents
Pryrolidine alkaloids, 8
Pseudomonas, 184, 191, 192
Psoriasis, 23
Pteroylglutamic acid, 177. *See also* Folic acid
Purine analogs, 236
Pyocyanine, 184
Pyrantel, 227–228
Pyrathiazine, 119
Pyrazinamide, 27, 180, 182
Pyrazines, sulfanilamide substitution, 175
Pyrazolidinediones, 43
Pyrazolinones, 151–153
Pyridazines, sulfanilamide substitution, 175

Pyridine, ring equivalents and, 29
Pyridoine, 23
Pyridonecarboxylic acids, condensed, 168–169
Pyridoxal, 23, 95
Pyridoxamine, 23
Pyrilamine, 110–111
Pyrimethamine, 202, 219
Pyrimidines, 30
    analogs of, 236
    sulfanilamide substitution, 175
Pyrivium pamoate, 227–228

Quantitative structure-activity relationship, 87–90
Quaternary ammonium salts, as antiseptic, 164–165
Quaternary carbon, sedative activity and, 136
Quaternary ions:
    antitrypanosomal activity, 201
    as ganglionic blocking agents, 32
Quaternary quinolinium compounds, 209
Quaternization, of solanaceous alkaloids, 108
Quinacrine, 207, 213–214, 226
Quinapyramine, 200
Quinidine, 10
Quinine, 10, 71, 217
Quinocide, 213
Quinoline analogs, 199–200, 213–217
Quinolinium compounds, quaternary, 209
Quinolols, 164, 207
Quinone system, mycobacteria and, 183
Quinuronium methosulfate, 209

Ranitidine, 55
Rauwolfia alkaloids, 10, 34, 39, 117
Reactivity:
    and bioisosteric replacement, 86
    and quantitative structure-activity analysis, 88
Receptors:
    acetylcholine, 55–56, 103–104
    drug, theories of, 12–14
    opiate, 64–65
Renin, 38, 192
Research:
    methods, 15–24
    organization of, 14
Reserpine, 34, 35, 39, 117–118, 134
Resochin, 214
Resorantel, 226
Rhabdomyosarcoma, 195
Ribavirin, 237
Rickettsias, 192
Rifamicin, 194
Rifampin, 183, 232

INDEX

Rifamycins, 179–180
Rimantidine, 242
Riminophenazine, 183
Ring equivalents, bioisosterism and, 29
RNA polymerase, 180
Ronidazole, 204
Rosaniline, 169

Salicin, 42, 43
Salicyl alcohol, 42
Salicylanilide, 164
Salicylates, 15, 42–43, 86, 150–151
Salicylic acid, 42, 171
Salol, 15
Saxitoxin, 59, 113
Scarlet red, 170
Schistosomiasis, 220–222
Schonhofer tautomerism, 214, 217
Scopolamine, 56, 57, 108
Screening, of drugs, 70
Secobarbital, 22
Sedative activity, substitution and, 136
Sedative hypnotics, 22
Sedatives, 134–135
  anticonvulsants, 139–145
  hypnotics, 22, 135–138
Senecio, 8
Serotonin, 20, 39, 52, 110, 117, 127, 134–135
Simarouba, 206
Sleeping sickness, 196, 198
Sleep states, drugs and, 134–135
Slow-reacting substance of anaphylaxis, 47, 110
Sn2 reactions, and biological activity, 82
Solanaceous alkaloids, 57–58, 108–109
Solubility, and bioisosteric replacement, 86
Somatostatin, 67
Sontoquine, 200
Sontoquinoline, 214
Spectroscopy, in peptide chemistry, 97
Spirochetes, 192
Spironolactone, 40
SQ 18506, 222
Staphylococcal infections, 173, 184
Steric factors:
  in bioisosteric replacement, 86
  in enzyme substrate complex stability, 83
Steroidal alkaloids, in amebiasis, 206
Steroids, 25, 148–150
  androgens, 154–156
  estrogens, 156–157
  progestational agents, 158–159
  see also Corticosteroids
Stibogluconate, sodium, 204
Stibophen, 204

Stilbamidine, 167
Stovarsol, 206
Streptococcal infections, 173
Streptomyces, 179, 189, 190, 192, 193, 194, 206
Streptomycin, 27, 31, 179, 190
Structural analogs, as lead compounds, 81
Structure-activity relationships, 21–22
Strychnos, 105
Substitution:
  and biological properties, 82–84
  enhancement factor in, 47
Substrate analogs, and irreversible inhibition, 75
Succinylcholine, 33
Suicide enzyme inhibition, 9, 73–74
Sulfachlorpyrazine, 202
Sylfaclomide, 176
Sulfadiazines, 175, 217
Sulfadimethoxine, 175, 202
Sulfamehoxydiazone, 175
Sulfamentoxine, 175
Sulfamerazine, 175
Sulfamethizole, 175
Sulfamethoxypyrazine, 175
Sylfamethoxypyridazine, 175
Sulfanilamides, 16, 26, 40, 80, 170–177
  bioisosterism and, 28
  bis-(4-aninophenyl)sulfone, 177–178
  mechanisms of action, 176–177
  N-substituted, 174–176
  sulfonamide dyes, 170–173
Sulfapyrazine, 210
Sulfapyridine, 174, 175, 217
Sulfathiazole, 174, 175
Sulfinpyrazone, 152
Sulfisoxazole, 175
Sulfonamides, 26, 169, 185
Sulfones, substituted, 177–178
Sulfonyl fluorides, 75
Sulfothiadiazoles, 181
Sulindac, 43, 44
Sulthiame, 141
Superlutin, 160
Suramin, 16, 166, 198, 225
Surfen, 200
Surra disease, 196
Suxamethonium, 106
Synthalins, 68, 166

Taft Es value, 88, 89
Tartar emetic, 198, 204, 221
Tautomerism, proton, 54, 55
Teclozan, 208
Teprotide, 41
Testosterone, 25, 154, 156

Tetrabenazine, 117, 128
Tetrachloroethylene, 223
Tetraethylammonium, 11
Tetrahydrocannabinols, 148
Tetrahydrouridine, 72
Tetramethylthiuram, 34
Tetramisole, 226, 229–230, 241
Tetrodotoxin, 59, 113
Thebaine, 63, 143–144
Therapeutics, foundations of, 10–12
Thiabendazole, 229
Thiacetazone, 180
Thiadiazoles, 68
Thiambutene, 147
Thiamine, 23, 202
Thiazetazone, 181
Thiazole, 29
Thiazothielite, 230
Thiazothienol, 229–230
Thienamycin, 189
Thiomides, 182
Thiobarbituric acids, 136
Thiogold compounds, 45
Thiophene, ring equivalents and, 29
Thioridazine, 120–121
Thiosemicarbazones, 31, 51, 181–182
Thioureas, 181–182
Thioxanthenes, 112, 121–123
Thromboxanes, 80, 153
Thymidine derivatives, 237
Thyroid hormones, 24, 92–95
Thyrotropin-releasing hormone, 97
Thyroxine, 24, 92–93
Thyroxine binding proteins, 93
Tibione, 181
Tilorone, 233–234
Timolol, 41
Tinidazole, 209
Tobramycin, 191
Tolazamide, 68
Tolbutamide, 68
Toxiferines, 105
Toxoplasmosis, 210
Trancyclopromine, 36
Transition-state analogs, 9, 72–73
Tranylcypromine, 73, 129–130
Treparsol, 206, 207
Triamcinolone, 149
Triazenoimidazoles, 50
Triazolam, 134
Trichloethanol, 135
Trichomoniasis, 168, 203
Trichuriasis, 227, 228, 229
Triclocarban, 166
Triclofos sodium, 135

Tricyclic antidepressants, 121–123, 126–127, 130–131
2-Tridecanone, 8
Triflubazam, 133–134
Trifluorothymidine, 238
Trifluperazine, 120
Triflupromazine, 120
3, 3′,5-Triiodothyronine, 92–94
Trimeprazine, 119
Trimethadione, 140
Trimethoprim, 217
Tripelennamine, 110–111
Triphenylmethane, 169
Triprolidine, 112
Trophoblastic tumors, 195
Trypan blue, 16, 197, 209
Trypanocides, 16, 165, 166, 170, 196–202
Trypan red, 16, 197
Tryparsamide, 198
Tuberculosis, 178–183, 190
Tumors, see Antiviral and antineoplastic agents
Turbocurarine, 11, 105, 106
Tyrocidin, 191
Tyrothricin, 191

Ureas, antibacterial, 165–167

Valinomycin, 191, 192
Valproic acid, 140
Vasoactive drugs, 32–33, 97
Vasopressin, 97
Veratrum alkaloids, 38
Vidarabine, 239
Vinblastin, 232
Vinca alkaloids, 232
Vinclofos, 226–227
Vincristine, 232
Vinylestrenolone, 161
Viomycin, 179
Viral infections, see Antiviral and antineoplastic agents
Vitamins, 22–24
  folic acid, see Folic acid
  D, 95–96
  K, 102
  niacinamide, 27, 182
  pyridoxal, 95
  thiamine, 202

Warfarin, 101–102
Wilms' tumor, 195
Wuchereria bancrofti, 224

X-ray crystallography, 91
X-ray diffraction spectra, 90